MEDICAL ENGINEERING

Projections for Health Care Delivery

CONTRIBUTORS

DON BAKER

T. GRAHAM CHRISTOPHER

JAMES S. COLE

JOHN T. CONRAD

COLIN H. DALY

A. W. GUY

J. HILDEBRANDT

LEE L. HUNTSMAN

CURTIS JOHNSON

JOHN LUFT

C. J. MARTIN

WAYNE E. MARTIN

GERALD H. POLLACK

CHARLES E. POPE, II

JOHN M. REID

ROBERT F. RUSHMER

SAM L. SPARKS

H. FRED STEGALL

D. E. STRANDNESS

GENE L. WOODRUFF

MEDICAL ENGINEERING
Projections for Health Care Delivery

Robert F. Rushmer
Center for Bioengineering
University of Washington
Seattle, Washington

 1972

ACADEMIC PRESS New York and London

ACADEMIC PRESS, INC.
111 Fifth Avenue, New York, New York 10003

United Kingdom Edition published by
ACADEMIC PRESS, INC. (LONDON) LTD.
24/28 Oval Road, London NW1

LIBRARY OF CONGRESS CATALOG CARD NUMBER: 77-182648

PRINTED IN THE UNITED STATES OF AMERICA

CONTENTS

Part I   BIOENGINEERING AND THE HEALTH CARE SYSTEM
Contributions, Complications, and Crises

Chapter 1 Engineering in Medicine and Biology
An Overview

Chapter 2 Health Manpower Problems
Potential Role of Biomedical Engineering

Part II   **ENGINEERING APPROACHES**
TO HEALTH CARE REQUIREMENTS

Contents   vii

Chapter 7 Medical and Biological Applications of Modeling Techniques
Lee L. Huntsman and Gerald H. Pollack

Chapter 8 Biomechanics and Biomaterials

Chapter 9 Technological Training of Medical Manpower

Chapter 10 Clinical Engineering
Future Engineering Requirements by Medical and Surgical Specialties
H. Fred Stegall (Editor)

Overall Summary
Future Options

LIST OF CONTRIBUTORS

Numbers in parentheses indicate the pages on which the authors' contributions begin.

DONALD W. BAKER (B.S.) Center for Bioengineering, University of Washington, Seattle, Washington (266)

T. GRAHAM CHRISTOPHER (M.D.) Harborview Hospital, Seattle, Washington (348)

JAMES S. COLE (M.D.)* Center for Bioengineering, University of Washington, Seattle, Washington (332)

JOHN T. CONRAD (M.D.) Departments of Biophysics and Obstetrics and Gynecology, University of Washington, Seattle, Washington (358)

COLIN H. DALY (Ph.D.) Department of Mechanical Engineering, University of Washington, Seattle, Washington (361)

A. W. GUY (M.D.) Department of Rehabilitation Medicine, University of Washington, Seattle, Washington (366)

J. HILDEBRANDT (M.D.) Department of Physiology and Biophysics, Institute of Respiratory Physiology, Virginia Mason Research Center, Seattle, Washington (340)

LEE L. HUNTSMAN (Ph.D.) Center for Bioengineering, University of Washington, Seattle, Washington (273,322)

CURTIS JOHNSON (Ph.D.) Department of Electrical Engineering, Center for Bioengineering, University of Washington, Seattle, Washington (253, 259,324)

*Present address: Department of Medicine, Baylor University, Waco, Texas

JOHN LUFT, Department of Biological Structure, University of Washington, Seattle, Washington (244)

C. J. MARTIN (M.D.) Institute of Respiratory Physiology, Virginia Mason Research Center, Seattle, Washington (340)

WAYNE E. MARTIN (M.D.) Department of Anesthesiology, Division of Bioengineering, University of Washington, Seattle, Washington (354)

GERALD H. POLLACK (Ph.D.) Department of Anesthesiology and Center for Bioengineering, University of Washington, Seattle, Washington (273)

CHARLES E. POPE, II (M.D.) Department of Gastroenterology, Veteran's Administration Hospital, Seattle, Washington (345)

JOHN M. REID (Ph.D.) *Center for Bioengineering, University of Washington, Seattle, Washington (263)

ROBERT F. RUSHMER (M.D.) Center for Bioengineering, University of Washington, Seattle, Washington (1,5,33,63,105,161,215,223,225,291,321)

SAM L. SPARKS (Ph.D.)† Center for Bioengineering, University of Washington, Seattle, Washington (369)

H. FRED STEGALL (M.D.) Center for Bioengineering, University of Washington, Seattle, Washington (257,267,326,331,354,356)

D. EUGENE STRANDNESS, JR. (M.D.) Department of Surgery, Peripheral Vascular Service, University of Washington, Seattle, Washington (337)

GENE L. WOODRUFF (Ph.D.) Department of Nuclear Engineering, Nuclear Reactor Laboratories, University of Washington, Seattle, Washington (249)

*Present address: Department of Physiology, Providence Hospital, Seattle, Washington
†Present address: Department of Sensory Engineering, University of Washington, Seattle, Washington

PREFACE

Rapid change is a characteristic common to all aspects of modern society. In recent years, medicine has been caught up in the accelerating pace of events. Great progress has been made through technology in improving the effectiveness of medical diagnosis and therapy, but many current crises in medicine are the unpredicted complications of technological success. Solutions to many of these problems will stem from further technological advances, but the time scale for achieving solutions is slower than the rate at which problems are developing. As a result there is a growing need for accurate, long-range forecasting of future requirements for technology in medicine.

The proper selection of objectives for biomedical engineering in the next decade depends on accurate prediction of the requirements for techniques and technology in the health care delivery system. The need for such forecasting is equally applicable to the kinds of basic bioengineering research which will be needed in support of these "applied" problems.

The demand for increasing the quantity and quality of health care services is mounting, and it will be accentuated by impending moves toward increased support by the Federal government of broadened coverage of health care for large segments of the population. The consequences of these trends can best be visualized by studying conditions in countries that have had large-scale health and welfare systems for some time now. For this reason, I devoted a period of study to current trends in health care delivery in Scandinavia, West Germany, the Benelux countries, and Switzerland, and gained an exposure to modern concepts of urban planning at the Athens Institute of Ekistics. With this experience as a background, the current crises in health care in the United States will be examined in this book in terms of manpower (Chapter 2), health care facilities (Chapter 3), and the distribution of health care (Chapter 4).

The flood of data stemming from the new types of information sources, such as automated laboratory techniques, requires a wholly new approach to the acquisition, editing, analysis, pattern recognition, storage, and retrieval of data in quantities beyond our wildest expectations of only a few years ago. Some of the implications of these data-processing techniques are considered in Chapter 5. Present developments and future prospects for new sources of diagnostic and research data are summarized in Chapter 6.

One basic biomedical engineering approach to complex systems in the utilization of simulation and modeling. These techniques are clearly applicable to research on physiological mechanisms, and they also contribute to the solution of very practical problems in analyzing health care delivery systems as a whole. The significance of these developments is suggested by three different examples described in Chapter 7.

Efforts to develop effective solutions to practical problems in clinical medicine depend on continuing research in biomedical engineering. For example, basic research in biomechanics and biomaterials holds promise of providing solutions to problems impeding progress in areas such as the development of new biocompatible materials, of nonthrombogenic surfaces, and of supportive substances for use in prosthetic devices. Some of these opportunities are considered in Chapter 8.

The impact on medicine of new techniques and technology calls for changes in the higher education and training of students in both medicine and engineering. In addition, there is growing recognition of the need for new types of personnel to occupy positions at various levels of responsibility in hospitals and laboratories. Such new breeds are required to develop, utilize, and maintain the sophisticated equipment which is crucial to the practice of modern medicine. Certain new training requirements for health manpower are discussed in Chapter 9.

A group of individuals affiliated with the Center for Bioengineering at the University of Washington, who are studying clinical problems in a diverse group of medical specialties, were asked to identify technological devices in common use in medical centers today. These descriptions and the prospects for new technological developments are described in Chapter 10.

Forecasting future technology is a difficult and somewhat foolhardy undertaking. There are thousands of ways of being wrong for each opportunity of being correct. I am extremely grateful for the courage and forbearance of the staff of the Center for Bioengineering, and many colleagues in medicine and engineering, who have been willing to contribute their ideas under these circumstances. Wherever possible, their contributions have been recognized in the text. Others have provided additional information, guidance, support, and editorial review of portions of the manuscript which deserve recognition, particulary Dr. Bertil Jacobsen and Dr. Gunnar Wennstrom of Stockholm, D. Tybaerg Hansen and Dr. C. Toftemark of Copenhagen, Dr. Constantinos Doxiadis of Athens, and Dr. Allan Hoffman, Dr. Paul Van Dreal, and Dr. Robert Leininger of Seattle. Thanks are also due many members of the Battelle Memorial Institute in Frankfurt, Geneva, and Seattle. It is impossible to acknowledge adequately the many contributions, large and small, from my many other friends and colleagues here and abroad.

Part I

BIOENGINEERING AND THE HEALTH CARE SYSTEM
Contributions, Complications, and Crises

INTRODUCTION

Bioengineering emerged as a complex multidiscipline during the past ten to fifteen years at a time when the health care system has been evidencing stresses, overload, and threats of breakdown. Paradoxically, dissatisfaction with medical care has reached a new high at a time when technology has provided vastly improved handling of many serious illnesses by new diagnostic and therapeutic techniques. A major criticism is the unrestrained upward spiral of medical costs when the principle stimulus is increased affluence and expanded sources of financial support for health care. The health care "system" is found to be a "nonsystem" with many overlapping hierarchies within individual institutions and facilities. Overutilization and inefficiency are assured by basic policies of principal purchasers of the services.

Like many other prominent features of modern society, changes are occurring with increasing rapidity, and to some degree appear to be out of control. Major problems appear to be unexpected complications of very great technological successes. The modern hospital is the central hub of health care and serves as the focal point for concentrations of techniques and technology of ever increasing sophistication. Engineering has contributed greatly to clinical research and practice. Biomedical engineering must help solve the resultant problems through improved long-range planning, design optimization, instrumentation, and management for which engineers are trained.

Effective interdisciplinary collaboration requires that engineers understand the operation and constraints implicit in the delivery of health care, and the nature and scope of bioengineering must be clear to physicians, health professionals and staffs. Biomedical engineering can be defined as a collaborative approach to problems of mutual interest by engineers and biomedical personnel. It encompasses virtually all disciplines of engineering and most, if not all, divisions of health care. In contrast, the health care delivery system cannot be simply defined, particularly if one is looking for the inherent opportunities and constraints imposed by its traditions and organization. In view of the importance of these considerations, the first part of this book is devoted to a very brief survey of the nature and scope of biomedical engineering (Chapter 1) and then to the principal problems facing modern medicine in terms of manpower, facilities, distribution of health care, and data processing.

ENGINEERING IN MEDICINE AND BIOLOGY
An Overview

The application of physical sciences, mathematics, and engineering to problems of medicine and biology has increased explosively during the past decade. Such rapid growth may be ascribed to the hybrid vigor resulting from cross-linking of engineers and life scientists in problems of mutual interest. Widespread application of the techniques and technology of engineering in basic medical science and clinical medicine is a relatively new development, but its origins can be traced far back in history (1, 2).

Within the last few years biomedical engineering has become one of the most rapidly growing fields in the country. This meteoric rise implies that biology and medicine are ripe for the application of engineering. The demand for this new multidiscipline results from the obvious need for quantitation in those areas of biology and medicine which have heretofore consisted primarily of large masses of qualitative descriptions, classifications, and correlations of various forms of life and their component organ systems. The keystone of future development in biology from fish to man will include quantitative analysis based on direct measurement of critical variables (3).

The fabric of medicine has been woven from patterns of signs and symptoms correlated over centuries with pathological processes observed during post mortem examinations. Medical practice has become highly structured despite the limited use of quantitative physical measurements in routine diagnosis and care of patients (4, 5). Holaday (6) summarized as follows: "The special problems associated with the practice of medicine do not excuse or justify the archaic methods we employ for collecting data, processing and interpreting it, and translating it into action." Rapid technological advances provide whole new systems for scientific research with new transducers, miniaturized amplifiers, telemetering equipment, large-capacity storage systems, high-frequency recorders, and high-speed computers. The medical community is stirring with interest in new approaches to research, diagnosis, and therapy based on new developments and new tools. For example, control systems analysis appears particularly applicable to the complex, interacting, biological control mechanisms.

Not long ago, mathematics, physics, and engineering were three distinct areas of interest. Mathematicians were engaged in providing new tools for quantitative analysis; physicists probed the frontiers of knowledge regarding the world about us, including the world of the atom; engineers were busy exploring the properties of materials for the benefit of mankind. Today, many scientists use mathematical tools to analyze physical principles with electronic computers. Such investigators function as mathematician, physicist, and engineer at one and the same time. The change in emphasis may also be recognized in the changing character of the faculties and fields of interests in these academic groups. Formerly, engineering faculties were largely made up of practicing engineers, many on a part-time basis. Today, the faculties of engineering are largely composed of full-time academicians with advanced degrees in engineering science or physics. They are generally engaged in research on subjects that were at the forefront of mathematics or physics only a few years ago (e.g., information theory, stochastic processes, solid state, and microwaves). The lines of demarcation between these three scientific disciplines are rapidly becoming obscured by a process of merging spheres of interest.

This trend toward unification of the sciences extends to biology and medicine through the rapid emergence of biophysics, bioengineering, and biomathematics. These new subdisciplines represent important steps in the conversion of biology and medicine into quantitative science (7); but the lines of distinction between biomathematics, biophysics, and bioengineering must be just as blurred as the borders of their basic disciplines.

The long-standing isolation of health sciences from quantitative sciences and engineering is particularly significant and regrettable in view of the diffuse interactions between many other segments of our modern society, over the past fifty years. A vast proportion of physicians were directly engaged in the practice of medicine having direct contact with patients and little responsibility for conditions outside the realm of medicine. The attention of the medical profession was focused on people suffering from various forms of illness as illustrated schematically in Fig. 1. 1 (A). Similarly, the members of other profes-

sional groups were concerned primarily with problems within relatively circum-
scribed areas. The legal profession was oriented toward disputes and crime with
some interested in politics and legislation. Philosophers were concerned with
ethics and morality, but their influence on the remainder of society was limited.
The sphere of politicians was limited to legislation and public works and the
economists were concerned with the sources and distribution of resources. As
our modern society became progressively more complex, the sphere of involvement of each group spread much more widely [Fig. 1.1 (B)]. For example, the
medical profession has become more actively involved in juvenile delinquency,
narcotics, alcoholism, abortion, and other aspects that formerly were regarded as
problems of law and morality. The subjects are now presented as integral parts
of medical training. Similarly, the problems of preventive medicine, highway
safety, pollution, smoking, have both political and medical overtones. The
growing utilization of insurance and of federal funds for financing medical care
increasingly involves economists and politicians. The greatly expanded sphere of
involvement and responsibility has served to greatly disperse professional
personnel, dilute their efforts, and extend their attention beyond limited and
well-defined professional goals.

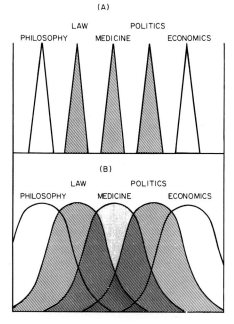

Fig. 1.1. (A) In the past, the various disciplines of science and society were more or
less discrete and isolated entities with obvious interactions but fairly discrete borders.
(B) Modern society is increasingly complex so the areas of concern have extensive overlap as
exemplified by such problems as population, pollution, drugs and health care.

EXPLOSIVE CHANGES IN PERIMEDICAL AREAS

Preparation for the practice of medicine has traditionally involved studies of zoology, heredity, chemistry, and physics. Much has been written and said about the desirability of including in the training of a well-rounded physician courses in sociology, anthropology, forensic medicine, ethics, history of medicine, and other cultural subjects. Despite these well-intentioned efforts, the subject areas peripheral to medicine received relatively little attention by medical students of the past. During the past quarter-century dramatic developments have occurred at the interface between many disciplines related to medicine (8) characterized by diversification or specialization in such fields of interest as comparative physiology, medical genetics, biochemistry, biomathematics, biophysics, and bioengineering as illustrated in Fig. 1. 2. The consequences of these developments can be visualized as impending explosions in other related fields.

Bioengineering can be regarded as the application of principles and the practice of mathematics and physical sciences to living organisms in health and disease. The opportunities for interaction between engineers and life scientists on problems of mutual interest are many and varied, since the concepts, knowledge, and approaches of virtually all engineering disciplines can be employed to study or analyze virtually all the organ systems of the body in various combinations.

The applications of engineering techniques to the practice of medicine have resulted in revolutionary changes in the organization, operational mechanisms, personnel interactions, and financial support of health care delivery. These, in turn, directly influence other aspects of modern society.

The skyrocketing costs of medical care stress the financial base of the total health care system, and the changing patterns of payment for health services and the disproportionate availability of medical care for different segments of the population have developed explosive pressures. They all contribute to the gathering storm of sociological problems, particularly among the disadvantaged in the central cities. The combination of very large expenditures, involvement of the federal government, and evolving social problems leads to political consequences in any industrialized culture or society. The new legislation developed in response to political pressures is an expression of legal involvement. The expanding threats of legal suits for malpractice represent a challenging problem. In addition, some dramatic therapeutic innovations pose legal problems and complex issues of ethics and morality. Consider the medical-legal-ethical problems that arise when the heart is intentionally stopped for significant periods during open heart surgery. The legal and moral rights and responsibilities of the

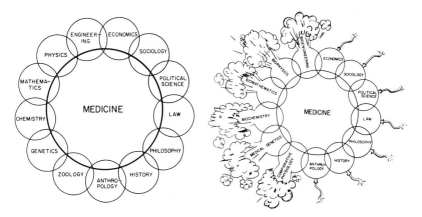

Fig. 1.2. (A) Medicine shares areas of common interest with many other subject areas such as zoology, genetics, chemistry, mathematics, physics—even more than can be represented on this schematic drawing. (B) In recent years, explosive changes have occurred at the interface between medicine and other disciplines, producing many new inter-disciplines such as biochemistry, biomathematics, biophysics and bioengineering. Accelerating progress presages spreading impact on other aspects of society such as economics, sociology, politics, law, and ethics. (Figure reproduced with permission of *Science;* see Ref. 8.)

patient, the donor, the family, and of the medical team involved in a heart transplantation are extremely complicated, essentially unexplored, and decidedly controversial. The solution of such unresolved problems will undoubtedly require all the wisdom we can invoke, past and present. We will be called upon to reconsider the concepts and traditions of the past as recorded by history and anthropology in light of the rapidly changing conditions of the next decade.

The various disciplines illustrated in Fig. 1.2 represent many, but not all, of the areas that overlap medicine. Additional examples might include psychology, architecture, business administration, and many other subjects of great interest. To assuage the reader whose favorite topic is not visible in Fig. 1.2 he is advised to visualize medicine as a sphere rather than as a circle with a wide variety of additional overlapping areas of interest in dimensions that could not be represented in the two-dimensional diagram.

INTERACTIONS BETWEEN ENGINEERS AND LIFE SCIENTISTS

The definition of an academic or scientific discipline is continuously changing since it is characterized primarily by the objectives and activities of men who claim affiliation with or acceptance as a member of the group. The boundaries of the new discipline of biomedical engineering have not been set because they are continually expanding as various categories of engineers and health scientists converge on new problems of mutual interest.

A realistic operational definition of biomedical engineering must have a very broad scope such as the following.

Biomedical engineering can be operationally defined as the constructive interaction of the full spectrum of the disciplines of engineering and life sciences for solving problems of mutual interest.

The opportunities for collaboration between engineers and life scientists are almost unlimited in their number and scope, but certain key problems are particularly susceptible to the application of the concepts and technologies of engineering to the problems of structure, function, and control of living systems. For this reason, the spontaneous development of collaborative or interdisciplinary projects tends to occur in certain major areas which are developing identities as subcomponents of the field of biomedical engineering. The identification of potential points of interaction between engineering and biomedical research can be achieved by listing some of the topics of common interest in living and nonliving systems as illustrated in Fig. 1.3. For example, fluid dynamics, the properties of materials, analysis of structure, heat, thermodynamics, mechanical waves, and vibrations are subjects that are commonly encountered in standard courses in colleges of engineering. These topics are covered, to varying degrees, in courses presented by the traditional academic units such as mechanical, chemical, aeronautical, or civil engineering. Topics such as dynamics and kinetics, instrumentation, computer applications, electrical circuits, control systems, communications theory, are topics of primary concern in electrical engineering, but with components found in the other standard engineering departments. Each of these terms corresponds to topics of prime importance in functional organ systems of living creatures. For example, fluid dynamics in engineering has a counterpart in hemodynamics (fluid dynamics of blood). The properties of materials correspond to the structural analysis and mechanical properties of various tissues under normal conditions (biological structure) or distorted by disease processes (pathology). Virtually all of these topics have implications for the nervous system, the cardiovascular system, the respiratory system, gastrointestinal, genitourinary systems, and the skeletal and muscular systems of the body. In view of the involvement of engineers in diverse biological problems, the ultimate scope of biomedical engineering extends far beyond its current levels of participation at this early stage of its development.

The cross-hatched areas in Fig. 1.3 are intended to imply that the concentration of effort in these various topics is far from uniform throughout the various disciplines of engineering or throughout the various organ systems of the body. The actual extent of involvement or activity is currently a subjective

TOPIC	Engineering Departments					MEDICAL CORRESPONDENCE	Functional Systems				
	Elec	Mech	Chem	Aero	Civil		Neuro	C - V	Resp	GI,GU	Mus-skel
Fluid Dynamics						Rheology Hemodynamics					
Properties of Materials						Mechanical Properties of Tissue					
Analysis of Structure						Anatomy Pathology					
Heat Thermodynamics						Metabolism Temp. Regulation					
Mechanical Waves, Vibrations						Heart Sounds Physical Therapy					
Dynamics, Kinetics Energy Work						Work Energy Trans.					
Instrumentation						Bioinstrumentation					
Computer Applications						Computer Applications					
Electric Circuits						Electrophysiology					
Control Systems						Biological Controls					
Communication Theory						Communication Theory					

Fig. 1.3. The topics of interest to students, faculty and professionals in various engineering specialties (left-hand column) are applicable to problems represented by the many different organ systems of the human body (right-hand column). The concepts and technology of the various engineering disciplines are applicable to a wide spectrum of biological and medical problems as indicated by the corresponding areas of interest.

impression. The spheres of interest among mechanical, chemical, aeronautical, and civil engineers have many common features, but each is capable of unique contributions to biology or medicine. Such a schematic representation provides an intuitive impression regarding some potential interaction between the engineering faculty and the life sciences. More substantial evidence of unequal distribution of effort is presented in Fig. 1.7.

THE CURRENT SCOPE OF BIOMEDICAL ENGINEERING

The opportunities for constructive collaboration between life sciences and engineering are assembled into a schematic representation of the broad spectrum of potential areas of interaction in Table 1.1. Just as biological structure is fundamental to studies of living organs or organisms, so biomechanics is the quantitative physical characterization of tissues and organs in terms of static and dynamic responses to imposed loads and stresses. Similarly, biomaterials is concerned with nonliving substances which contact, supplement, or replace

TABLE 1.1

Basic bioengineering		
Biomechanics	Transport mechanisms	Simulation
Properties of tissues	Mass—transfer	Mathematical modeling
Stress—strain relations	Hydrodynamics	Systems
Viscoelastic properties	Diffusion	Sequences
Tensile strength	Active transport	Interactions
Compliance	Secretion	
Contraction—relaxation	Excretion	Biological models
Damping	Digestion	
	Absorption	Comparative
Biomaterials		Anatomy
Support materials	Energy transmission	Physiology
Artificial joints	Electromagnetic waves	Bionics
Bone substitutes	X-ray	Sensors
Artery—vein substitutes	Ultraviolet	Networks
	Visible light	
Dialysis membranes	Infrared	Control system analysis
Nonthrombogenic surfaces	Microwaves	Neural controls
	Radiowaves	Neuromuscular
Artificial skin	Mechanical waves	Autonomic
	Subsonic	Visceral organs
	Sonic	Glands
	Ultrasonic	Temperature
		Blood pressure
		Hormonal Controls
		Metabolic controls
		Psychological responses

Potential Areas of Interaction between Life Sciences and Engineering

Applied bioengineering			
Instrument development	Therapeutic techniques	Health care system	Environmental engineering
Research tools	Physical therapy	Organization	Pollution
Physical meas.	Surgical	Medical economics	Air
Chemical comp.	instruments	Longrange planning	Water
Microscopy	Respiratory		Noise
Isotope	treatments	Methods	Solid waste
	Radiation	improvement	Human fertility
Clinical	therapy	Support functions	Population control
instruments		Service functions	
Neurology	Monitoring	Nursing	Aerospace
Cardiology	Intensive care	Facilities design	Environment
Respiratory	Surgical, postop	Medical care	control
Gastrointestinal	Coronary care		Closed ecologic systems
Genitourinal	Ward supervision	Operations	Physiological adaptation
Musculoskeletal		research	
	Artificial	Optimization of	Underwater
Diagnostic	organs	laboratories	Compression effects
data	Sensory aids	Support functions	Heat conservation
Automation	Heart–lung mach.	Personnel	Communication
Chemistry	Artificial kidneys	Processing	
Microbiol.	Artificial	Scheduling	
Pathology	extremities		
Multiphasic	Arms	Cost-benefit	
screening	Legs	analysis	
		Cost accounting	
Computer		Evaluation of results	
applications		Beneficial economy	
Data processing			
Analysis			
Retrieval			
Diagnosis			

living tissues. Biomechanics and biomaterials are closely related because the identification or development of substances which are chemically and physically compatible with tissues for implantation in the body demands knowledge of corresponding characteristics of the tissues which they contact or replace. These problems are discussed in more detail in Chapter 8.

Transport mechanisms of living organisms can be considered under mass transfer and energy transmission. Mass transfer encompasses fluid dynamics, mixing, diffusion of dissolved ions, and active transport as occurs in secretion, excretion, digestion, absorption, and similar vital processes. Transmission of energy through tissues is of importance in understanding both functional mechanisms (i.e., electrical currents in excitable tissues) and also as the basis for utilizing and improving access to information about the body from both intrinsic sources (heart sounds or electrocardiography) and external energy probes. These topics are of crucial importance in developing new types of diagnostic instruments (see Chapter 6).

Simulation and mathematical modeling are common approaches to gain greater insight into complex situations or complicated machines. Mathematical or computer models are simplified representations of more complicated systems (see also Chapter 7). Similarly it is possible to identify biological models in lower forms of life which can serve as experimental prototypes of more complicated versions found in higher forms and man. Bionics is the term employed to represent a process by which mechanisms found in living organisms are incorporated into engineering designs to provide improved solutions to technical problems. Considering the diversity of effective mechanisms found in phylogenetic development, this approach has contributed surprisingly little to modern technology as indicated later in this chapter.

Living organisms can be likened to complicated chemical processing plants with a vast array of interesting control systems functioning to maintain balance and smooth functioning of the component parts. The analysis of control systems responsible for regulation of neural reflexes, involuntary function (e.g., blood pressure, and body temperature), or metabolic processes is fertile ground for engineering approaches.

Practical applications of biomedical engineering often take the form of new types of instruments or apparatus, initially for research applications and later adopted for routine use on human patients if sufficiently safe and beneficial. A rich field of endeavor is the development of clinical instruments for use on the various organ systems (see also Chapter 10). Automation in clinical laboratories and efficiently organized diagnostic sequences (multiphasic screening) can provide vast stores of information with great speed. The growing mass of clinical data which can be accumulated from patients requires computers to store, analyze, and retrieve the information in more efficient ways. Engineering contributions to patients are exemplified by artificial organs, prosthetics, physical therapy, and patient monitoring systems designed to follow the course of disease and evaluate therapy. The health care delivery system is being scrutinized and analyzed by methods improvement, operations analysis, cost-benefit analysis in quest of ways to improve the organization and functioning of the system for the benefit of the patient and society as a whole. Some additional examples of these various components of biomedical engineering are presented to provide a more comprehensive picture.

REPRESENTATIVE EXAMPLES OF BASIC BIOENGINEERING

The applications of the concepts and techniques of mathematics, physics and other engineering sciences to fundamental mechanisms of biology or medicine can be regarded as basic biomedical engineering science just as biochemistry, anatomy, physiology, or microbiology are basic sciences for medicine. The description of structure and function of living systems in terms of the physical and physicochemical properties is surely not new, but the recent involvement of engineering concepts and technology to such problems has greatly accelerated progress in many of these efforts. Some examples of engineering approaches to our understanding of the structure, function, and control of tissues, organs, and organisms provide insight into the potential contributions we can expect from the constructive merger of the disciplines involved.

BIOMECHANICS

Movement and mobility are essential characteristics of many living creatures and the many diverse mechanisms of ambulation reflect successful designs emerging progressively over eons of evolutionary history. The motility of amebas, angle worms, fish, birds, and man represent adaptations to the environment by mechanisms of widely varying complexity, yet even the simplest of these defies our ability to fully understand or duplicate. A most humbling experience is to compare the limitations of the most complex substitute for a human hand with its original counterpart in terms of the power sources, the speed, discreteness or accuracy of movement, and the precision of control. In such a comparison, modern technology developed over a hundred years cannot compete with natural evolution over billions of years.

The forces developed by contracting skeletal muscle are very large in relation to the size of the force generator. For example, at a moderate rate of walking, the force exerted on a tendon by a single muscle in the leg (triceps surae) is equivalent to about 4 times the person's weight. The gluteus muscle may exert a force of 1400 lb. If all the muscles of the body combined their tension in one direction, they could develop a force of 25 tons. This may be impressive but consider also the relative forces developed by a jumping grasshopper or a weight-lifting ant. The coupling and interaction of these forces on the various kinds of levers represented by the bony skeleton of man or the exoskeleton of insects constitute problems that can challenge any mechanical engineer.

In addition to the familiar form of muscle which produces skeletal movement in man, vital functions are subserved by more primitive smooth muscle. Such nonstriated muscle is responsible for a variety of completely automatic (nonvolitional) activities such as changing the size of the pupil in the eye, adjusting the caliber of very small blood vessels to regulate blood flow to various tissues, or to propel liquids along various kinds of channels.

FLUID DYNAMICS

The body contains many different kinds of pumping systems serving diverse functions such as digestion of food, excretion of wastes, propulsion of lymph, delivery of secretions, and maintenance of the immediate environment of tissue cells. For example, the urinary bladder has a wall composed primarily of smooth muscle capable of accommodating a progressively increasing volume of urine without increased internal pressure until time for expulsion when it can deliver a urinary stream with considerable force. Another type of smooth muscle expulsion system is the uterus which delivers newborn infants to the outside by rhythmic contractions.

Engineers interested in fluid dynamics have tended to concentrate their attention on the blood circulatory system which is the most obvious hydraulic system in the body. Formidable problems must be overcome to study and analyze the rheology of the pulsatile flow of a non-Newtonian slurry through distensible, branching, tethered conduits. The convergence of engineering talent on this problem has greatly accelerated progress but much remains to be accomplished. Similarly, a major effort at developing an artificial mechanical heart is sharply focusing attention on the remarkable attributes of the normal heart. At a heart rate of 70 strokes per minute, the normal heart beats over 100,000 times a day without any pauses longer than a few seconds for the lifetime of most individuals. It is a self-contained power generator developing energy from chemical stores. Its performance characteristics are difficult to match. For example, each synchronous contraction of the heart muscle ejects blood into its outflow artery with such force that peak velocities range from 100 to 200 cm/sec and the acceleration ranges from 6000 to 10,000 cm/sec^2. We have been unable to locate a mechanical pump that will duplicate this performance so that we can calibrate dynamic response of flowmeters used to study the function of the heart. In the design of artificial mechanical hearts, it is not certain even now whether incorporation of an impulse generator capable of such high acceleration is necessary or desirable for the distribution of blood to the tissues.

BIONICS

The wide diversity of living creatures represents many successful evolutionary experimental solutions to many different problems encountered by modern engineers. For example, the flight of birds served as the prototype for early experiments on machines for manned flight. The differences in power sources rendered the movable wing an inappropriate approach to the problem. In contrast, aquatic mammals and fishes can pass through water with remarkably little drag and appear to represent design characteristics which might well be emulated by marine engineers. A wide variety of biological adaptations were presented at a symposium organized by Jackman and Von Gierke (9). Unfortunately, the advantages of living systems as prototypes for industrial design have generally been recognized only in retrospect. The FM sonar system that bats use to avoid obstacles or catch their prey is a splendid system but did not lead to the development of either radar or sonar. Bats emit very loud ultrasonic pulses (as high as 100 dyn/cm) at 10–20 pulses per second. As he closes range with his prey, this rate increases to 200 per second, improving his resolution on target. His ears are extremely sensitive so that he must suppress the sensitivity of his receiver while he is sending. He can successfully navigate and hunt in a very noisy environment. Moths which are the prey of bats have their own defense mechanisms which include extremely sensitive ears responding to a wide range of frequencies (3000–200,000 Hz). The moth wings are covered with sound-damping fine hair and some are capable of jamming the bat radar with their own ultrasonic emissions. The navigational ability of some birds extends beyond current theory. Among recorded feats is the case of a shearwater, which was flown by plane from England to Boston, and found its way back to its nest (3000 miles) in only 12 days. The migration of birds from nesting places in the arctic to the southern hemisphere and back, in flocks or singly, is difficult to explain. Caged warblers have been seen to orient themselves in the direction they would take during migration no matter how many times the cage is turned. They also tend to orient their bodies in accordance with pattern of celestial stars rotating in a planetarium.

The bombadier beetle is reported to be able to discharge a pungent liquid with great accuracy up to 20 inches using a high explosive rocket fuel (hydroquinone and hydrogen peroxide) which passes through a muscle valve to a combustion chamber of hard chitin where the material explodes. These two chemicals explode spontaneously when mixed in a test tube. Rocketry would have been greatly facilitated had this living prototype been thoroughly studied fifty years ago. Fireflies produce a characteristic glow based on a substance called luciferin which emits light in the presence of an enzyme. Each species of firefly has unique flash intensity, duration, and pattern, which serve to distinguish mates from members of another species. The origin of this cool light is not understood.

Living creatures monitor their environment with many sensors having exquisite sensitivity. The infrared sensors of a pit viper respond to the warmth emitted by a mouse at 5 or 6 feet distance. A breed of moth on Trinidad has such keen chemical detection that the male is reported to respond to a female as much as a mile away; its sensitivity computed to approach detection of a single molecule at a time. The human ear can sense a sound of such slight intensity that the ear drum moves no more than the diameter of a hydrogen molecule. Light receptors in the eye may be capable of responding to a single photon.

Processing of the information picked up by these sensitive biological transducers is highly developed in the nervous systems of animals large and small. Living systems display adaptive controls which are difficult to reproduce by our most complex computers. The pattern recognition capability of living systems cannot be duplicated by existing computer techniques. The quantity of information which can be culled at a glance from a single picture defies analysis. We must learn to take advantage of this remarkable ability in the display of data to facilitate recognition of diagnostic patterns signifying normal or abnormal function (see also Chapter 5). An enormous quantity of information regarding the external environment is processed by the nervous system to produce appropriate responses of various types. In addition a very large quantity of information about the function of internal organs is picked up by sensors within the body which serve as the input to control systems regulating activity of many different organisms in an integrated fashion.

MODELING

The human intellect boggles at visualizing more than two or three simultaneously changing variables in any system, but mathematical expressions can be effectively used to keep track of many interacting factors. A powerful engineering approach to understanding the function of complicated mechanisms is to develop mathematical expressions which incorporate existing knowledge and relationships in a system. The mathematical model can then be inserted into an analog or digital computer in such a way that variations in the component variables produce interactions and change in output which can be checked against the behavior of the original system. The mathematical model is characteristically a simplified version of the original system so the model generally displays deviation from that of the target function. The mathematical model may be used to predict changes which are tested by inducing appropriate changes in the original system so that suitable changes can be introduced into the model to improve its approximation in a series of sequential cycles. Thus utilized, mathematical models help provide insight into the behavior of the system and also give strong clues regarding what measurements and experiments must be conducted on the living system to provide progressively improved understanding. This approach is particularly valuable in assessing and understanding the kinds of control system which are encountered in living systems. Modeling and simulation as quantative approaches to analyzing complex systems are presented in more detail in Chapter 7.

CONTROL SYSTEMS

The ability of man to stand or execute discrete voluntary movements depends upon multiple control loops by which sensors within muscles, tendons, and joints continuously feed into the nervous system information about the tension in the muscles and the spatial localization of the extremities. We accomplish complex and coordinated actions so easily that the intricacy of the necessary control loops is obscured. The problems are thrown into proper perspective by current efforts to develop artificial extremities. Consider the problem of developing an artificial arm capable of driving a nail, threading a needle, and playing the piano. The controls required for coordinated movements are readily observed and appreciated. Less obvious are the controls over the functions of internal organs.

The regulatory mechanisms which maintain the arterial blood pressure within a relatively narrow range have been studied extensively. The basic principle is precisely that which would be employed in a hydraulic system like that illustrated in Fig. 1.4. Sensors on the high-pressure side of the circuit provide an input into a control box with an output which regulates both the pump output (rpm) and the setting of the valves which collectively determine the hydraulic impedance to outflow of liquid. The body is replete with such control loops, but a vast majority remain largely unexplored. The extent to which this is true can be suggested by considering how many different kinds of controls are necessary to maintain the composition of blood and body fluids. The body weight tends to remain relatively constant over periods of days or weeks and the relative quantity of water in the body and its chemical composition also remain within relatively narrow limits despite rather wide changes in the intake of water, salts, and foodstuffs of various sorts. This means

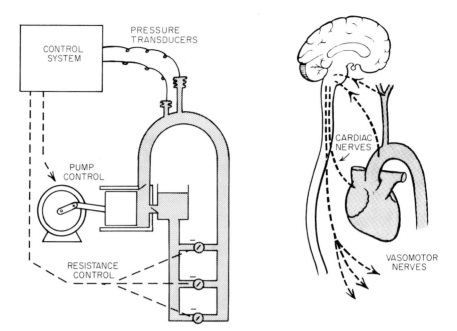

Fig. 1.4. The control of the pressure head in a mechanical hydraulic system by changes in the output of the pump and the resistance of outflow valves has a close correspondence to the mechanisms which adjust the output of the heart and the peripheral vascular resistance to regulate arterial blood pressure.

that the total quantity of each of these substances in the body is being regulated. The fact that the composition of the body liquids tends to resist changes indicates the need to postulate some sort of sensors, integrating mechanisms, and outflow of information to appropriate effector mechanisms. These could take the form of mechanical, chemical, hormonal, or neural controls. Some of these are becoming clarified by intensive investigation of mechanisms like endocrine control over certain electrolytes. The opportunities for further elaboration are widespread and of very great importance in understanding how the body functions (see also Chapter 7).

APPLIED BIOMEDICAL ENGINEERING

The factors which control the blood pressure, heart rate, body temperature, respiratory rate, body weight, body fluid composition, and other elements of vital function are of particular significance in the diagnosis of disease states which induce changes in these controls. The measurement of the significant variables is an essential step in reaching appropriate diagnoses. When patients are acutely and dangerously ill (i.e., from injury, blood loss, coronary occlusion, following surgery, or other serious conditions), these vital functions are monitored to maintain surveillance over the changing condition of the patient. For such purposes, instrumentation must be developed and continuously improved in terms of reliability, convenience, safety, and cost. In addition, new types of instruments serve as research tools opening up new fields of investigation. Bioinstrumentation is one of the major routes by which engineering can favorably influence both the research productivity and the care of patients by the medical community.

BIOINSTRUMENTATION

"The great advances in experimental physics have come from men who have designed specific apparatus to provide an answer to a question formed by a background of theory." This view was expressed by Morowitz (7) who emphasized that to expect unmodified commercial equipment to serve biologists for their research is "an admission that the instrument manufacturer has a better understanding of biological problems than that possessed by the instrument user." Commercial apparatus for amplification of electronic signals, recording, storage, retrieval, or analysis of data, designed and constructed for general scientific application, is also likely to be applicable to medical research. However, the transducers used in industry or in other scientific fields are not usually applicable to biological systems. For example, devices employed to measure pressure, flow, dimensions, or viscosity in an industrial plant or physics laboratory are not necessarily appropriate for research on the heart and blood vessels. One limiting factor in current research is the lack of appropriate transducers. As new research techniques or tools become available and accepted, great numbers of basic science and clinical laboratories converge and rapidly exploit their research potential. Such was the sequence with respect to cardiac catheterization, vectorcardiography, radioisotopes, indicator dilution techniques, electromagnetic flowmeters, artificial heart-lung machines, and so on. A very large research potential is embodied in hundreds of medical research laboratories poised and ready to take advantage of new research tools. The conclusion seems justified that the medical research capacity of this country greatly exceeds the supply of new research techniques and new ideas. The current slow rate of development of scientific instruments constitutes the single most important obstacle in the progress of biology and medicine toward becoming a quantitative science.

SOURCES OF PATIENT DATA

The information required to arrive at diagnosis of specific illnesses stems from a number of sources, including the past history, current symptoms and their development, the physical examination of the patient by the physician, and a wide variety of laboratory tests. An important source of quantitative information stems from the analysis of samples obtained from the patient and subjected to analytical procedures. These samples represent gases exhaled from the lungs, blood, body fluids removed from particular sites, urine, feces, and biopsy specimens of tissues. These samples are tested for chemical constituents or analyzed for cell counts and identification, examined for signs of cancer or cultured to determine the presence of specific bacteria (Fig. 1.5A). To an increasing degree, engineering is contributing to improvements in the testing procedures and in automation of procedures which promises to provide increasingly accurate information at lower cost per determination.

Additional information regarding the location, size, structure, function, and control of internal organs can be acquired by direct exposure (i.e., in surgery) or by inserting needles, catheters, or sampling devices through the skin or into body orifices. These methods involve slight but real hazards to the patient in most instances. A preferred approach would be to analyze intrinsic energy originating within the body or by directing beams of energy into the body and detecting the changes induced during propagation of the energy

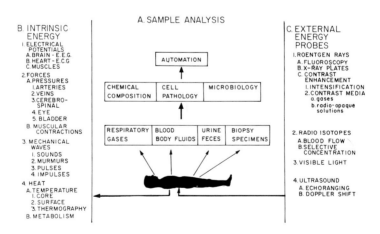

Fig. 1.5. Principle sources of information for diagnostic purposes include (A) analysis of samples from the body such as chemical composition, cellular structure or presence of bacteria; (B) analysis of intrinsic electrical, mechanical or thermal energy generated within the body; and (C) external energy sources which provide information regarding structure and function of internal organs.

through the tissue of interest. Examples of these two approaches illustrate the diversity of opportunities available. There are numerous sources of energy generated within the human body which can be sampled and analyzed to detect changes in the function of internal organs as summarized in Fig. 1.5B. Alternatively, beams of energy can be directed into or through the body from external sources and serve to probe for information regarding the size, shape, position and function of internal structures in the body. These two approaches represent avenues of development by which engineering can be expected to contribute to diagnosis of disease states.

The intrinsic energy sources (Fig. 1.5B) have been extensively studied and exploited as sources of diagnostic information. The widespread utilization of electrical potentials from the heart, brain, and skeletal muscles are generally familiar. The forces developed by contracting muscles are easily tested and provide valuable information regarding their function and control by the nervous system. The pressures recorded in the arteries of the body are important indicators of the status of cardiovascular function and control. The clinical measurement of the arterial blood pressure by sphygmomanometer (cuff) method is dependent upon audible detection of vibrations in the artery below the cuff produced by turbulence in the blood or acceleration transients in the walls. The changes in body temperature indicate the varying balance between heat production and heat elimination of the body.

Extrinsic energy probes provide opportunities to extract information about the location, size, displacement, velocity, and acceleration of structures within the body, but without mechanical penetration of the skin. A common example of a familiar energy probe is the use of X rays to provide fluoroscopic images or photographic plates displaying the size and shape and radiodensity of internal organs. Increased contrast is often obtained by injecting radioopaque dyes into the bloodstream to bring into sharper relief the borders of the heart chambers and blood vessels. Similarly, air can be introduced into spaces in or around the brain to emphasize tumors or displacement caused by them. Iodinated oils may be introduced into the bronchial tubes to show their branching patterns more distinctly. Such techniques represent important technological contributions to accurate and definitive diagnosis antidating biomedical engineering as a source of many more nondestructive diagnostic techniques.

Many tissues transmit light and can be examined directly in a darkened room. For example, the presence of inflammation in sinus cavities can be detected because they fail to transmit light like a normal air-filled sinus. Similarly light transmission through the head of a newborn infant can be directly observed in a search for tumors or evidence of enlarged liquid-filled spaces as occur in hydrocephalus. A beam of light can also be used to evaluate the gas content of the bowel in infants. The blood content of the skin or other accessible tissues of the body (i.e., mucous membranes of mouth or nose) can be estimated or recorded by a photocell responding to changes in reflected or transmitted light by a technique called photoelectric plethysmography. All of the techniques employing light beams can be rendered more convenient, more consistent and more objective by modern technologies including photoelectric cells, various light sources, and fiberoptics for directing the beams.

The distribution of heat over the surface of the body has been shown to reflect the existence of "hot spots" under the skin due to local inflammation or certain types of tumors. Instead of exploring areas of skin with a thermometer or thermocouple, it is now possible to directly display the temperature distribution over the surface by applying thermochromic materials (i.e., liquid crystals of cholesterol esters) which change color over the whole spectrum as the temperature is changed over a narrow range such as $3°-4°C$. Alternatively, infrared cameras produce a picture of the body in terms of temperature distribution over its surface and directly indicate the areas of high temperature (see also Chapter 6).

Ultrasound is a form of energy which can be easily generated and controlled; it travels as a beam or pulse through tissues and is scattered or reflected from particles or interfaces, such as red blood cells, arterial walls, heart valves, heart walls, and a midline structure of the brain. The velocity of ultrasonic transmission through tissues is well known so that the time required for a pulse to travel from the transmitter to a reflecting surface and back to a receiver is an accurate indication of its distance from the skin surface. This information can be directly displayed on the face of a cathode ray tube. The process is very similar to the use of sonar to detect the range of a vessel from a submarine by timing the reflection of sonar pulses from the hull. Ultrasonic beams are now increasingly common as means for locating the midline of the brain, the movement of heart valves, the velocity of blood flow in accessible arteries and veins through the undamaged skin. Such applications and the future prospects are discussed in more detail in subsequent chapters.

CLINICAL ENGINEERING

The direct application of engineering techniques and technology to patient care can be considered in two main categories: data acquisition and therapy. The sources of patient data illustrated in Figs. 1.5 and 1.6 may contribute to diagnosis either directly or by means of computer processing. In addition, many of these sources of patient data can be utilized for evaluation of therapy and following the progress of the illness. Various combinations of recording equipment are being incorporated into monitoring systems for use during surgical operation, postoperative care, intensive care following coronary occlusion or traumatic injury. Monitoring equipment has not been widely accepted as an aid to nursing supervision of patients during general ward care.

The principal contributions of engineering to definite patient therapy has been organ supplements or substitutes of various types as illustrated in Fig. 1.6. In addition, wave energies are commonly employed in physical medicine. Clearly, the contribution of engineering to clinical diagnosis and therapy is at a very early stage of development (see also Chapter 10).

FERTILE FIELDS FOR BIOMEDICAL ENGINEERING

The wide variety of opportunities for applying engineering to pure and practical problems in biology and medicine described above may give the erroneous impression that the coverage of potential applications is fairly complete. The examples illustrated in Fig. 1.5 and other illustrations in this chapter might well support such a conclusion. A closer examination reveals a very pronounced tendency for engineers to gravitate toward particularly attractive areas where they can most readily identify problems and feel relatively comfortable. As a result, very large proportions of the total effort have been confined to relatively few areas such as electrocardiography, computer applications, modeling and simulation, control systems analysis, biomechanics and biomaterials; other areas of great importance have received little or no attention.

Unequal distribution of effort among the many opportunities has an historical basis. Many of the biomedical engineering programs have emerged from within academic departments of electrical engineering in affiliation with receptive members of departments of physiology, or medicine, stemming from

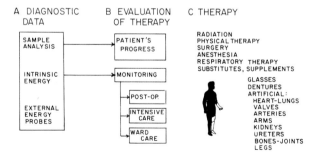

Fig. 1.6. Engineering applications to patient care include the sources of diagnostic data illustrated in Fig. 1.5, many of which are utilized for monitoring the patient's progress or response to therapy. In addition engineering can be applied to therapy in a variety of ways (C) including the development of artificial organs, tissues, and supplements.

the obvious applicability of systems analysis and instrumentation development to the complex living systems. Mechanical engineers and materials scientists have tended to team up with faculty in biological structure, orthopedics, or physical medicine. Chemical engineers find common interest with physicians interested in the lung or kidney function and in the development of artificial substitutes for these organs. As a result of these spontaneous affinities in addition to the availability of research support from specific granting agencies, major concentrations of effort can be identified in a few key areas. One way to analyze this distribution of research and development effort is to obtain samples of current research projects and arrange them into various categories in an organized fashion(8). For this purpose, 1067 abstracts of research projects were obtained from the Science Information Exchange (SIE) by computer retrieval in three main categories, namely, biomaterials, instrumentation, and radiation. The abstracts were then examined for content and assigned to subgroups under these three main headings. In addition the biological target of each was identified in terms of materials, cells, tissues, specific organs and tissues as indicated across the top of Fig. 1.7. Despite the subjective nature of the process, the number of abstracts falling in each square gave an indication of the distribution of effort among the diversity of opportunities. Obviously no two squares represented opportunities for research of equal importance, but the very heavy concentration of effort on diagnostic and therapeutic instrumentation for the heart reflects both an area of wide mutual interest plus support for the development of an artificial heart. In contrast, many squares are remarkably devoid of activity in this particular sampling. Obviously, certain squares may not represent significant or reasonable opportunities for study. These and other tabulations indicated a sharply biased distribution of effort concentrated on the cardiovascular system with a somewhat lesser convergence on the functional aspects of the nervous system. Criteria are lacking to determine what the total amount or distribution of biomedical effort should be, but it seems clear that the many blank squares in Fig. 1.7 contain many potential opportunities for productive collaborative efforts. For example, it seems clear that the contributions of engineering to gastroenterology, urology, dermatology and neurology do not compare in amount or diversity to the concentration of resources being applied to the cardiovascular, respiratory, and muscle–skeletal systems, judging from available data or impressions. Some of the specific opportunities represented by the empty squares will be identified and expanded in subsequent chapters.

The unequal distribution of research effort displayed in Fig. 1.7 indicates that the priority system for problem selection by the investigator or the availability of research and development support is based on criteria other than that of achieving widespread coverage. In fact, the priorities for medical research are very frequently established upon evaluation of the importance of a problem in terms of the incidence of the disease, the mortality of ailments, and to some

Category	Subcategory	MATERIALS	Plant, Micro	Animal	CNS	Heart	Arteries	Microcirculation Lymphatic	Veins	Gas Exchange	Airways	Mouth, Teeth	Gut	Kidney	Urinary Tracts	Reproductive System	Bones, Joints	Muscle	Prosthetics	SKIN	WHOLE MAN
BIOMATERIALS	Structure	30	1	2	7			3									3			1	
	Mechanics	5	2		2	2	6	1		1		20			1		17	1		1	2
	Compatibility, Usefulness	18			5	6	28	1				39	2		1		5		2	5	
INSTRUMENTATION	Research	7	5	30	17	6	11	2		1	3	3	1				1			4	
	Diagnostic	5		9	7	31	11	1	1	4	1	3	1		3			2		1	9
	Therapeutic	8		8	23	145	24		7	28	4	10	4	22	4		3	1	31	15	
RADIATION	EM Long wave	1	3	7	3	1	1	1									1			2	1
	EM Visible	28	25	22	48			2												16	2
	Ionizing – Short wave	35	49	24	6												1			7	4
	Sound	1	2	2	9	1															1
	Ultrasound			6	2		1			2	2						2				

Fig. 1.7. The distribution of research and development projects in the fields of biomaterials, instrumentation and radiation are distributed over many targets extending from materials and cells to tissues and organs, but major efforts are concentrated in a few very prominent areas, leaving many additional opportunities in areas which are receiving little or no attention. (Reference 8 reproduced with permission of A.A.A.S.)

degree the popular or emotional appeal of a disability. A common result is the expenditure of large sums of money in pursuit of research goals for which the fund of basic knowledge and technology have not been adequately developed. Another rewarding approach is to identify those opportunities to apply to problems in biology or medicine techniques or technology which have been fully developed in other fields such as engineering.

NEGLECTED OPPORTUNITIES

The cost analysis of medical service is extremely precarious and requires consideration of many different factors and assumptions. As a result of one such survey (10), the distribution of national health expenditures was expressed in terms of the specific types of service and diagnostic categories related to the different organ systems (Table 1.2). The relation between the costs of medical, nursing, and hospital care is of particular interest. The distribution of money and resources for the care of patients with diseases of these various organ systems bears no obvious relation to the distribution of effort in biomedical engineering research and development in Fig. 1.7. Largest expenditures were for diseases of the digestive tract. This is extremely important to the future of biomedical engineering involvement, which is highly concentrated on the cardiovascular system and negligible in the gastrointestinal system. The very large expenditures for ailments involving the gastrointestinal tract, nervous and mental disease, and injuries are scarcely acknowledged at all in the research orientation of the biomedical engineering community.

One of the major problems in modern society is the skyrocketing cost of medical care. The historical reasons for this are discussed in some detail in Chapter 2. There has been a widespread tendency to attribute these increased costs to the increasing complexity and sophistication of medicine resulting from applied technology. The enormous benefits of scientific approaches to medicine cannot be disregarded or discouraged in order to stem the rising costs. Instead we must employ new techniques to help solve the complications which have been created by undeniable medical progress. The applications of engineering science and technology to medicine are in their infancy; enormous opportunities remain unexplored or unexploited. It is becoming clear that many of the problems confronting hospitals and the care of patients are amenable to engineering solutions. In many instances, the state of the art of engineering or of medicine may not be sufficiently developed for such solutions. If the solutions require commercial development of instruments or equipment, we must be prepared for a lag time of several years before items can be made available in numbers sufficient to make a widespread impression. The most immediate prospect of identifying remediable problems and solutions in hospital and health care facilities is by means of reorganization, optimization, and improved efficiency through modern methods.

HOSPITAL INDUSTRIAL ENGINEERING

In the past, industrial engineering suffered from a very unfavorable image because of the so-called "efficiency expert." A common attitude in medicine is that its purpose is to save the patient's life, not to save his money. This attitude presupposes that the available resources are ample to meet the requirements of the patients who need medical care. This unstated premise is manifestly erroneous. Priorities must be established whenever demands exceed the available resources. Recently, the entire medical community is suddenly confronted with the need to justify or counter the progressively rising costs. It is difficult to assess the costs and impossible to evaluate the benefits since "no monetary value can be placed on human health or life." It is no longer possible to hide behind this philosophical shield. Currently, hospital administrators are accepting the challenge of attempting to establish an equitable balance between the "quality" of care and its cost. For this purpose, the entire medical care delivery team will

TABLE 1.2

National Health Expenditures: Distribution of Expenditures for Specified Health Services by Selected Diagnoses and Type of Service, 1963 [a]

Diagnosis	Total[b] $	Hospital care $	Nursing home care $	Physicians' services $	Dentists' services $	Nursing care[c] $	Other professional services[d] $
Amount, total (in millions)	22,530.0	11,579.0	825.0	6867.0	2369.0	460.0	430.0
Percent total	100.0	100.0	100.0	100.0	100.0	100.0	100.0
Neoplasms	5.7	8.7	3.3	3.0	—	8.6	—
Mental, psychoneurotic, and personality disorders	10.7	17.8	3.6	4.1	—	4.3	2.6
Diseases of nervous system and sense organs	6.3	5.9	21.6	7.4	—	10.0	—
Diseases of circulatory system	10.1	11.0	25.1	10.4	—	15.9	—
Diseases of respiratory system	7.0	6.5	—	11.7	—	5.8	—
Diseases of digestive system	18.5	11.5	1.1	5.8	100.0	10.2	—
Diseases of bones and organs of movement	6.3	4.3	6.3	6.6	—	8.5	89.3
Injuries	7.6	8.6	8.8	8.8	—	6.6	—
All other[e]	28.0	25.7	30.2	42.2	—	30.0	8.1

[a]From D. P. Rice, Estimating Cost of Illness, *Amer. J. Public Health* **57**, 424–440 (1967).

[b]Totals may not add due to rounding in all tables in this report.

[c]Includes the services of private duty professional nurses in the hospital and home, private duty practical nurses, and visiting nurses.

[d]Includes the services of podiatrists, physical therapists, clinical psychologists, chiropractors, naturopaths, and Christian Science practitioners.

[e]The breakdown for each major diagnostic category included in "all other" may be obtained from the author for all tables.

29

ultimately be involved. The hospital is commonly regarded as a totally inappropriate place to utilize techniques commonly viewed as appropriate to mass production in large industry (11) (see also Chapter 2).

The distinction between efficiency and effectiveness are frequently misunderstood particularly with reference to hospitals. Efficiency is frequently expressed as a percentage when it is a ratio between input and output in the same units (dollars, hours, etc.). If the output and input are in different units, the corresponding ratio may be expressed in units per hour, items per dollar, and is termed productivity. In a hospital productivity could be evaluated in terms of the number of meals served per dollar, the number of prescriptions filled per hour, the number of patients admitted per admitting clerk, etc. It is obviously possible to be effective in achieving objectives without being efficient or attaining high production rates without favorable productivity. Discrepancies between these determinants often represent wastage which can be corrected without loss of quality of health care through proper analysis and constructive action. Hospital industrial engineering techniques are readily applicable to the support functions of hospitals as will be described in more detail in Chapter 3. Most of these functions do not involve direct or unique hazard to the patient and can be readily compared to similar functions outside hospitals (food services, laundries, pharmacies, accounting, records, etc.). This responsibility should be included in the field of biomedical engineering. Improved effectiveness, efficiency and productivity in the sensitive realm of direct patient care will require more sophisticated techniques and analysis for which biomedical engineering can make unique and invaluable contributions. Many of the defects or deficiencies in the functioning of a hospital, uncovered by analytical studies, will be amenable to engineering solutions. Design of new processes and operations will be required for which engineers will be required to draw on the extensive knowledge of engineering mechanisms, electronics, computers, and many other segments of engineering science and technology.

There is a growing tendency for hospitals to gain the benefits of hospital industrial engineering by several mechanisms, including consultants, hospital associations, or systems analysts as staff specialists, many being engaged in methods improvement and optimization in utilization of the facilities and services.

Well-established analytical techniques have been developed to probe procedures, sequences and facilities to improve their utilization, productivity and cost-effectiveness to the benefit of all concerned. For this purpose, the process of medical care must be studied in an organized fashion to determine how the individual members of the health team are spending their time, interacting with the available facilities and patients and how their efforts could be rendered more effective and less frustrating. By identifying and correcting activities that waste time, dissipate effort, or contribute little or nothing, the quality of care can be improved with reduced effort on the part of the individuals in the health care system. Of prime importance is the recognition of ways by which the valuable time of the professionals can be spared routine activities and allowed to concentrate on the essential features of care for which their training and experience are uniquely required.

METHODS IMPROVEMENT AND OPTIMIZATION OF HEALTH CARE

The waste of talent and under utilization of highly trained medical personnel in hospitals is displayed whenever procedures are carried out by the most highly qualified person available instead of the individual with an optimal degree of training and experience (12,13). The extent to which physicians, nurses, and other highly trained professionals engage in activities for which their training and experience are not required has been documented by a few studies using well established techniques designed to identify methods of improving procedures of many different types. The need for such studies in hospitals, wards, clinics, pharmacies and service functions is becoming increasingly apparent but the total quantity of solid information remains scanty in comparison with the scope of the problem. Very little is known regarding the distribution of time and effort by physicians in private or group practice, but the few studies which have been conducted reveal startling information. For example a time and motion study of four pediatricians in active practice demonstrated that they devoted 48% of their 8.3-hour working day with about 20 patients most of whom were healthy or had minor illness which did not require the attention of such highly trained physicians (14). In another study a variety of tasks being performed by highly trained pediatricians were identified as being much more appropriately carried out by "medical auxiliaries." The present economic crisis from rising costs of medical care provides a powerful stimulus to engage in more extensive and intensive studies of medical and surgical practice to determine how the valuable time of highly trained professionals can be spared for the tasks and decisions for which they were trained at such enormous expense of time and effort. These opportunities are considered in greater detail in subsequent chapters.

SUMMARY

Applications of engineering to biological and medical problems represent a most promising and challenging opportunity to accelerate progress toward improved care of patients. The potential interactions between engineers and life scientists are extremely varied since they may involve virtually all the disciplines of engineering and the life sciences. Because biomedical engineering is in an early phase of development, its ultimate potential cannot be assessed. One of its most important roles will be to accelerate the transformation of our vast store of knowledge about living systems in health and disease into effective mechanisms for recognizing disease states, administering therapeutic measures and evaluating their success. Early examples of new technologies have already been adopted in hospitals and contribute to their effectiveness but also produce complexity and complications. Biomedical engineering can contribute tools and techniques to patient care and methods for improving the effectiveness of health care facilities. The ultimate role of biomedical engineering must be based on projections into the future. The extent of its potential contribution can be conveyed considering the manpower problems (Chapter 2), health care facilities (Chapter 3), and distribution of health care (Chapter 4).

References

1. Kisch, B. The medical use of scales; an historical remark. *Amer. J. Cardiol.* **5**, 262 (1960).
2. Mendelsohn, E. I. The pre-history of biomedical engineering. *In* "Dimensions of Biomedical Engineering" (E. Salkovitz, L. Gerende, and L. Wingard, eds.). San Francisco Press, San Francisco, 1968.
3. Davis, F. H. Patterns in the distortion of scientific method. *Southern Med. J.* **53**, 1117-1121 (1960).
4. Page, I. H. Physiological thinking and your hypertensive patient. *Physiol. Physicians* **1**, 1-4 (1963).
5. Stark, L., Arzbaecher, R., Agarwal, G., Brodkey, J., Hendry, D. P., and O'Neill, W. Status of research in biomedical engineering. *IEEE Trans. BioMed. Eng.* **15**, 210-231 (1968).
6. Holaday, D. A. Where does instrumentation enter into medicine? *Science* **134**, 1172-1177 (1961).
7. Morowitz, H. J. The relationship between the process of measurement and instrumentation in biology. *Ann. N.Y. Acad. Sci.* **60**, 820-828 (1955).
8. Rushmer, R. F., and Huntsman, L. L. Biomedical engineering. *Science* **167**, 840-844 (1970).
9. Jackman, K. R., and Von Gierke, H. The future of bioengineering in our daily lives. 15th Annual Technical Meeting and Equipment Exposition, Anaheim Convention Center, Institute of Environmental Sciences, April 1969.
10. Rice, D. P. Estimating cost of illness. *Amer. J. Public Health* **57**, 424-440 (1967).
11. Smalley, H. E., and Freeman, J. R. "Hospital Industrial Engineering." Reinhold, New York, 1966.
12. Kissick, W. L. Effective utilization: the critical factor in health manpower. *Amer. J. Public Health* **58**, 23-29 (1968).
13. Bennett, A. C. "Methods Improvement in Hospitals." Lippincott, Philadelphia, 1964.
14. Bergman, A. B., Dassel, S. W., and Wedgwood, R. J. Time-motion study of practicing pediatricians. *Pediatrics* **38**, 254-263 (1966).

Chapter 2

HEALTH MANPOWER PROBLEMS
Potential Role of Biomedical Engineering

The health professions of the United States are confronting a group of crises of unprecedented proportions. Since the turn of the century, the accuracy of diagnosis and effectiveness of therapy have been enormously increased, but patient satisfaction has progressively diminished and mutual confidence between doctor and patient has become strained. Hospitals equipped with increasingly complex equipment and staffed by an ever expanding array of professional and technical personnel have evolved as the center of patient management but they have become overutilized, inefficient, and uncontrollably costly. Despite the largest expenditure on health in the world, the health care system in the United States is inferior to that of several other countries in terms of overall quality and distribution of health care to the total population.

At the center of this controversial maelstrom is the medical practitioner who has persistently defended his unique and sole responsibility for providing for the health needs of the nation. The medical profession which once enjoyed universal respect and confidence is now subjected to widespread complaints and criticism from all sides. The general public wishes access to effective medical care when and where it is needed, administered with sympathy and individual concern and provided at reasonable cost. Paradoxically, this is precisely the kind of service which the medical profession desires to deliver, but no one has developed mechanisms that would make this goal attainable in present circumstances. The factors that prevent attainment of these mutual objectives are very complex and not under the direct control of the individuals or groups concerned; the patient, the doctor, and health profession, the hospital adminis- trator or governmental officials. Some of the problems which created current crises can be attributed to the increased influx of technology. Many of the solutions must come from applications of engineering knowledge, techniques, concepts, and approaches. Rapid scientific and technical progress in medicine have outstripped ability to manage the resulting complexity, or respond to the requirements and demands of society as a whole. Industrialized societies have become such a complex system of interlocking components that changes occurring or induced in one segment have unpredicted effects on others. It is no longer reasonable or effective to confine attention to one ailing portion of the system. Defects emanating from new technologies can be alleviated only by a careful consideration of the various facets of the scientific, engineering, eco- nomic, political, social, legal, and other aspects of the issues. Many of the most important contributions of engineering to society have required detailed plan- ning and interactions with many different components of our increasingly complex social, economic, and political institutions. In recent years, the magnitude of problems facing all of us has prompted some to mount an appeal for a return to the "good old days," even if it means abandoning beneficial products of progress. The clamor for the return or renovation of the old family doctor has been so widespread and persistent that it needs to be placed in proper perspective.

HISTORICAL PERSPECTIVES

Current mechanisms and attitudes in health care are based on long tradition, extending back centuries but undergoing extremely rapid changes in recent years. The peculiarities of modern medicine stem from the distant past and yet they greatly influence both the current conditions and available options for the future. The soothing mists of time tend to obscure the harshness of the conditions in the past. We can learn from history only by retaining a realistic view of past events and conditions.

"OLD DOC" REVISITED

The most vocal critics of modern medicine call for a warm, sympathetic ministration by a doctor who is readily available night or day, with compassion and concern for the patient, his family, and his pocketbook. This combination

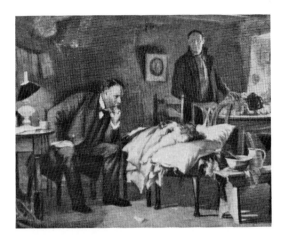

Fig. 2.1. The traditional view of the sympathetic family doctor on call day or night and personally involved with the problems of the patient and his family as portrayed in this familiar painting. From the modern point of view, this physician is watching helplessly as a child dies of diphtheria, something which should never happen today.

of characteristics epitomizes the romanticized version of the old country doctor or the family physician at the turn of the century Fig. 2.1.

Medical practitioners were once well distributed throughout the country, even in isolated places. According to Fein (1) the number of physicians was greater in relation to the population in 1900 (157/100,000) than it was in 1962 (136/100,000). Rural areas had even better distribution of physicians than urban America (2). Hospitals were scarce and most medical care was delivered at home, including much of the childbirth and surgery. Most nursing requirements were met by the extended family. Chronic illness and aging were not such great problems because only an exceptional few lived to a ripe old age; the life expectancy was around 40–50 years. Patients and their families did not expect their doctor to be able to cure most of the ills that beset them. They depended upon him to relieve their minds, shoulder some of their responsibilities, ease their symptoms, sagely announce either that "the crisis is past" or the patient has "passed away." He was able to affect the course of only a few illnesses by his ministrations. His bag contained an assortment of compounds and extracts (mostly of vegetable origin) and of such limited effectiveness that only a few derivatives are in use today (e.g., quinine and digitalis). He could ease some symptoms such as pain or fever but was essentially powerless when confronted by the common illnesses of the day such as pneumonia, tuberculosis, diabetes, typhoid fever, malaria, or the epidemic scourges such as cholera, plague, or other communicable diseases. Surgery was undertaken with hesitation, the outcome was uncertain, and the risks were very great. Thus the mean level of competence in the medical profession was depressingly low at the beginning of the twentieth century. The family doctor was most effective in his handling of personal problems and psychological disturbances. A warm and sympathetic health counselor in combination with a scientifically competent physician would be an asset which we have failed to provide in adequate number (3).

The signs of success in modern medicine are distinctly different from those of most human endeavors. The industrial segment of society produces tangible evidence of their activities in the form of buildings, products or material goods. The value of their contributions can be readily appreciated. In contrast, the crucial criterion of success in medicine is the disappearance or absence of disease. For example, the contagious diseases that used to decimate or threaten large segments of population are largely under control in developed countries such that epidemics of smallpox, typhoid, plague, cholera, malaria, tuberculosis, and poliomyelitis have become rare and are no longer an immediate threat to the public. Pneumonia, which used to terminate the lives of so many weak, aged, or disabled people is now largely controllable with antibiotics. Modern medicine is now capable of directly influencing the course of a large proportion of infections and many metabolic disturbances. The ultimate success of medical care is the elimination of diseases or signs and symptoms so that the product of truly successful care is a kind of void rather than a tangible product. As a result of this peculiarity of medicine, evidence of its most magnificent successes disappears from view and tends to be forgotten by the general public. The failures of medicine or its incomplete successes remain highly visible to all. It is worth remembering that science and technology have been responsible in large measure for the disappearance of many previous threats to health through medical research, active programs of public health, and sanitary engineering. In addition, many technical procedures have become part of the fabric of medicine, including electrocardiography, radiography, and many laboratory tests. In general, these have been produced commercially as an industrial response to a growing market.

INFLUX OF TECHNOLOGY INTO MEDICINE

Roentgen rays were explored for medical applications within a few years after their discovery in 1896. Facilities for roentgenography, fluoroscopy, and radiation therapy have become incorporated into most hospital facilities, large and small, despite the very large expenditures involved. Similarly, electrocardiographs and electroencephalographs and other more sophisticated diagnostic equipment has also become widely used. Batteries of equipment for clinical laboratories, microbiology, and pathology have become standard. Following World War II, the development and acceptance of medical technology accelerated with the commercial production of electronic transducers for pressure measurement, oximeters for blood oxygen saturation, heart-lung machines for open heart surgery, radiation sources and detecting devices for therapeutic and diagnostic applications. Additional wide varieties of new materials and equipment, permanent and disposable, have been made available for hospital, clinic, and office. These developments have resulted from commercial development in response to spontaneously emerging medical markets. Although the nature of medical practice has been drastically affected by progressive adoption of new technologies, the rate at which they have been developed and accepted has been extremely slow in comparison with other segments of society: The engineer does not know what is needed; the physician does not know what is possible.

Lacking a mechanism for identifying the technological requirements of the rapidly developing medical and surgical specialties, the rate of development of commercially produced equipment has been spotty and haphazard. The emergence of biomedical engineering as a recognized university function provided a channel by which engineering techniques and technology could be applied more readily to biomedical research and practice. Active collaboration between engineers and physicians promises to serve a vital role in recognition of problems susceptible to engineering attack, to develop specifications for new and objective techniques, design and evaluate prototypes and channel them into industrial production for widespread utilization.

DIVERSIFICATION OF HEALTH MANPOWER

The progressive and accelerating influx of more sophisticated diagnostic and therapeutic equipment and techniques has caused a major expansion of the numbers and diversity of manpower for health care. This trend developed gradually before World War II but progressively more explosively since that time. In the 10-year period between 1955 and 1965, increase in general population (17%) and of dentists (13%) and physicians (22%) were in about the same proportion while the nonprofessional nurses and medical auxiliary personnel increased by 63%. (See Fig. 2.2 and Reference 4.) Accompanying this very great percentage increase in the numbers of allied health personnel (5) is an expansion in the diversity of their activities (Table 2.1). The very large numbers of professional nurses, practical nurses and aides and orderlies, are supplemented by many different categories of technicians employed to use various types of equipment in clinical laboratories (e.g., ECG and EEG) chemical laboratories, radiology laboratories, histology laboratories, microbiology laboratories, and a wide variety of therapeutic facilities (pharmacy, occupational therapy, physical therapy, speech therapy, inhalation therapy). The many opportunities to provide new and improved equipment in support of these diagnostic and therapeutic

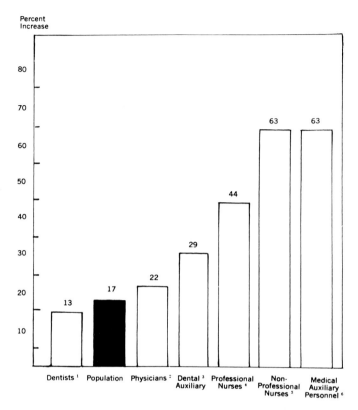

Fig. 2.2. The number of dentists and doctors increased at about the same rate as the general population (black bar) but while nonprofessional nurses' aides, technicians, and clinical laboratory personnel increased more than three times as rapidly between 1955 and 1965 (4).

activities represents an important potential contribution of engineering to health care. These varied activities have called for greatly expanded medical record facilities and personnel (see Table 2.1). Such an impressive array of allied health personnel serves to illustrate the degree and type of change in medical practice since the family doctor engaged in solo practice provided most of the care at home or in the office.

SPECIALIZATION

Only thirty years ago, general practitioners constituted more than 80% of the practicing physicians; only 17% of physicians in private practice were full-time specialists (6). Today, more than 70% of active physicians (including

TABLE 2.1

Personnel Needs in Hospitals, 1966[a]

Categories of hospital personnel	Present
Total professional and technical	1,380,800
Professional nurses	370,200
Licensed practical nurses	148,500
Aides, orderlies, etc.	373,100
Psychiatric aides	156,200
Medical technologists	52,900
Radiologic technologists	23,700
Laboratory assistants	14,400
Medical record personnel	
Professional	6,200
Technical	10,000
Surgical technicians	17,400
Dieticians	12,600
Food service managers	5,300
Cytotechnologists	1,600
Histologic technicians	3,900
Electrocardiograph technicians	5,900
Electroencephalogram technicians	1,900
X-ray assistants	5,700
Occupational therapists	4,600
Occupational therapy assistants	5,600
Physical therapists	8,000
Physical therapy assistants	5,200
Speech pathologists and audiologists	1,000
Recreation therapists	4,600
Inhalation therapists	5,500
Pharmacists	9,500
Pharmacy assistants	5,500
Medical librarians	2,900
Social workers	12,100
Social work assistants	1,500
All other professional and technical	105,900

[a]Estimates for all registered hospitals based on United States Public Health Service–American Hospital Association survey. [column 1] based on 4600 returns. From Reference 5.

those in training) are full-time specialists. At the same time the number of general practitioners has rapidly declined. Specialists in internal medicine and pediatrics provide service related to that of the general practitioner, but the total numbers in all three groups still represent a net reduction in primary care physicians. The momentum of this trend is sufficiently great that powerful forces or incentives will be required to reverse it. In vitually all lines of endeavor,

TABLE 2.2

Specialty Practice in the United States[a]

	Number of physicians in full-time specialty	
	1963	1931
1. Medical specialties		
American Board of Dermatology	3,261	638
American Board of Internal Medicine		
Allergy	831	(1)[b]
Cardiovascular disease	1,703	(1)[b]
Gastroenterology	561	(1)
Internal medicine	34,334	4,003
Pulmonary disease	1,137	465
2. Pediatrics		
American Board of Pediatrics	14,077(5)	1,568
Pediatric allergy	72(6)	(2)
Pediatric cardiology	105(6)	(2)
3. Psychiatry and neurology		
American Board of Psychiatry and Neurology		
Child psychiatry	518	1,401
Psychiatry	15,569	1,401
Neurology	1,788	1,401
4. Surgical specialties		
American Board of Anesthesiology	7,623	152
American Board of Colon and Rectal Surgery (Proctology)	670	195
American Board of Surgery (General Surgery)	25,331	4,320
American Board of Neurological Surgery	1,817	
American Board of Obstetrics and Gynecology	15,683	1,418
American Board of Ophthalmology	7,839	6,410
American Board of Otolaryngology	5,166	6,410
American Board of Orthopaedic Surgery	6,791	609
American Board of Plastic Surgery	990	(3)
Board of Thoracic Surgery	1,291	0
American Board of Urology	4,630	1,346
5. Other specialties		
American Board of Administrative Medicine	3,329	0
American Board of Pathology	7,321	518
American Board of Physical Medicine and Rehabilitation	910	0
American Board of Preventive Medicine		
Aviation medicine	688	0
Occupational medicine	1,732	(3)
General preventive medicine	708	
Public health	1,550	778
American Board of Radiology	8,725	1,005
Part-time specialists	(4)	

TABLE 2.2 (Continued)

	Number of physicians in full-time speciality	
	1963	1931
General practitioners	85,157	125,599
Retired or not in practice	14,747	5,981
Total	276,477	156,406

[a]Data From Reference 6.

[b]The figures in this column *do not* denote the number of doctors certified by the boards.

[c]Key to numbers in parentheses: (1) Included in Internal Medicine; (2) included in Pediatrics; (3) included in General Surgery; (4) the concept of part-time specialist has changed, and this category was not included in recent years; (5) includes pediatric specialties; (6) these data are for 1964 and are not included in totals.

scientific progress is accompanied by specialization, partly because it becomes increasingly difficult for one individual to encompass and assimilate the rapidly expanding fund of knowledge and technology. The number of official and certified specialties is continuously changing and evolving, but a comparison of the number of physicians in various full-time specialties in 1931 and 1963 indicates the nature and extent of the trend (6). The first specialty board, ophthalmology, was established in 1915; many more were added in the 1930's. Subspecialties representing the requirement for special skills and knowledge are emerging within many of the specialties (Table 2.2, reference 6).

As physicians shifted attention from patients as people to "cases" with illnesses, the faculties of medical schools became engrossed in the "scientific" elucidation of problems. Students flocked into scientific training and concentrated in the vicinity of medical institutions rather than in the areas where people live. The extended training required for specialization tended to reduce both the numbers of available physicians and correspondingly shortened their effective span of medical service. The integrating functions of the general practitioner became fragmented into the multitude of specialties, making it necessary to consult several physicians and specialized facilities to handle illnesses that had previously been the province of a single physician. This process greatly expanded the requirements for more doctors, at a time when their numbers were barely keeping up with population growth (2).

Confronted by the wide spectrum of specialties, the patient requiring medical care is understandably confused, particularly with the progressive extinction of general practitioners to coordinate the process of diagnosis and therapy. In effect, patients are being forced into a position of identifying the nature of their own illness with sufficient accuracy to decide what kind of physician in this vast array they should consult.

THE DOCTOR–PATIENT RELATIONSHIP

The unquestioning confidence in the wisdom and judgment of the physician was particularly important when a very large proportion of medical problems could be treated only symptomatically. The beneficial aspects stemming from the doctor–patient relationship must not be discounted because relief can be very real even if the mechanism is psychological. When the physician assumes professional responsibilities for a patient with an illness that cannot be effectively treated by surgery or medicine, he must achieve full confidence of the patient that "everything possible is being done." The less effective the available therapy, the greater the reliance on this intangible bond between the doctor and his patient.

Many factors in our rapidly changing society tend to disrupt the close personal relationship between physicians and their patients. This rapport is most readily established and maintained where the population is stable and the selection of doctors in the vicinity is limited. The major changes in life styles wrought by land, air, and sea transportation plus immigration from rural to urban settings have greatly impaired the maintenance of this relationship. The progressive trends toward specialization in medicine and the widespread development of partnership and group practices have imposed further obstacles. Acceleration of the process occured during World War II when a large proportion of the population moved or migrated and many became dependent upon military medicine. In battle or on shipboard, the primary contact with the medical corps was with a medical corpsman or pharmacist mate. The military personnel had no choice of physician and the system was very largely impersonal compared with the concept of a family doctor.

The military physicians and surgeons were responsible for large groups of young and healthy men who rarely required prompt attention outside of battle areas. Medical officers discovered that life could be more pleasant when night calls did not disturb sleep. Many of them were separated from their practices for several years and took advantage of this break to go to some other location or to change their type of practice. Meanwhile, the civilian doctors were desperately overburdenced and found they could not survive and still respond to calls day and night. They could decline to make night calls with impunity and the patients generally understood. Obviously much larger numbers of patients could be seen either in the office or in hospitals and they could be served far better with the existing doctor shortage. The net result of these and other factors produced a basic change in medical practice after the war was over. Both military and civilian doctors broke with the long tradition of constant availability of doctors and generally declined to be constantly on call day or night. While a large segment of the medical professional community discarded the central theme of the doctor–patient relationship, they remained unwilling to delegate this role to anyone else. Instead, organized medicine fought to maintain a semblance of the traditional doctor–patient relationship by insisting that the patient must pay his doctor directly rather than through a third party. This commercial symbolism proved a poor substitute for a noble philosophy. It tended to convert the public image of a physician from the kindly old family doctor to the source of the healing art by a businessman not too different from any other commercial service.

Rapid improvement in the effectiveness of therapy has undoubtedly played an important role in reducing the need for an effective doctor–patient relationship. The advent of the "sulfa" drugs during World War II, soon followed by the wide spectrum of antibiotics, placed in the hands of physicians weapons to directly combat a very large number of the most troublesome problems, namely bacterial infections. If a patient with an infection visits a physician and receives a bottle of pills that rapidly cures it, he might not be so deeply concerned about who made the diagnosis or who provided the pills so long as the result was prompt and satisfactory. However, the increased effectiveness of medical care has not eliminated the patient's desire for human warmth and understanding of his personal problems.

The impact of these changes on the recipients of health care was subjected to a controlled study in 1955 (7). At that time there was surprisingly little complaint regarding the cost of medical care. (Complaints of high fees were received from only 19% of respondents distributed among all three economic categories.) In contrast, 51% criticized physicians for being unwilling to make house calls or for requiring that sick individuals be brought to the office or hospital. A similar percentage (47%) objected to the poor organization of physicians' practice causing long waits in the office after appointments made for a specific time. The incidence of greatest criticism (64%) centered upon the nature of today's physician-patient relationship. The modern technique-centered practice lacked the warmth of the old-time practitioner (who probably knew less about medicine but more about his patients). This response was not limited to the older age groups but was even more definite among families with husbands under 40 years of age. The patient no longer sees the physician as a special person (Fig. 2.1) but as a member of a team (Fig. 2.3).

These observations can be interpreted to mean that the patients who receive health care are generally more satisfied with its quality than with the manner in which it is presented (7). Applications of new technology to medicine

Fig. 2.3. Patients may view the physician as a member of a team that may be far more effective in preserving or restoring health but lacking in the warmth and sympathy which characterized the old style family physician portrayed in Fig. 2.1.

thus far may have contributed to the effectiveness of health care, but may have contributed to both patient dissatisfaction and the overall cost.

Mechanisms by which patients can gain access to forms of medical service appropriate for the particular illness must be developed either by resurrecting family physicians, developing primary care physicians or by alternative acceptable options (see Chapter 4). The widespread trend toward specialization is reflected in both the organization and function of hospitals, oriented to provide settings in which patients are surrounded by concentrations of specialized equipment and trained personnel.

MOBILIZATION AND UTILIZATION
OF HEALTH MANPOWER

A wealth of opportunities for improving the efficiency of medicine exist in virtually all the medical areas and specialties. They have not been generally recognized because few people have looked. The most obvious examples involve a greater delegation of responsibility to individuals with more appropriate levels of training and the addition of new techniques and technologies to extend the capabilities and effectiveness of members of the health team at all levels.

PHYSICIAN-DIRECTED SERVICES

The widely recognized shortage of physicians has already stimulated mobilization of more allied health personnel. The National Advisory Commission on Health Manpower (4) reported a very significant expansion of the extent to which physicians are being assisted by rapidly increasing numbers of complementary personnel. (See Fig. 2.2 above) The net result has been a very great increase in the services directed by dentists and physicians. They both depend increasingly on trained personnel. The hospital services also increased greatly (65%) so that there was a substantial rise in the availability of health services available on per capita basis. It was predicted that "this recent rapid expansion of health services can be expected to continue through 1975."

Some examples of the numbers and diversity of health manpower are presented for selected occupations in Table 2.3 (6). This list is impressive but far from complete. The rapid growth rates in the various categories are readily apparent by comparing numbers for 1950 and 1960. Note the relatively slower growth rate in the number of physicians (i.e., about 8000 per year in a total of some 275,000). The column listing the number of years of training indicates that physicians generally require a minimum of 8 years after high school and commonly require 12 years as a minimum for specialization. Certain specialties require even longer. Health manpower can be divided into three main groups on the basis of the duration of their training as illustrated in Fig. 2.4. Between these groups are wide gaps in the personnel ladder which tend to exaggerate the differences in the amount of responsibility and authority assigned to the individuals. The exaggerated stratification of health manpower has impeded the delegation of responsibility for patient care by the medical profession. However, an expanding role for specially trained nurses and the training of new types of health personnel (e.g., physicians' assistants) have helped fill some of these gaps. This process may be of great importance in helping to relieve the shortage of physicians.

The development of new positions of importance in health care delivery is proceeding rapidly and must be greatly expanded because there is no possibility of training enough physicians to meet all the current and expected demands on the health care system.

EXPANDING ROLE FOR NURSES

The largest single component of the Health Manpower pool is in the category of registered nurses. In the not-too-distant past, nurses were paid

TABLE 2.3. *Estimated Manpower in Selected Health Occupations*[a]

Profession or occupation	Estimated number of persons in occupation			Approximate number of graduates per year	Years of training required for qualification (after high school)	Average hours of work per week	Average annual income in dollars
	1950	1960	(1962 or 1963)				
Administrators, hosp. (1933)	8,600	12,000	12,500	500	5 to 7	40 (est.)	8,000 to 35,000
Dentists (1859)	87,000	101,900	105,500	3,200	6 to 7	42.8	15,500 (1961)
Dietitians (1917)	22,000	26,000	28,000	n.a.	5	39.5	5,400
Medical record librarians (1898, 1953)	4,000	8,000	9,000	n.a.	5	39.5	5,600
Nurses							
Registered (1896)	375,000	504,000	550,000	30,000	3 to 4	39.5	4,500
Practical	137,000	206,000	225,000	16,600	1	39.5	3,450
Nursing aides and orderlies	221,000	375,000	410,000	n.a.	None	39.5	n.a.
Pharmacists (1852)	101,100	117,000	117,400	n.a.	4	39.5	8,100
Psychologists (clinical, etc.)	3,000	8,000	8,500	n.a.	7	40 (est.)	9,000 to 15,000
Rehabilitation counselors (1957)	1,500	3,000	5,000	n.a.	5 to 7	40 (est.)	7,000 to 13,000
Social workers (1922)	6,200	11,700	15,000	n.a.	6	38.5	6,100
Technologists							
Medical laboratory (1933)	30,000	68,000	n.a.	2,700	4	39.5	4,900
X-ray (1920)	30,800	70,000	n.a.	2,800	2	39.5	4,300
E.E.G. (1949)	n.a.	n.a.	n.a.	n.a.	½ to 1	40 (est.)	3,500 to 4,300
Cytotechnologists	n.a.	n.a.	700	400	3	40 (est.)	5,000 to 7,000
Therapists							
Occupational (1917)	2,000	8,000	8,000		5	40 (est.)	5,000 to 7,000
Physical (1921)	4,600	9,000	12,000		4	39.5	5,600
Speech (1925)	1,500	5,400	10,200		4 to 7	40 (est.)	6,000 (B.A.)
Hearing (1925)	1,500	5,400	10,200		4 to 7	40 (est.)	7,500 (M.A.)
							12,700 (Ph.D.)
Inhalation (1949)	n.a.	n.a.	n.a.	n.a.	1	40 (est.)	5,000 to 8,000
Physicians (1847)	220,000	260,500	276,500	8,000	8 to 12	59.6	25,050

[a]From R. M. Magraw, reference 6.

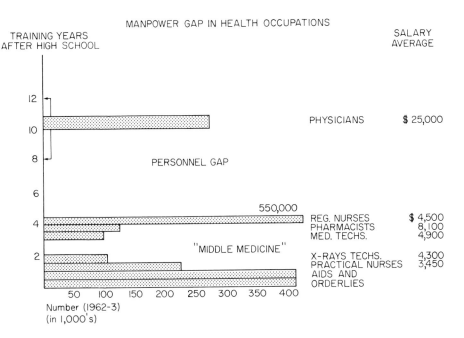

Fig. 2.4. Health manpower concentrates in three groups on the basis of the duration of their training after high school as indicated in Table 2.3. Intensive efforts are being directed toward filling in these personnel gaps. Salary levels have also greatly increased since 1962–1963, contributing greatly to the total costs of health care delivery.

slightly more than janitors and were expected to perform tasks that bore little relation to the extensive training they had received. The average salary for nurses in 1962–1963 was $4500 (Fig. 2.4). This figure has increased greatly, possibly as much as 50% in less than 10 years. In many hospitals nurses have been relieved of many menial tasks and awarded more responsibility. This downward shift in responsibility is but a token of the changes needed. Comments are frequently made that up to one-third of nurses' time is still spent on nonnursing duties (i.e., clerical and administrative functions). An exciting example of the potential contribution of nurses has been brought to light by recent experience in coronary care units which displayed great variability in the mortality rates of patients receiving such intensive care. In the coronary care units where specially trained nurses were given authority to proceed on their own initiative with administration of drugs and defibrillation, the survival rate was greatly improved over the units in which the nurses were required to obtain the approval or help of a physician for such life saving measures. This makes very good sense because a physician cannot always be reached on a moment's notice these days. It is but one example of greatly expanding horizons for appropriate use of nursing talents.

A project has been initiated in which patients attending an ambulatory clinic (i.e., for hypertension, atherosclerotic heart disease, arthritis, obesity, etc.) were divided into two groups, with one receiving all their medical care from a nurse (8). They reported that the nurses were accepted by patients as a primary source of care. The patients appreciated the improved adherence to schedules and lower cost. Similarly, nurses serve as anesthetists in many hospitals. The development of improved monitoring devices to provide more precise and reliable indication of the patient's condition could greatly increase the proper utilization of nurses and others with special training to serve this vital function. A variety of trials are underway to evaluate expanded roles for nurses in handling patients with chronic disease, including long term anticoagulation, diabetes, tuberculosis, alcoholism (9). The role of the nurses in prenatal and well-baby care has been mentioned (10). One follow-up clinic for rheumatic fever utilizes nurses rather than physicians as the major patient contact.

The total pool of trained nurses is very much larger than the 550,000 listed in Table 2.2 because a very large percentage of them are not working at their profession for many reasons. It seems entirely possible that a major upgrading of the role of nurses could draw substantial numbers of nurses back into active practice. Nurses appear to be a most effective source of sympathetic care and concern in the area where this is needed most, namely, the medically disadvantaged in central cities and in areas where care by physicians is seriously deficient. Such opportunities can be readily explored under the direction of qualified physicians.

Until about 15 years ago, the infant mortality in the United States was among the lowest in the world. In the decade of 1950, the decline in mortality rate slowed down and now is about 25 per 1000 live births (11), substantially worse than many other countries in Western Europe. Like most of these other countries where deliveries are characteristically conducted in the home (about 80%) midwives give prenatal and natal care as independent practitioners (11). The infant mortality rate is low (about 15 per 1000 live births in The Netherlands) with corresponding conditions in Scandinavia, England, Wales, and many other well-developed countries. A study of 174 countries published in 1966 noted 600,000 midwives in 153 countries covering 75% of the world's population (12). In contrast, midwives have been very rare in the United States where obstetricians deliver large numbers of babies in hospitals. "Thus do many obstetricians become the highest paid midwives in the world" (13). Obstetricians generally ascribe the higher mortality rates in this country to the disadvantaged peoples who have little or no medical care during pregnancy. However, there remains a distinct possibility that excessive ministrations by medical doctors contribute by the acceleration of labor, the use of anesthesia and the occasional application of forceps. These are hazards to infants not found in the usual uncomplicated delivery handled by midwives or without any attendance. In the

United States there may well be fewer than 500 midwives, with probably less than ten programs training nurse-midwives. Barnes (14) drew the following distinction between the obstetrical nurse that remains an assistant to a physician and a nurse midwife who assumes the responsibility herself."She shakes hands with the patient, looks her in the eye, and assumes responsibility for this woman's health and welfare, and the health and welfare of her baby during the coming time." Documented experience in many countries has clearly demonstrated that the common reliance on physicians for obstetrics in the United States may represent a substantial waste of highly trained medical manpower. However, there is growing likelihood that both the medical profession and the many maternity patients will readily accept modifications of such traditional patterns.

PHYSICIANS' ASSISTANTS

A variety of programs are rapidly evolving to fill a major gap in the manpower pool with new types of personnel trained specifically to fill various roles implied by the term physicians' assistants. The number of these is not accurately known because they are developing spontaneously and rapidly in many different settings. Some twenty programs in as many institutions were mentioned in a recent report (15). These included a four year baccalaureate program, child health assistants training, medical specialty assistants for medical and coronary intensive care units, ophthalmic assistants, anesthesia assistants, emergency medical technicians. Many ophthalmic assistants learn their duties "on the job" without benefit of previous training (16). One particular advantage of ophthalmology is its firm basis in physics of optics and the availability of quantitative tests which provide numerical data subject to specific verification. This feature may have been responsible for the well-established role of the optometrist in the role of testing and prescribing for defects of visual acuity (see below). Presumably, progressive development of quantitative measures of function in other fields of medicine and other organ systems will provide a correspondingly firm base for the careers of other types of physicians' assistants (see also Chapter 6).

New health personnel are expected to join in new combinations with traditional personnel and with new machines, such as autoanalyzers, computers, and other instrumentation. New types of assistants include those performing functions now usually performed by physicians. Three categories of physicians' assistants have been identified (17). Type A is conceived as being capable of approaching the patient, collecting historical and physical data, organizing these data and presenting them so clearly that the physician can visualize the medical problem, and arrive at diagnostic and therapeutic decisions. They might be called physicians' associates. Usually functioning under medical direction, they might function without immediate surveillance on specific types of problems. Type B assistants are not expected to be so fully prepared with general medical knowledge; they would be exceptionally skilled in some clinical specialty, beyond that possessed by Type A but not well enough trained to attain the degree of independence described above. Ten different programs were summarized by an ad hoc committee of the   National Academcy of Sciences (17).

An important source of manpower for some programs is the pool of trained medical corpsmen who leave the armed forces each year, many of whom have had extensive training and experience under stressful conditions requiring good judgment. Many of these corpsmen have had course work in basic medical sciences and invaluable on-the-job training and experience in hospital and outpatient settings. Most of the military medical corpsmen are totally lost to the health professions when they return to civilian life. The extent to which medical corpsmen can extend the services of physicians is being actively evaluated in the Medex Program under the direction of Richard Smith (18). Selected corpsmen with extensive experience have been given intensive training and then paired with interested physicians mostly engaged in active practice in isolated rural areas where the demands are very heavy. Under direct supervision of the physician the Medex trainee assumes those responsibilities delegated to him and mechanisms are being set up to provide immediately accessible communication. The general practitioner is committed to hire the Medex trainee at the conclusion of his preceptorship. Under these rather optimal conditions, the program appears to be highly successful and points the way toward one possible remedy for situations in which medical care is sorely deficient or unavailable because of inadequate health manpower. If the program is successful under these conditions, its applicability to other types of medical practice and service will undoubtedly follow quickly. It must also facilitate the downward transfer of functions to nurses and other types of health team members who can be trained for specific purposes and function under direction. The manpower pool of medical corpsmen has been variously estimated at around 6000 per year (18). Clearly, the civilian educational system must be geared to provide corresponding or necessary education for a broad spectrum of men and women trained specifically to relieve the overburdened physicians of tasks which do not require such extended training. A challenge to bioengineering is to provide technical supplements to physicians' assistants which will allow them to play increasingly valuable roles.

PHARMACISTS AS COUNCILORS ON THERAPEUTIC AGENTS

The profession of pharmacy was established for compounding medicines, powders, lotions, pills, and elixirs prepared specially for each patient. The effectiveness of many of these concoctions is now highly suspect and in many cases they have been largely replaced by a bewildering array of pills, capsules, and formulas distributed ready for consumption by large drug companies. Information about these drugs is presented to physicians by highly biased advertising and detail men, who encourage physicians to use samples of trials. The relative merits of various preparations are difficult to glean from the scientific literature, and most physicians are too busy to search out critical evaluations of the many types of drugs required for treatment. It appears that the original role of the pharmacist has been largely transferred to the mass producers of drugs and remedies. The training in colleges of pharmacy does not reflect this changing role or prepare its graduates for a more significant role in the health care system. This training includes courses in basic medical science (physiology, anatomy, pharmacology) and knowledge about a wide spectrum of therapeutic agents. Furthermore they directly dispense enormous quantities of remedies that do not require prescriptions.

In situations where physicians are scarce or not available, the pharmacist may be the most accessible representative of the health care system. The pharmacists numbering over 120,000 could play a more significant role in the delivery of health care. At present they dispense remedies for what the patient believes is a minor illness. He is also a kind of councilor in therapeutic materials, but his advice is rarely sought by physicians. The pharmacist is far more likely to be acutely aware of the enormous discrepancy between the costs of the same proprietary drugs sold under different labels than either the doctor or patient. A study of the current role of pharmacists might well be conducted to determine the current and future prospects for more significant position in the scheme of medical care. It would probably indicate that pharmacists are overtrained for their current function of dispensing prescriptions, and are undertrained for the role of advising physicians regarding therapeutic effectiveness, or patients regarding treatment of minor illnesses with over-the-counter drugs. Modifications of the curricula in the colleges of pharmacy might be recommended to build additional competence upon the current base of course work. For example, the University of California has developed a curriculum for Doctor of Pharmacy with significant expansion of the basic medical sciences and hospital pharmacy including clinical applications that could provide a nucleus of a new breed of clinical pharmacist. If the pharmacist could achieve a status of therapeutic consultant for both physicians and patients while enjoined to refer the more serious kinds of illness, a substantial increment in the health care system might well be covered with minimal distortion or expansion of current programs.

OPTOMETRISTS

Medical specialists trained in the management of diseases of the eyes prescribe glasses on the basis of quantitative tests for the nature and severity of refractive errors. The number of these ophthalmologists is grossly insufficient to fulfill this role and their efforts are supplemented by some 16,000–20,000 optometrists who are trained for testing and fitting of glasses in a four-year undergraduate course sequence in colleges of optometry. Their importance can be visualized by imagining the consequence of suddenly legislating optometrists out of existence. Although any physician is legally entitled to examine eyes, it is clear that the medical profession could not readily assume this additional role. One reason for the widespread success of optometrists in this paramedical role is the fact that they are equipped with optical testing devices that can measure quantitatively the refractive errors of the eyes in a way which can be checked and verified.

They can elicit numerical measurements of the visual acuity and the ability of the eye to accommodate to near vision. Refractive errors of the eye can be specified quantitatively, in terms of spherical or astigmatic corrections. Deviation of the eyeballs because of imbalance of the extraocular muscles can be specified. The measurements of the nose and interpupillary distance required to fit the glasses are directly measured. It is very likely that success and widespread acceptance of optometry is related at least in part to its quantitative basis. This observation might serve as a stimulus to bioengineering development of quantitative methods for others serving the role of physicians' assistants or paramedical personnel.

EMERGENCY CARE BY POLICE, FIREMEN, AND AMBULANCE DRIVERS

Accidents are the most common cause of death and disability in very large segments of the population. In 1965, 104,000 people were killed by accidents (46,000 by motor vehicles); 52,000,000 were injured (3 million in motor vehicles).

Three groups of individuals are most commonly in direct contact with a large fraction of these injuries, the ambulance drivers, police, and firemen. They represent identifiable groups who could save many lives if properly trained and equipped with appropriate equipment. Opportunities for improving emergency medical care are considered in more detail in Chapter 3.

ROLES OF BIOMEDICAL ENGINEERING

The greatly expanded diversity of health occupations indicated in Tables 2.2 and 2.3 reflect a progressively wider utilization of physician-directed services, many of which result directly from adoption of new technologies. A major goal of biomedical engineering must be the continued development and improvement of the many diagnostic and therapeutic tools of health care to optimize their effectiveness, reliability, safety, and overall benefit. In this way, the time and effort of the corresponding manpower can contribute maximally to the welfare of patients with minimal cost and risk. Increasingly complex and comprehensive testing devices are being added to the clinical laboratories. The medical and surgical specialties continue to expand their funds of knowledge and technical capability by progressive adaptation of newly developed tools for research, diagnosis, monitoring, and therapy. Biomedical engineering is an appropriate mechanism to contribute to these processes.

IMPROVED UTILIZATION OF HEALTH PERSONNEL

With the expanding role of technology in medicine, the number and diversity of personnel and organizational relationships have expanded greatly. Compare the function of the solo family physician in Fig. 2.1 with the medical team schematically illustrated in Fig. 2.3. With the increasing complexity of health care facilities, there is growing need for appropriate applications of proven techniques of management engineering, or operations analysis of these groups for two main purposes—(a) to determine the most appropriate and effective organizational and spatial relationships for achieving and maintaining high levels of effectiveness, and (b) to identify opportunities for use of new techniques or technology to improve performance and job satisfaction. This process is in such an early stage of development that it is not yet easy to find many specific examples. However, one or two will suffice to indicate the potential rewards of this application of engineering analysis of complex functions.

The physician is the highest paid member of the health team. In recent years, interest has been aroused in the improved utilization of the physician's time. The doctor's training is specifically designed to prepare him to collect and analyze data, identify disease states, select appropriate therapy and follow the course of illness. The medical profession has effectively resisted any considerations of cost-effectiveness or operations analysis designed to improve the efficiency of medical practice. The kinds of time-motion studies which have proven of value in other segments of the economy (19) have rarely been applied to aspects of the health care system (see Chapter 1), but their techniques can provide most revealing information when applied to a branch of medicine. For example, Bergman et al. (20) studied the activities of four pediatricians during their normal practice to determine the distribution of their time among various activities. It was found that 50% of the pediatrician's working day was spent with well children and 22% on patients with minor respiratory illnesses. "Intellectual understimulation" resulted from spending a majority of time with children who did not require the talents or training of a physician. The results of this simple study confirm the experience of many others that the vast bulk of the pediatrics practice revolves around supervision of well children and minor respiratory tract infections. Other demands for time and attention include advice regarding formulas, diet, behavioral problems, allergies, immunizations. Many of these activities can be accomplished as well or better by an experienced, mature woman with limited formal training but with medical supervision. The logical reaction to such studies is to delegate responsibility for the nonmedical aspects of practice to individuals with a more appropriate level of training, relieving the physician of the need to dissipate his efforts at routine and menial activities and conserving his talent for dealing with problems for which he is uniquely trained, namely the care of the sick.

A Pediatric Nurse Practitioner program has been established by Silver et al. (21) to provide specific training in interviewing techniques, counseling of parents, assessing and managing many minor problems including upper respiratory infections, gastrointestinal upsets, skin eruptions, and common contagious diseases. These problems represent a very large proportion of current pediatric practice and have been shown to be well handled by nurses, to the mutual satisfaction of patients, parents and physicians (see below). The nurse–patient relationships can be just as satisfactory as a doctor–patient role. By relieving physicians of these aspects of pediatrics, his skill can be directed to the more complex and serious illnesses for which he must take direct responsibility.

Critical evaluations of the physician's role have been rare and largely unnoticed in the past. Pediatrics is a particularly obvious choice as an opportunity to greatly increase the efficiency of the physician by sharing responsibility with other members of the health community. Obstetrics is a prime prospect as discussed in preceeding sections. Other examples of inappropriate utilization of physician's time are easily discovered in other medical and surgical specialties. They present opportunities to divest highly trained physicians of unnecessary and undesirable routine activities to the mutual benefit of all concerned. This may take the form of simple devices. For example, Drui et al. (22) conducted a time and motion study on radiologists which revealed that they expended a substantial proportion of their time manually removing X-ray plates from envelopes and positioning them on view boxes as they were reading and interpreting them. By designing a mechanized display system, the radiologist could merely press a button operating a series of simple racks which emerged from a container carrying sequences of X-ray plates for viewing (Fig. 2.5). By this means, the time of the radiologist was conserved by eliminating unnecessary manual routine and directed to the interpretation of X rays for which he was specifically trained. This is an example of a technological advance stimulated by a systematic study of a process leading to the design or procurement of specified equipment to solve the problem. The device

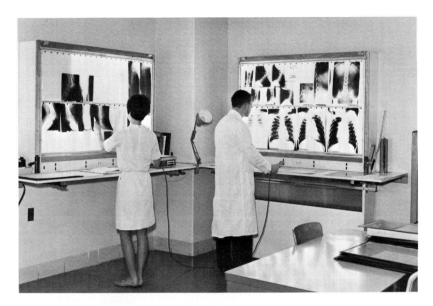

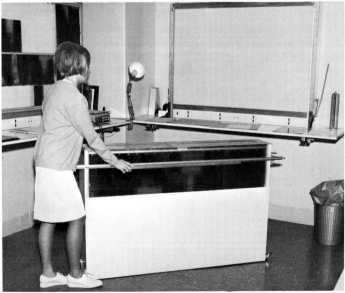

Fig. 2.5. A radiologist can greatly improve the speed and ease with which he interprets X-ray films which are placed for viewing manually by a technician (above left) or by an automated viewing screen as an example of how technology can increase manpower utilization. (Photograph reproduced through the courtesy of Professor Albert Drui.)

illustrated in Fig. 5 is a local counterpart of a device produced by the Philips Company (AutoAlternator) which has been enthusiastically utilized in Sweden.

A very large portion of current technology is concentrated within hospitals and health care facilities containing many opportunities to improve the utilization of personnel, equipment, facilities and organizational relationships.

OPERATIONS RESEARCH

Operations research has been carried on for centuries under the name of "organized common sense" or the "scientific method," but it was first awarded this name by M. S. Blackett in Britain about 1938. It can be defined simply as a study of man–machine systems that have a purpose but more accurately as a miltidisciplinary analysis of processes in relation to a stated purpose and a measure of merit. Originally the process was introduced to evaluate effectiveness and optimize methods of warfare and the measure of merit was frequently the fraction of approaching aircraft shot down by an antiaircraft system (23). The relevance of operations research to health care delivery systems is not immediately obvious until examples are encountered like the problems of scheduling, optimal utilization, computer simulation of processes like those found in medicine. The human engineering problems faced by pilots in aircraft are also being encountered in the displays of information for the surgeon and anesthetist during major surgery or the team operating an intensive care or coronary care unit. Experience in methods for acquiring, editing, analyzing, storing, and retrieving information of many different types in industry or government may be extremely relevant to the increasingly complex problems posed by the flood of diagnostic data that currently accumulates during routine handling of patients in major medical centers.

Flagle (24) identified three major points of contact between operations research and health care: (a) systems analysis of scheduling patterns of hospital resources; (b) decision-making in medical diagnosis and therapy; and (c) work study and methods improvement (see also Chapter 3). For example, the flow of patients in erratic and random fashion through clinics and wards was recognized early as conforming to classical concepts of queueing as they were analyzed for the patiently waiting lines of Britons, and subsequently for multiple services. Studies of wards and clinics at Johns Hopkins Hospital led to the concept of progressive patient care (intensive, intermediate and self-care). Although one might picture a hospital as containing orderly processes of observation, analysis, decision, action, and feedback, such a picture is unrealistic. Rather, many different physician-patient relationships simultaneously generate a wide variety of demands on the hospitals services. Mechanisms are set up by which the hospital resources respond to the physicians orders without any high-level administrative decision, but by an extremely costly communication process accounting for about ¼ of the hospital operating costs.

The decision-making process is central to the screening, diagnosis, or therapy of individual patients. In one sense the physician is engaged in a game with nature as his adversary. He has knowledge and experience concerning a wide variety of disease patterns and a battery of screenings and diagnostic tests that are generally imprecise and with somewhat uncertain relevance. The most crucial deficiency is any acceptable yardstick or measure of merit on which to base judgments regarding cost of errors or the value of success. The enormous increase in the volume of quantitative data from the many sources described in subsequent chapters should provide opportunities and develop more effective guidelines regarding the efficiency of various approaches and the cost-benefit of various aspects of medical care (see also Overall Summary, page 375).

SYSTEMS ANALYSIS IN HEALTH CARE DELIVERY

The next ten or fifteen years will undoubtedly witness many necessary changes in mechanisms for administering medical care. Various approaches must be weighed very carefully before decisions are made to avoid unpredicted complications. Undesirable choices are extremely expensive because of the enormous investment in facilities equipment and manpower. Systems analysis is a valuable process for identifying in advance the possible consequences of various choices. The essential tool of systems analysis is the mathematical model or simulation, which is described in more detail in Chapter 7. The construction of a mathematical model requires an interaction of individuals with intimate knowledge of the system under study, administrators who must make it function and analysts competent in the process of transforming the essential and known components into a mathematical expression which can depict their behavior under various relevant conditions. Murray (25) described how models may be created in a form that is intended to provide appropriate answers to questions of two types—(a) decision, and (b) innovation.

An ambulatory medical clinic operated in a limited number of examination rooms is an important component of a university teaching program. A study was conducted in anticipation of construction of new facilities to modify (optimize) the clinic's teaching function, staffing, and method of operation. One question called for a decision regarding the consequences of linking the emergency room to the medical clinic so that patients were screened for previously untreated illness in the emergency room and referred to the medical clinic. The decision hinged upon the question of whether the additional service would overload the clinic staff. The question did not dictate a choice but provided additional information so that various alternatives could be considered.

The implications were pervasive, including effects on the number of new patients, follow-up patients, laboratory facilities and other factors which rendered the problem complex.

A second question stemmed from the interest in increasing the educational experience of medical students in the ambulatory clinic. This called for innovation and required projection of the consequences of the changes in clinical facilities and organization on the number of medical students that could be effectively handled. When innovation is the prime objective, the model can be used to search various alternatives for the most promising approach.

Systems analysis sometimes leads to mistakes or disappointments attributed to one of the three following types of inappropriate applications (25)— (1) The problem assumed to exist is actually bogus; (2) the model's outputs are tangential to the issues of the problem; and (3) the model is relevant in output but is so comprehensive and tedious to build or use that it collapses under its own weight. The potential applications of simulation to many different basic and applied problems are indicated by examples described in Chapter 7.

TECHNOLOGICAL VOIDS IN MODERN MEDICINE

Thirty or more years ago patients sought the aid of a physician for quite different reasons than are prevalent today. Medical care and training tended to be geared to handle serious illness of threatening nature. Hospitals tended to be filled with a much larger proportion of patients with overt and well developed signs and symptoms. With expanded knowledge, improved technology, and higher incidence of successes, the ability to "save patients' lives" became the hallmark of the physician to an increasing degree. During that period, patients with minor illness or annoying symptoms such as low back pain, dysmenorrhea, indigestion, headache would rarely consult a physician. They relied mainly upon home remedies, patent medicines and stoicism. With the heightened reputation of physicians as healers, the expectations of the general population increased so that growing numbers anticipated corresponding benefit for minor illnesses that are usually not influenced or improved by a visit to a physician. Health care delivery now involves a much broader spectrum of illness than previously confronted the physician extending all the way from brain tumors to headaches, broken bones to blisters or intestinal bleeding to acne. Increasing proportions of physicians' time (probably more than 50%) are devoted to patients who expect improvement or help for minor disturbances but are not likely to experience any benefit.

The most spectacular and successful recent endeavors of both medicine and biomedical engineering are to be found in handling certain life threatening emergencies (i.e., major surgery, artificial heart-lung assists, intensive care, shock

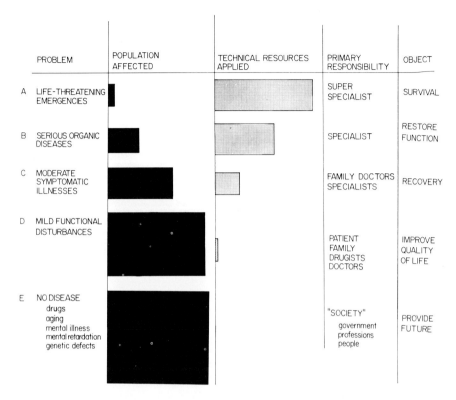

PROBLEM	POPULATION AFFECTED	TECHNICAL RESOURCES APPLIED	PRIMARY RESPONSIBILITY	OBJECT
A LIFE-THREATENING EMERGENCIES			SUPER SPECIALIST	SURVIVAL
B SERIOUS ORGANIC DISEASES			SPECIALIST	RESTORE FUNCTION
C MODERATE SYMPTOMATIC ILLNESSES			FAMILY DOCTORS SPECIALISTS	RECOVERY
D MILD FUNCTIONAL DISTURBANCES			PATIENT FAMILY DRUGISTS DOCTORS	IMPROVE QUALITY OF LIFE
E NO DISEASE drugs aging mental illness mental retardation genetic defects			"SOCIETY" government professions people	PROVIDE FUTURE

Fig. 2.6. The most extensive application of technology is for life threatening emergencies with superspecialists working for survival of a very small segment of the population. Technology plays a decreasing role in the more common lesser illnesses which affect the large bulk of the population.

centers, coronary units) as illustrated schematically in Fig. 2.6. Extensive technical resources (equipment and people) are mobilized for the benefit of a relatively small proportion of the population at extremely great expense. The responsibility for such care is on "superspecialists," men who are highly trained in a component of a medical specialty, and their objective is survival of the patient with only secondary concern for his future functional status, or the cost. A realistic cost-benefit analysis of such efforts would bring into sharper focus key questions regarding the value of human life. Serious organic disease (Fig. 2.6B) is the main province of the various types of specialists who deal with a much larger proportion of the population benefitting from technology to a major degree through clinical laboratory testing and sophisticated therapeutic measures. A large proportion of the symptomatic illness of moderate severity is handled by family physicians or by specialists depending upon circumstances or on patients' choice. The objective is full recovery of function and the success rate is very high in some categories such as infections or replacement therapy for metabolic diseases. Many patients receive relatively little benefit from consulting physicians. In either case, the number of patients is very large and the utilization of technology very small.

The mild functional disturbances account for an extremely large proportion of patients seeking medical advice and help, particularly in recent years of increasing affluence. Many of these ailments, real or imagined, may interfere with the quality of work, play or living but do not pose threat to life or employment. The patient or his family decide whether to ignore the annoyances, purchase proprietary drugs or seek medical advice. A very large proportion of such patients will obtain little relief from the physician whose major role is to ascertain that the symptoms are not signs of more serious illness. The mild and moderate illnesses (Fig. 2.6D) represent an extremely large and growing medical clientelle whose expectations from medicine are often not realized. These groups are responsible for an enormous surge of new demand from expanded access to medicine (i.e., Medicare, Medicaid, and imminent expansions of subsidized insurance plans). The medical profession would be well advised to undertake an educational process for the entire population to explain more candidly what the symptoms mean and what patients can expect from medical care of patients with a very large spectrum of such symptoms (low back pain, dysmenorrhea, headache, fatigue, muscular pain, etc.).

It is time that the medical profession rebuilt its credibility by disclaiming common and misleading promissory statements (e.g., that donations to charity will cure cancer, prevent birth defects, eliminate heart disease etc.,). To avoid being totally engulfed by the deluge of patients in the future, intensive effort must be directed toward new and more effective mechanisms for identifying, channeling and handling the mass of individuals with forms of illness for which a visit to a physician is unrewarding.

Currently all aspects of the health care system are engulfed by an influx of patients with real or imagined problems for which the physician has no ready answers and cannot turn away. As a result, patients may regard the physician as unsympathetic or incompetent; the physician may regard the patient's problem as a trivial intrusion on a busy schedule. There appears to be need for new definitions and categorization of sickness with an altered distribution of responsibilities so that the bulk of the minor ailments for which medicine appears ineffectual could be handled to a greater degree by individuals with much lower levels of training. Provisions must be made to assure that patients with significant and treatable diseases are consistently referred to physicians.

Masses of people without disease in the formal sense (drug abuse, aging, mental illness) pose severe problems for which medicine in conjunction with the rest of society must seek answers (Fig. 2.6). This is an expression of the extensive overlapping responsibility between the many disciplines and professions as illustrated in Figs. 1.1 and 1.2. The fact that these problems are not and cannot be the sole responsibility for the medical profession should be definitively established through mobilization of all the available talent to tackle these vexing issues which confront us all. The current reliance on general hospitals or doctors' offices for the administration of most medical care must be broadened to include effective utilization of a broader selection of facilities designed to accommodate individuals with illness of various types and degrees of severity (see Chapter 3).

Additional medical manpower can be mobilized by upward mobility of personnel and downward assignment of responsibility. By such techniques, nurses, physicians' assistants and technicians can absorb some of the important but routine activities of physicians without his having to relinquish overall responsibility for the patients' health and welfare. By development of automated multiphasic screening sequences, information can be gathered utilizing the technicians in efficient organizations and presented to physicians for evaluation and decisions. The contributions of all members of the health care community should be examined to see if they can be upgraded through additional training or supervised responsibility without diminishing the quality of the service.

Systems analysis and hospital industrial engineering techniques may help to identify opportunities to improve the organization and procedures in various settings. Systematic approach to evaluation of even the most mundane aspects of the system may reveal opportunities to significantly improve the overall performance. The need to develop techniques for assessing the "quality" of health care has never been greater. The process will require positive contributions by the medical profession for the establishment of criteria and specifications by which such judgments can be rendered more consistent and more objective.

References

1. Fein, R. "The Doctor Shortage: An Economic Diagnosis. Studies in Social Economics." The Brookings Institution, Washington, D. C. 1967.
2. Silver, G. A. American medicine; technology outruns social usefulness. *Bull. N. Y. Acad. Med.* **46**, 148-160 (1970).
3. Knowles, J. H. Where doctors fail. *Sat. Rev.* 21-23 (1970)
4. "Report of the National Advisory Commission on Health Manpower." Volume I. U.S. Gov. Print. Off., Washington, D.C., November 1967.
5. Social Forces and the Nation's Health; a Task Force Report. U.S. Dept. Health, Education, and Welfare, Health Services and Mental Health Administration, Bureau of Health Services, Washington, D.C., May 1968.
6. Magraw, R. M. "Ferment in Medicine: A Study of the Essence of Medical Practice and of Its New Dilemmas." Saunders, Philadelphia, 1966.

7. Friedson, E. "Patients' Views of Medical Practice: A Study of Subscribers to a Prepaid Medical Plan in the Bronx." Russell Sage Foundation, New York, 1961.
8. Lewis, C. E., and Resnik, B. A. Nurse clinics and progressive ambulatory patient care. *New Eng. J. Med.* **277**, 1236-1241 (1967).
9. Connelly, J. P., Stoeckle, J. D., Lepper, E. S., and Farrisey, R. M. The physician and the nurse—their interprofessional work in office and hospital ambulatory settings. *New Eng. J. Med.* **275**, 765-769 (1966).
10. Ford, P. A., Seacat, M. S., and Silver, G. A. The relative roles of the public health nurse and the physician in prenatal and infant supervision. *Amer. J. Pub. Health* **56**, 1097-1103 (1966).
11. International Conference on the Perinatal and Infant Mortality Problem of the United States. National Center for Health Statistics, Series 4, No. 3. U.S. Department of Health, Education, and Welfare, Washington, D.C., June 1966.
12. Speert, H. Midwifery in Retrospect. The Midwife in the United States. *Trans. 3rd Josiah Macy, Jr. Conf., 1968.*
13. Kissick, W. L. Effective utilization: the critical factor in health manpower. *Amer. J. Public Health* **58**, 23-29 (1968).
14. Barnes, A. C. Training Programs for the Nurse-Midwife in the United States, The Midwife in the United States, *Trans. 3rd Josiah Macy, Jr. Conf. 1968.*
15. Can doctors' aids solve the manpower crisis? *Med. World News* **11**, 25-30, (1970).
16. Stein, H. A., and Slatt, B. J. "The Ophthalmic Assistant." Mosby, St. Louis, 1968.
17. New Members of the Physician's Health Team: Physicians' Assistants. Report of the Ad Hoc Panel on New Members of the Physician's Health Team of the Board of Medicine of the National Academy of Sciences, 1970.
18. Smith, R. Medex. *J. Amer. Med. Assoc.* **211**. 1834-1845 (1970).
19. Bennett, A. C. "Methods Improvement in Hospitals." Lippincott, Philadelphia, 1964.
20. Bennett, A. B., Dassel, S. W., and Wedgwood, R. J. Time-motion study of practicing pediatricians. *Pediatrics* **38**, 254-263 (1966).
21. Silver, H. K., Ford, L. C., and Stearly, S. G. A program to increase health care for children: the pediatric nurse practitioner program. *Pediatrics* **39**, 756-760 (1967).
22. Drui, A., and Gunderson, J. How to measure the work of professionals. *Ind. Eng.* **1**, 42-43 (1969).
23. Page, T. The nature of operations research and its beginnings, *In* "New Methods of Thought and Procedure" (F. Zwicky and A. G. Wilson, eds.). Springer, New York, 1967.
24. Flagle, C. A decade of operations research in health. *In* "New Methods of Thought and Procedure" (F. Zwicky and A. G. Wilson, eds.). Springer, New York, 1967.
25. Murray, G. R. Systems analysis in health care delivery: Homeopathy or medicine? *Pharos* **34**, 23-27 (1971).

Chapter 3

HEALTH CARE FACILITIES
Concentration of Competence and Technology

In earlier times, the majority of medical care was administered in the patient's home or the doctor's office. Hospitals were avoided as pest houses or poor houses. Now a physician without access to a well-equipped hospital is greatly limited in the scope and quality of care he can deliver. The scene of action has shifted from the home to the hospital where specially trained personnel and sophisticated equipment and procedures have become concentrated. Patients derive obvious benefit from ready access to modern technology and the physician finds it convenient to have patients concentrated in a single geographical location. The physician can personally visit his patients and at the same time share his responsibilities with the professional, technical, and administrative staff of the hospital with no cost to himself. Since professional fees are independent of hospital charges, physicians benefit greatly from having concentrations of patients in such close proximity. Both the patient and the doctor have come to view the hospital as the most obvious and convenient way to meet the problems of sickness, particularly when the costs are being covered by third parties (i.e., insurance companies or the government).

In spite of its obvious attractions, the hospital is a strange and unsettling environment for a patient. No patient gains much relief or comfort from the admissions procedures, laboratory testing, and other ancillary features of hospital activity. The typical nurse is too harried to resemble the sympathetic lady of the lamp. Many nurses have administrative responsibilities for a cadre of practical nurses, aides, and other helpers. To the patient, she is but one of a crowd composed of pharmacists, technicians, radiologists, serving as a buffer between him and his physician (Fig. 3.1). The patient is vaguely aware that behind the scenes are many people probing his inner self, recording and assembling information about him in this strange environment where both the culture and language are unfamiliar. This picture is a far cry from the old country doctor (Fig. 2.1) and is partly responsible for the widening schism between the medical profession and the public.

TYPES OF HOSPITALS

The word hospital generally conjures up a picture of private nonprofit or government owned hospitals established to provide short-term care for rather acute illnesses. Most of the admissions are to voluntary nonprofit (65%) or government (19.3%) hospitals (Fig. 3.2). It is frequently forgotten that less than half of the 1,700,000 hospital beds in this country are set up for care of acute illness. Fully one-half of the total beds are in long-term hospitals, including mental and tuberculosis hospitals, supported primarily by the states, and fully 80% are occupied by the mentally ill (1). One-tenth of the beds are operated by the federal government for its special charges (e.g., military, veterans, or public health). Proprietary hospitals, operated for profit, are generally rare, but their number is increasing rapidly in a few metropolitan areas (i.e., in Los Angeles and suburban New York City). In addition proprietary owners operate 70% of the 362,000 nursing-home beds in the nation. The principle biomedical engineering effort in support of hospital care has been directed toward the patients receiving short-term care, primarily provided by voluntary (nonprofit) hospitals containing 70% of the short-term beds and admitting 65% of the patients. Technology in support of patients undergoing extended care in long-term hospitals, in nursing homes and in private homes and other facilities must be developed to provide comprehensive coverage of the health needs of the country.

ANOMOLOUS ORGANIZATIONAL RELATIONSHIPS IN HOSPITALS

Virtually all hospitals share a common characteristic that complicates administration and obstructs constructive changes; administrative responsibility and power are shared among several individuals or groups such that none is in a position to set goals or induce change. Power does not reside in any one individual and many employees serve more than one master. As a result, each group in the hospital structure tends to exhibit considerable autonomy (2). The hospital administrator is expected to be an organizational manager, responsible

Fig. 3.1. Patients see a hospital as a concentration of technology, technical competence, and services—a far cry from the age of the family physician (cartoons by Phyllis Wood, included with permission of Carole Profant).

for housekeeping and business functions and general responsibility for nurses, but he can exercise limited power. For example, the nurses are not only responsible for carrying out physicians' orders, but are also responsible to the nursing supervisors. The hospital bears reponsibility for the quality of care; shared with that of the physician over whom it has very little control. The chiefs of the various services also have autonomous power, individually over their services and collectively as a medical board, which is rarely challenged by the hospital administrator. The board of trustees of the hospital is commonly composed of businessmen and community leaders with authority over both the hospital administration and the medical board but without either the background or technical knowledge to exercise its power. The specialist groups also tend to have cohesive forces that supersede the authority of the hospital organization. In addition, the individual physician retains a high degree of autonomy which can be threatened only in response to rather clear-cut

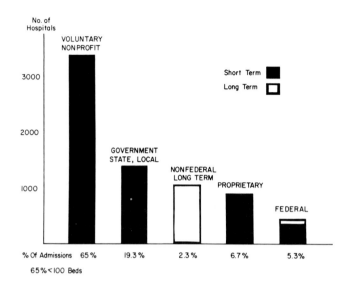

Fig. 3.2. A majority of patients are admitted to small voluntary, nonprofit hospitals designed for short term care (for explanation, see text).

negligence or misbehavior. For these reasons, the hospital may be realistically viewed as a rather loose affiliation of autonomous groups that interact for the care of patients but are not engaged in organized action toward any but the most general goals. The autonomy of the groups is so great that there exists abundant veto power, but little tendency to engage in joint action (2). This peculiar organization has extremely important connotations for the future. Decisions or adaptation to changing conditions cannot be expected to occur promptly or efficiently because responsibility and authority are divided. Instead, the resolution of issues occurs by a process of political maneuvering without a mechanism for resolving controversial issues or conflicts between components. Such a nonorganization imposes extremely serious obstacles in the path of anyone interested in introducing innovations or improvements because it is so unrewarding to assume initiative in such a setting. Pressures are building up to institute changes which necessarily involve participation by biomedical engineers.

The ultimate responsibility for the quality of care is actually shared between the doctor and the governing board of the hospital acting on behalf of the community. This is but another example of the common tendency of personnel in the hospital to wear multiple hats. The legal responsibility for negligence or errors of omission or commission is similarly dispersed and obscure, contributing not only to caution in accepting innovation but also an apparent reluctance to restrain skyrocketing costs if there is the remotest chance that an economic measure might increase the hazard to patients, even in unforeseen ways (see the section on skyrocketing costs, page 69).

The hospital administration cannot function adequately without the participation of an effective staff of physicians. They must provide services and facilities that will attract a competent staff even if they are operated at a loss. As a result virtually all hospitals, large and small, have very similar basic services and facilities for a variety of clinical specialties including maternity, pediatrics, radiology, operating even at a deficit. The basic designs of hospitals in general reflect a widespread lack of innovation or imagination. The interiors of hospitals are too much alike, lacking total planning for effective organization and operation as a necessary prerequisite to good design.

Until recently hospitals were regarded as doctors' workshops, containing concentrations of the tools of medical practice. Staff physicians wielded great power and the administrators managed the bookkeeping, housekeeping, and the grossly underpaid personnel. Since World War II hospital administration has become a more widely accepted profession altering the unassailable position of physicians as the only agents in the field of medicine who could assume responsibility for health care.

ECONOMIC SIGNIFICANCE OF NONPROFIT HOSPITALS

The prevailing pattern of ownership and control of health facilities in this country is that of a nonprofit agency. In many cases this is the most appropriate mechanism but there are penalties associated with the nonprofit approach. Such health facilities are not required by any pressure to achieve economies or accomplish high levels of efficiency. In fact, hospitals are granted nonprofit status to free them of considerations of economy found in competitive profit-making enterprises. They are forced to de-emphasize economies and may be roundly criticized if they are caught cutting corners, even if they are asked to cut costs. The controlling motivation in hospital design is to be safe rather than sorry; extra capabilities and stand-by facilities are built in and maintained in order to meet any possible problem rather than to accommodate to the most probable requirements of the patient. The calculated risk is avoided in handling patients.

Since hospitals are under no obligation to show a profit, their management has had little incentive to keep a lid on costs. A profit-making organization will eliminate paper work, including substitution of computers for clerks if a cost analysis shows that greater efficiency will be achieved. The standard response to an unmet need was to hire additional employees. The wages for hospital employees have risen very rapidly in the past few years, having been sorely depressed in the past. In general hospital workers are still not paid handsomely compared with other workers of comparable skills and training. An important element in rising costs is the lack of efficiency in utilization of personnel. A generally accepted principle of sound management is to assign tasks to the person with the least amount of training qualified to perform the job. In hospitals, one often finds highly trained individuals performing routine or menial chores. The rising wages of hospital employees is enormously compounded by the greatly increased number of employees. The number of employees per patient was 1.48 in 1947 and 2.11 in 1957 (3). This enormous expansion in hospital manpower has been continuing ever since. There are now approximately 2.5 employees per patient.

INCREASED SPECIALIZED SERVICES

It would seem reasonable that the larger the hospital, the lower the cost per patient per day. Actually costs are generally higher in large hospitals presumably because they provide a wider diversity of extremely expensive services which outweigh the economy of large scale (3). Advances in medical knowledge or techniques lead to higher cost because more can be done for the patient. The criteria for good medical care are continuously being revised upward, often with a large number of procedures per day. A much larger array of clinical laboratory tests available to the physician are being regularly utilized. If the improved and diversified facilities and services had served to improve productivity or efficiency, their increased costs would surely have been countered by shorter illness and briefer hospital stays. The average duration of stay in the hospital ceased its decline in the middle 1950's (1) and has actually increased in some cases. This may reflect improvement in the mortality rate of the seriously ill and aging patients who may require extended care.

THE QUALITY GAP

The quality of medical care provided a particular patient is extremely difficult to estimate since no really objective criteria have been established as guidelines. The basis for most clinical judgments are subjective and individual; retrospective evaluation is frequently biased. The medical profession has undertaken self-policing by peer review mechanisms usually by committees of hospital staff or medical association, but they can really detect only overt evidence of errors of judgment or commission. Occasionally groups of "experts" evaluate the care of groups of patients to assess the adequacy of care. Judging by such studies the advance in medical knowledge and technique has increased the potential for preventing or alleviating illness but the potential for improvement has not been fully realized (3).

Medical records of a random sample of 420 patients admitted to 98 different hospitals in New York City during May 1962 were reviewed by expert clinicians. According to competent evaluation, only 57% of all patients and only 31% of general medical cases received "optimal" care. In voluntary hospitals affiliated with medical schools, 80% of all patients were judged to have received optimal care while only 47% of patients in proprietary hospitals. The accuracy of clinical laboratories has also been questioned with surveys indicating erroneous determinations in as many as 25% of the samples.

Among studies of the quality of health services in the United States from two community hospitals and two hospitals affiliated with medical schools the figures shown in the following tabulation are representative.

	Patients receiving fair or poor care (%)	
	Community hospital (%)	Medical school affiliates (%)
Medicine	46	74
Surgery	39	60
Obstetrics gynecology	50	74

According to Lee (4) "It is a tragedy—for which all must share in the responsibility—that the lives of millions of the poor could be quickly and dramatically improved without a single major addition to our knowledge of the science of medicine. The barriers here have been the barriers of delivery of services, in organization, in financing, in communications, and utilization."

SKYROCKETING COSTS OF HEALTH CARE

The inflationary spiral of wages, goods, and services is a source of great concern in the United States and elsewhere. The steeply rising expenditures for health care appear to be soaring progressively and uncontrollably as indicated in Fig. 3.3. Between 1955 and 1965, the expenditures for health services and supplies more than doubled (from $17.1 billion to $37.3 billion) and the projection to 1975 indicates more than doubling again to $94 billion. All of the categories of expenditures participated in this flood of money but the largest are the expenditures for physicians and general hospital services. The costs of health care are increasing much faster than inflation in other areas of the economy. The

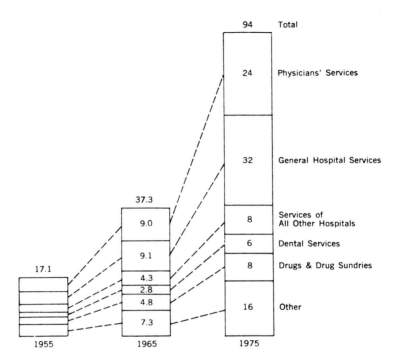

Fig. 3.3. Expenditures for Health Services and Supplies in billions of dollars. The total cost of health care has risen sharply in recent years with the largest increments in physician's fees and general hospital services (from Report of the National Advisory Commission of Health Manpower. Washington, D. C., Nov. 1967).

economic characteristics of health care are distinctly different from other portions of our nation's enterprises.

DISTINCTIVE FEATURES OF MEDICAL ECONOMICS

The health field is sufficiently different from other segments of our economy that few economists have engaged in analysis and many of the basic principles are generally not applicable. Klarman (1) summarized some of these peculiarities.

1. Sliding Scales

Both physicians and hospitals have a tradition of adjusting their charges in accordance with the patient's economic status, not only for truly destitute but also the medical indigent and the individual who can support himself except when confronted by large and unexpected expenses. The sliding scale tends to aggravate the unequal distribution of physicians who concentrate in suburban areas and avoid rural areas and central urban districts. Charity medicine and a sliding scale have been regarded by the medical profession as proof of benevolent motives and professional standards of behavior. It is possible because medical care is not transferable between persons. It enables the poor as well as the wealthy to receive medical care and to this extent tends to equalize income. The sliding scale has diminished or disappeared under many circumstances as a result of factors such as specialization, group practice, and standardization of fees by insurance companies or government.

2. Unexpected Expense

Unpredictable occurrence and incidence of illness is an important factor in encouraging the pooling of costs by insurance or by prepayment plans. The patient cannot effectively plan or save toward this unknown eventuality. In addition, illness tends to reduce earning capacity.

3. Varying Value

Medical care is of extreme worth at the time it is administered but is not repossessible by the seller as would be the case of durable goods. When a patient regains his health, the value of the service in retrospect may not appear as valuable as it did at the time of its administration.

4. No Criteria for Consumer's Comparison

Lack of knowledge by the consumer is another characteristic of health care. He cannot judge the quality even after he has received the services. He chooses a physician but he does not determine the type and amount of care he will obtain. It is the physician who decides what kind of services will be rendered. The consumer's ignorance places a heavy responsibility on the integrity and judgment of the physician. The patient is in no position to engage in comparison shopping—he is rarely aware of the exact nature of the services he has received—to say nothing of their value.

5. Predominantly a Personal Service

A major portion of medical care represents personal service which means that it must grow by an increase in labor force rather than by increased productivity as can occur in industrial segments of the economy. This fact is largely responsible for medical costs rising more steeply than for goods and services produced by industry.

6. Mixed Objectives

Health care is a mixture of consumption of goods, services, and an investment in health and productivity. A program reducing or preventing illness among workers has a different economic aspect from services provided men or women retired from the labor force.

In summary, the distinctive economic characteristics of health services include the irregular, uncertain and occasionally communicable nature of illness; special attitudes of the public toward health and medical care, unusual character of the input—output relationships; the unique forms of organization of these services. Any economist may have conflicting views of the health care system. He is generally impressed with technical advances but tends to view conservative approaches or the absence of treatment as a deprivation. The criteria for appropriate quantity and quality of health care are not established. Such intangibles tend to thwart objective analysis of costs and benefits.

RISING COSTS OF PHYSICIANS' SERVICES

The demand for medical care has increased greatly because increased numbers of people have sought the services of physicians more frequently. The increased size of the clientele results from the increased affluence of the population as a whole. Further the total population of the country is increasing and the various health insurance plans are more widely used. The physicians are seeing more patients and also have raised fees at an accelerating rate. For example, their fees increased at an average of 3% per year from 1960—1965 and 8% per year in 1966. This is about two times the rate of increase of consumer prices. The supply of physicians has not responded promptly to increased demand and rising prices. Limitations in the number of spaces in medical schools and the prolonged training required to prepare physicians restricts entrance into the field and sustains an imbalance between the supply of medical services and their requirements.

FACTORS CONTRIBUTING TO INCREASING FEES AND CHARGES

In the American economy, doctors and their representatives have insisted on fees for service so that rising demand produced higher incomes. But physicians react differently from most other groups in that increased demand produced increased service through longer hours and by more visits per hour. The value of each visit diminished and the income increased even though the fee for each visit may increase more gradually.

To some extent the rise in doctors' incomes may be associated with increased productivity. In the early postwar periods, prices were generally rising and physician fees increased much less, proportionately. This indicates that physicians passed on the gains of rising productivity to the patient. However, this has not been evident in the 1950's since total costs have risen. Further, the increased productivity may be due to improved facilities and therapeutic effectiveness, concealing to some extent the deterioration of service from hurried visits, reduced home calls, and the like.

THOROUGHNESS

The traditional criterion for the quality of a physician's services is thoroughness. Generally accepted standards call for a comprehensive search for disease and disability demanded by the patient, the patient's family, by his colleagues and the community in general. From his earliest training in medical school a physician is roundly criticized if he misses a diagnosis or has a discouraging therapeutic result because he failed to take advantage of opportunities to test or treat. He is rarely subject to criticism for having excessive zeal in his testing or retesting, examination or reexamination, prolonged therapy, hospital stays, or follow-up. The threats of law suits for malpractice are multiplying rapidly. The only protection against the growing threat of exorbitant judgments in malpractice suits is to have covered all eventualities. When the cost of the extra tests or examinations is borne by the patient or his insurance, the physician has little to gain and much to lose by showing restraint in the use of hospital facilities or the number and frequency of tests.

THE COST OF REASSURANCE

A very large percentage of patients have minor ailments or symptoms which are not functionally significant. The physician is reluctant to dismiss these symptoms and reassure the patient unless he has become quite convinced he is not overlooking something more serious. Similarly, the patient is likely to be unconvinced or dissatisfied by a physician's reassurance without a fairly extensive examination. Thus both the patient and the physician frequently require reassurance through comprehensive examinations even in the absence of significant dysfunction. Since the incidence of minor complaints is so very high, the overall cost of this kind of reassurance is astronomical. The development of highly organzied and inexpensive multiphasic screening facilities may ultimately have great value in reducing the cost of uncertainty and the expenditures for reassurance. Allied health personnel may also play a greater role in supplementing this function.

The traditional requirement for thoroughness evolved ages ago when available tests were few and cheap. At that time comprehensive examination merely provided a longer personal contact between the physician and his patient. The beneficial results of more careful examination and prolonged deliberation were obvious. However, unrestrained thoroughness and a comprehensive approach to medical care at present is imposing an enormous drain on the resources of the medical community. The common practice of ordering a large variety of complex and expensive tests repeated at frequent intervals is extremely costly and has not been critically examined to determine the extent to which it actually improves health care. When a patient is referred to another physician, the consultant frequently repeats all the tests and adds more.

Only the most secure and self-confident physician can muster the courage to be content with minimal or optimal diagnostic tests or exhibit the patience required to allow the course of a disease to render the diagnosis more evident. Both of these attitudes by competent physicians can conserve funds without jeopardy to the patient but they do not necessarily avoid criticism. A pressing need exists for a greatly expanded foundation of reliable information regarding the realities of health care delivery so that the distribution of our resources can be channeled into the most effective uses for the benefit of all concerned.

New technology brought into the health care system may be capable of increasing the quality of patient care at lower cost. This potential for improving the cost-effectiveness of the system can only be realized if the innovations are utilized appropriately. With the development of automated clinical laboratory facilities, the cost of each test may be greatly reduced and significant savings could be achieved for those who pay the bills. However, the exorbitant use of these tests have so greatly increased the total number being run that the total cost to the community is certainly no less and may be more.

FACTORS FOSTERING EXCESSIVE HOSPITALIZATION

At any particular moment, a large proportion of the patients occupying the beds do not really require prompt or immediate access to the facilities and service of a modern hospital. Depending upon criteria used, as many as half the patients are receiving little service or care which are necessary for their health and well being, that could not be provided by other and cheaper means. Excessive hospitalization has been directly encouraged by payment policies such that physicians may not be allowed to receive fees from insurance companies or government sources unless his patient is hospitalized. Patients tend to expect or demand hospitalization if they do not feel well, especially if they do not have to pay the bill directly.

The duration of hospitalization may depend upon factors completely unrelated to the severity of the patient's illness. The physician's decisions regarding the appropriate duration of hospital stay may be unduly influenced by hospitalization coverage of insurance rather than by the cost of health care or the most effective utilization of hospital facilities. With the steeply rising cost of expanding specialized services, hospital administrators have little incentive to insist on prompt discharge of recovered patients who require minimal care. Their "over" charges can be used to help defray expenses in more specialized units which commonly run at a deficit (i.e., obstetrics, intensive care, coronary care units, etc.). A physician caring for patients may regard an empty hospital bed as a sign of success but to a hospital administrator it stands as a "failure to fill."

RISING COSTS OF HOSPITALIZATION

The most challenging problem in the health care delivery system is the care of the sick in modern hospitals. The skyrocketing cost of hospitalization is only one aspect of a complex problem, with an increase in daily charges by some 380% from 1940 to 1961 (3). In addition, the costs of more specialized auxiliary services have further increased. There is no way of judging whether this increased cost could be justified in terms of increased value received because there is no generally accepted yardstick for the product. The rapidly rising prices have been accompanied by shorter average stays in the hospital because of increased effectiveness of therapy. More services are provided for patients and a larger spectrum of diseases and pathological states are being effectively treated. On the other hand, the cost of hospitalization is widely variable among institutions providing apparently equivalent services. For example, costs per patient day varied from $46–$96 among hospitals selected from a set of "distinguished" institutions (see Reference 4, Chapter 2).

In all this complex organization framework, there exists no implicit mechanism or pressure to restrain costs or introduce economy in the care of hospitalized patients. The physician charges fees which are independent of the unit costs of hospital stay or special tests and services ordered. The hospital makes charges which are designed to defray expenses no matter how large they may be. This is a form of "cost-plus" economics which has proved so costly in military contracts and has contributed significantly to the rapidly spiralling costs of patient care. In general, hospital administrators tend to preserve high rates of cash flow paying less attention to economy. Indeed, an increase in productivity or efficiency which would cause empty beds or reduce the use of services would tend to further imbalance a precarious financial state of the hospital. The small voluntary hospitals with a full spectrum of services and facilities are economically unsound in most instances. Extreme pressures for restraining the upward spiral of expenditure for medical care are at least beginning to stimulate a trend toward the development of group practice and prepayment plans with diverse types of sponsorship.

MEDICAL–SURGICAL SPECTACULARS

In recent years, the general public has been stirred and impressed by a series of innovations in patient care that achieved world-wide interest. The development of techniques for open heart surgery for repair of defects under direct observation with the heart action intentionally arrested was a dramatic and impressive achievement. This was followed by the development of artificial kidneys, first used for short-term therapy of transient kidney failure, later for chronic therapy of patients with advanced kidney malfunction and also for the support of patients awaiting availability of donors for kidney transplants. A massive effort for development of aritificial hearts stirred the imagination, expanded hopes and awakened fears for a future in which men might live on with implanted vital organs. Intensive care facilities have been established to monitor and treat patients who are seriously and acutely ill, particularly for patients with coronary occlusion and myocardial infarction. Each of these developments represents substantial contributions of engineering techniques and technology. At the same time they represent new opportunities to provide care and hope for the future to a very small segment of the population but at extremely high cost. For example, if every patient needing an artificial kidney were provided one at current rates, the total bill would become $200 million per year at the end of 10 years. The general public may get the impression that the medical profession in general and the biomedical engineering effort in particular are directing their attention toward such medical spectaculars and tending to ignore the enormous problems which confront the health professions and the rest of the nation to provide necessary solutions to the massive array of health requirements and their steeply rising costs.

Thus, the medical profession has a long tradition of rewarding thoroughness and applauding extensive, even excessive use of resources. The physician has little incentive to exhibit restraint in use of testing and therapeutic facilities. The threat of lawsuits with large judgments is a powerful incentive for extreme thoroughness or even excesses in costly tests and treatments. Medical care is geared to deal with serious, life-threatening illness, but is not nearly so effective in managing the common annoying health problems which constitute the bulk of modern medical practice.

BIOMEDICAL ENGINEERING INVOLVEMENT IN A GROWING SPECTRUM OF HEALTH CARE FACILITIES

The options available to most patients at present are extremely limited. Medical care is generally administered either in the doctor's office or in a hospital (Fig. 3.4). A vast majority of these hospitals are very small with some 65% being smaller than 100 beds (see Fig. 3.2).

Typically, a particular bed in such a hospital might be occupied by a patient dying of leukemia, and one or two days later be occupied by a patient who had been admitted for laboratory tests or for a minor illness. If the services and facilities which are readily available to a specific hospital room are optimal for a patient who is seriously ill, they could be exorbitantly expensive and costly for the ambulatory or convalescent patient. Hospital administrators generally regard an empty bed as a sign of failure and a source of financial deficits. Beds filled with patients who require minimal care help defray the very large expenditures of the more seriously ill patient. This procedure tends to overcharge for minor illnesses in hospitals and to obscure the very high cost of caring for the seriously ill. It also tends to increase the total cost of health care by maintaining very expensive facilities which are utilized for only a fraction of the patient census. Most of the small hospitals are attempting to provide precisely the same diversity of services which can be found in the larger medical centers. These efforts are exorbitantly expensive and frequently produce overlap or redundancy wasteful of important resources. A more rational approach would be to develop specialized facilities for the care of specific categories of illness depending upon the severity of illness, with separate facilities for providing minimal nursing care and services for the convalescing and

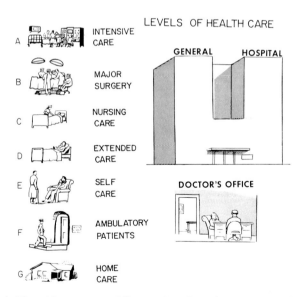

Fig. 3.4. The wide spectrum of illnesses from intensive care to ambulatory medicine is treated either in general hospitals or in physician's offices.

ambulatory patient. We should anticipate a greatly expanded spectrum of health care facilities for patients having different types and different degrees of illness as illustrated schematically in Fig. 3.5.

COORDINATED COMBINATIONS

The development of components or subsystems for health care on an individual basis ultimately requires coordination to achieve full effectiveness. For example, the finest mobile coronary care unit or fleet of emergency vehicles cannot function effectively without the coordination and availability of appropriate facilities set up to deal with emergencies of various degrees of severity. The need for local and regional planning is most clearly evident with respect to provisions for emergency care. Each facility must be considered as a component of larger systems. For example, a trauma center and shock control station cannot function in isolation but must have ready access to laboratories, radiology, general surgery, neurosurgery, thoracic surgery, and other clinical specialties. With modern means of high speed transportation, it is not always necessary to have all these essential ingredients in immediate proximity, but the mechanisms for communication and interchange must be planned, organized, and rendered effective. These principles are applicable to all components of the

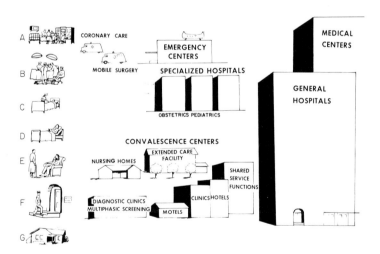

Fig. 3.5. The many levels and types of illness may be more effectively handled in specialized facilities equipped and organized for illness of specific types or severity with improved provisions for ambulatory or convalescent care, combinations of clinics and housing, diagnostic centers, specialized treatment centers, and large fully equipped general hospitals and medical centers.

health care facilities in a region, an aspect of the problem which is just now gaining recognition and attention (see also Chapter 4).

The traditional groupings of wards, departments and specialties in hospitals are no longer optimal. It is quite possible that new combinations can be demonstrated more effective by innovative approaches. For example, the Thorax Centrum in Rotterdam incorporates concepts of Paul Hugenholtz which include provisions for both cardiology and thoracic surgery integrated in the same highly sophisticated environment with the most modern equipment and full provisions for intensive care, either coronary or postoperative. The center is supported by carefully designed computer capabilities for extracting maximal information from available data sources, and a research program to develop new transducers. In addition, the facility is designed to provide three levels of care—*maxicare, midicare, and minicare*—to accommodate progressive stages of treatment as patients recover from the initial threat but need access to the most intensive care if complications unexpectedly supervene. This is but one example of a common tendency to develop health care facilities designed to specifically accommodate the specific needs of particular groups or categories of patients.

Maxicare

Maxicare can be characterized by trauma centers, stroke control stations, coronary care facilities and postoperative recovery rooms, as indicated above. In each of these instances, patients face extreme hazards that call for sustained therapy, frequent monitoring, and undivided attention of highly trained personnel. The gravity of the situation warrants the most sophisticated equipment available, integrated into well organized systems. Patients undergoing major surgery are also in sufficient jeopardy to warrant instrumental monitoring equipment to supplement the traditional clues currently utilized by anesthesiology. Biomedical engineering contributions to surgery have been extremely valuable in a few isolated instances (i.e., artificial heart-lung machines, anesthetic machines, etc.) but have produced very limited changes in routine surgical operations. Emergency care, intensive care and their bioengineering implications are discussed in greater detail below.

Although the need for intensive care is generally associated with dramatic examples like trauma centers, shock-control stations, major surgery, postoperative recovery rooms, many other medical problems call for extensive monitoring, therapy, and nursing service. Acute episodes or advanced stages of disorders affecting the various organ systems also call for "maxicare," including respiratory diseases, metabolic diseases (e.g., diabetic coma), nervous disorders, gastrointestinal emergencies, severe skin burns, and many others. The cost of such care is so catastrophic that there is good reason to consider providing these services by local government with federal subsidy as required in accordance with carefully conceived regional plans (see Chapter 4).

Midicare

Intermediate levels of health care are less dramatic but extremely important in the spectrum of medical services. These forms of medical care can be provided in medical centers, general hospitals and specialized hospitals as illustrated in Fig. 3.4. At present a very large proportion of beds in such facilities are occupied by patients whose illnesses do not require the full complement of facilities, services, and personnel available, unless complications develop. The costs of medical care are unnecessarily inflated when hospital beds are occupied by patients because they *might* need prompt attention or treatment. The options available for providing various levels of health care, suggested by Fig. 3.5, should be enlarged to take advantage of modern techniques of telecommunication so that a prompt response can be available on a contingency basis for patients recovering from serious illness who are not yet ready for full discharge.

Minicare

A significant reduction in the average duration of hospital stay has been one of the most encouraging and important recent results of increased therapeutic effectiveness. By developing alternative housing facilities (motels, hotels, or nursing homes) in close proximity to concentrations of physicians in hospitals or diagnostic facilities, contingency care should be provided at greatly reduced cost. The only requirement for such a mechanism is effective organization and telecommunication between the grouped facilities as suggested in Fig. 3.5. Provision should be made within hospitals or in adjacent buildings for facilities in which ambulatory patients can take care of their own personal needs for food, bathing, toilet, and mobility.

If sound regional planning (see Chapter 4) results in the location of major medical centers or effective trauma centers at strategic locations near freeways or other transportation channels, a clinical cluster of hotels or motels adjacent to medical facilities would serve patients on self-care regimes, families of patients, emergency patients under observation, and casual travelers. If the clinical complex is appropriately planned and organized, such a multifunction approach would encourage not only increased utilization, greater efficiency through shared services (see below) and high cost-benefit ratio but could also be a sound economic venture. For example, the laundry and food services could be organized to handle the entire clinical cluster and motel facility together.

TECHNOLOGICAL REQUIREMENTS FOR IMPROVED HOME CARE

Patients who are bedridden but not in need of immediate access to doctors or the nursing functions of a hospital may be more comfortably managed in the familiar environment of their own home. This problem is particularly acute for those individuals of an advancing age who are somewhat disabled but could take care of themselves with some external support. For example, the wheelchairs in common use today are little changed over the past 100 years (5). Technology could significantly alter this picture. For example, there are a number of designs in the patent office for various combinations of beds, chairs, toilets and bathtubs which could be incorporated into functional units. One can conceive of a combination bed and wheelchair which could be moved about fairly readily and which could be converted into a bathing facility by a waterproof overlay on the bed and a proper shower head on a flexible hose. Modern developments in chemical toilets might also be incorporated into such a unit. Various combinations of beds, chairs, toilets and baths, and bedside units of different sorts could be made available from hospital supply rental agencies (Fig. 3.6). The cost of such a unit per day should be considerably cheaper than the cost of a hospital room. For those individuals who are debilitated to the point where they could not prepare their own food, bedside units could include frozen food compartments, a microwave oven to warm the food, a source of water and of beverages, and even an automatic device for presenting the patient with capsules and pills at regular intervals in accordance with their prescriptions. Incorporated into such a unit might be a communications system with access to not only an intercom within the household but also a connection to a source of medical advice and help. Such a home care unit is schematically represented in Fig. 3.6. It does not now exist but all the necessary technologies have been developed to make it work.

Various programs have been tried experimentally to provide meals on wheels to patients at home. Mobile units could bring drugs, therapeutic equipment to the home, and deliver samples to the laboratory without utilizing

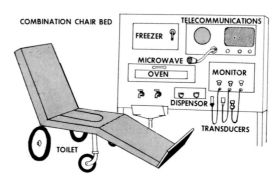

Fig. 3.6. Engineering innovation is badly needed to facilitate effective care at home, by delivered goods and services, combination units to serve as beds, chairs, toilets, baths, and rental bedside units with improved communication with health care centers, and possibly mechanisms for food service for the bedridden but not acutely ill.

the expensive time of highly trained physicians. Another alternative would be to develop a small trailer house which is fully equipped as a home-hospital unit and could be driven into the driveway of a house or apartment and utilized there as a temporary hospital facility.

The widespread utilization of standard plumbing in the bathrooms in hospitals seems unimaginative. Heinz Wolff is one of the few who questioned whether bathtubs, basins or toilets of radically different design might be preferred for patients, particularly those with immobilizing disabilities (see Reference 22, Chapter 5). The contributions of engineering to improving the prospects of self care or home care are negligible to date. The ultimate demand for well designed and engineered combined facilities for self-care should be great and growing because of the aging population. They would find use not only in private homes but also in nursing homes, retirement, and extended care facilities. The potential demand should provide an incentive for commercial development without subsidy.

TRAUMA CENTERS AND EMERGENCY SERVICES AS SUBSYSTEMS

A prime example of an integrated service in desperate need of improvement and modern technology is the mechanism for dealing effectively with traumatic injuries, on the highways, in industry and in the home, many of which are tragic complications of technological progress (i.e., the automobile). "Accidental death and disability" has been presented as the neglected disease of modern society in a report of committees of the National Academy of Sciences (6). In 1965, the accident death toll was approximately 107,000, including 49,000 from motor vehicles, 28,000 at home, and 14,000 at work. The tragedy of the high accidental death rate is that thousands are killed who could otherwise have long life expectancy and productive contributions to society, whereas those afflicted with malignancy, heart disease, and many other chronic diseases die late in life and are past their productive years. Automobile accidents are the leading cause of death for persons between ages one and thirty-seven. In 1965, the disabling injuries numbered over ten million and included 400,000 that resulted in some degree of permanent impairment. The cost of accidents totals some 18 billion dollars, approaching the current national annual appropriation for conducting the war in Vietnam. In 1965 more than two million victims of accidental injury were hospitalized and occupied one out of eight beds in general hospitals in the United States. Despite the magnitude of the problem and its general recognition, the response has been sluggish and ineffectual. Expert consultants returning from both Korea and Vietnam have publicly asserted that the chances of survival of a man with a particular injury would be better in a zone of combat than on the average city street. The problem is compounded by the fact that 70% of motor vehicle deaths occurred in rural areas and in communities with populations under 2500. Thus the bulk of accidental injuries on the highways occur at the sites which are remote from concentrations of health care facilities.

A review of ambulance services in the United States indicates a paucity of firm data, a diversity of standards (often very low), and equipment which is frequently poorly designed, unnecessarily expensive, and generally inadequate. Few communities provide financial support for ambulance services. When provided they usually are maintained by the fire or police departments. Approximately 50% of the country's ambulance services are provided by 12,000 morticians. In many instances, the vehicles are unsuited for active care during transport and the attendants are not properly trained. Communication is seldom possible between the ambulance en route and the emergency department that it approaches. There can be no doubt that the provisions for medical emergencies represent a segment of health facilities which is badly in need of a new and comprehensive approach.

TYPES OF EMERGENCY FACILITIES

The limited information now available indicates that most emergency departments across the country must be classed as advanced first aid stations without full time professional staff. Only modest first aid equipment is available so they are not prepared to care for a patient who is critically injured. Somewhat superior is the limited emergency facility which is found in many hospitals whose emergency departments function twenty-four hours a day as out patient clinics. They are frequently confronted with major emergency care beyond their capabilities. In sparsely populated areas or small communities, facilities of this type are essential and by proper sorting, large numbers of medical and surgical patients can be adequately handled at these limited emergency facilities. Mechanisms must be made, however, for transport of patients with critical illness whose management requires specialized facilities in more advanced hospitals incorporating major emergency facilities.

PROTOTYPES FOR AMBULANCE AND EMERGENCY SERVICES

Although the discouraging picture which has been painted above is extremely widespread in the United States, first-class ambulance service does exist in a few cities. For example, Baltimore's highly trained, full-time ambulance attendants are equipped with up-to-date vehicles and equipment as a separate mission of the fire department. There is central screening and dispatching to ensure open traffic lanes and communication with the ambulance attendants en route to aid distribution of casualties to assigned hospitals. In other parts of the United States, enormous numbers of critically injured patients die from crashes on our highways for lack of the most fundamental requirements. For example, some 50 different telephone numbers are listed to call the police in Los Angeles. In St. Louis there are 32 numbers to call the police and 57 to call for fire emergencies.

THE RUSSIAN VERSION

Emergency numbers are the same throughout Russia; in any city in the country one can dial 01 for fire, 02 for police, 03 for health emergency, and 09 for information (7). This relatively simple change greatly streamlines access to emergency care. The Russian polyclinic is a neighborhood group-practice type of resource that serves ordinary health needs. The physicians and nurses of the polyclinic respond to any emergency call from home or factory between 8:00 am and 9:00 pm daily. Any emergency in the street by day or anywhere after 9:00 pm is the responsibility of a well-organized city-wide emergency service in urban Russia. The telephone to a central dispatcher conveys the necessary information about the nature of the emergency and this information is immediately relayed to an emergency substation where an emergency brigade is waiting. Each emergency brigade consists of a physician, two specially trained nurses, and a chauffeur. The buildings housing emergency services are strategically located on or near principle streets in the central part of the district which they serve.

AN INTEGRATED SYSTEM IN FRANKFURT, GERMANY

An emergency and rescue service which could serve as a prototype for many large urban centers of the world is to be found in Frankfurt, Germany. Under the direction of Ernest Achilles (an engineer serving as Oberbrand or director of the fire department of the State of Hesse) an ambulance service has been organized as a function of the fire department, which epitomizes the fundamental nature of cooperation required in such complicated services. The fire department is manned by 850 professionals. Each professional member of the fire brigade has training in emergency rescue and rotates through service driving ambulances or as the attendant in ambulances. There are 35 ambulances located at strategic sites associated with seven fire departments in different regions of the city. The fire trucks are diversified and contain an exceedingly wide variety of equipment to meet all types of dangerous circumstances including toxic gases, oil spills, radioactive materials, and so forth. When either a fire or an emergency call is received, a card is selected from the card file which lists the route from the designated fire station to the street address and also includes any special health hazards which are located on that street including stores of noxious chemicals or radioactive substances. This card is placed on a pneumatic tube and is delivered immediately to the proper vehicle which contains the necessary equipment to combat fire or an emergency in that particular environment. It also gives street-by-street directions for driving. All fire fighters have training by a doctor and his staff, two weeks in hospital emergency wards, two weeks in anesthesia, two weeks in maternity wards, and back to emergency wards again. Each fire fighter is then assigned on a rotation basis to an ambulance accompanying a man with extensive experience. From this large pool of highly trained firemen, additional training is provided for a mobile surgical team which extends for an additional six to nine months. The mobile surgical team is assigned to one of three mobile surgical trucks which contain all the necessary facilities for handling seriously ill patients including intubation, tracheotomy, thoracotomy, and amputation. Two personnel from the fire department are assigned to each mobile unit and they rotate weekly between the emergency room and in the operating room of the hospital to which the mobile surgical unit is assigned. In this way they are working closely with the three physicians who are assigned to the emergency ward. The surgeons in the emergency room rotate on duty in terms of priority 1, 2, and 3. When a call is received by the central dispatcher that indicates a serious accident, both a regular ambulance and a mobile surgical unit are automatically sent to the scene. The calls for mobile units average three to six times a day. The ambulance is fully equipped even including a small portable electrocardiograph which can be directly placed on the thorax and immediately displays the pattern on the face of a cathode ray tube. In two and one half years, open heart massage has been carried out 15 times in the mobile unit by the attending physician. The most important feature of this program is the intimate relationship between the mobile surgical units and the hospitals to which they are assigned. The physicians declare that the emergency attendants are among the most desirable and highly trained assistants in the hospital.

Disinfection procedures have been established to handle ambulances and their attendants which have conveyed patients with infectious diseases. Whenever such a patient has been delivered to the hospital, the ambulance is ordered to drive directly to a disinfection unit where it enters directly through special doors into a room equipped for fumigation. The attendants leave this room and enter a special washroom where they remove their clothes and under supervision, wash and shower while their clothes are going through a complete disinfection process. The ambulance remains in the fumigation unit for several hours depending upon the severity of the illness and the nature of the infection. The ambulance driver and attendant are kept in isolation for a period of hours depending upon the nature of the infective agent. Every ambulance is exposed to a spray disinfectant at least once per week.

In the last two or three years, helicopters have been obtained on loan from the United States Air Forces or the German Air Forces in case of serious emergency on Autobahns where injury accidents are commonplace. Extensive tests have been carried out using the Bell helicopter for this purpose. They have also tested a German helicopter, the BOCO 105 and the Vertol H21. An effort is being made currently to obtain two helicopters and helicopter pads are already available at the three mobile surgical center bases.

In the United States there is a recognized need for sophisticated emergency facilities to render complete care to the severely injured in strategically located sites; few such facilities exist. Most emergency departments of large hospitals have inadequate space and personnel to carry out their mission. Indeed even these hospitals tend to be staffed on a rotating basis so that a critically ill patient may first be seen by a nurse, or medical student who must be expected to take proper action and make proper judgments under circumstances of great stress. Provision should be made for 24-hour staffing by highly competent medical and paramedical personnel thoroughly trained in resucitation and other life-saving measures and also fully aware of the facilities and location of the necessary instruments within the emergency room. Such facilities are rare indeed and are not necessarily chosen by the ambulance drivers. Since the ambulance drivers usually have no mechanism for communication with physicians, hospitals, or other sources of information, they decide, with little basis for judgment, where they should deposit their patients. Since emergency facilities are not accredited in relation to their capabilities, patients are commonly deposited in facilities which are inadequately equipped to care for them. This means unnecessary delay which can be crucial in many cases.

Emergency care for the critically injured or acutely ill is a chain for which each stage in the process must be effective or the entire effort can be meaningless. Easy access to organized emergency facilities requires rapid and accurate communication (i.e., standardized telephone numbers). Appropriate vehicles must be dispatched with necessary equipment and adequately trained personnel. Patient handling and transport must be both expeditious and nontraumatic. Supportive therapy and effective monitoring should be reinforced by communication with the emergency facility to which the vehicle is routed to provide essential advice and guidance during transit and in preparation for arrival. The emergency vehicle must arrive at an emergency facility appropriate to the type and severity of the injury. The facilities, staff and organization must be designed and geared for streamlined function. Among the most effective mechanisms for identifying objectives, developing appropriate plans and evaluating the results is by a common engineering technique based on input-output relationships and commonly termed *systems analysis.*

A SYSTEMS APPROACH TO EMERGENCY CARE

Complex systems can be subjected to detailed analysis by assessing the relationships between the input and the output of the individual components of the system. The basic principles of such systems analysis can be simply represented as follows.

The output represents the objective of the component of the system, and the input corresponds to the resources which are available for its attainment, including materials, personnel and facilities. The process should be tailored so that the available resources can be most effectively used to approach the stated objectives. Modern engineering science involves not only research and development of new technology but also the optimization, management engineering, methods improvement, and hospital industrial engineering. Clearly, biomedical engineers are capable of contributing much of value to the deliberations and analysis of local and regional planning groups addressing the complicated problems of providing comprehensive health care for all the people who need it. The objectives, requirements and specifications for emergency care can be concisely and explicitly stated, including rapid responses, effective communication, reliable instrumentation, streamlined organization. An innovative approach is needed to provide both the necessary facilities and integration to produce a smoothly functioning system. It can serve as an example of the ways in which biomedical engineering can contribute to stringent requirements of one component of health care delivery.

BIOMEDICAL ENGINEERING APPLICATIONS FOR EMERGENCY CARE

As a representative example, the systems approach to evaluating the engineering requirements of emergency care can be displayed in simplified form as an input-process–output relationship as illustrated schematically in Fig. 3.7. Essential ingredients in successful operation of such a system would include (1) a simple, unambiguous telecommunication link with a dispatcher, (2) ready access to medical consultation by both the dispatcher and the ambulance attendants to aid in selection of facilities and for the proper procedures en route to an appropriate facility for definitive care. None of these are generally available in the United States. The anticipated output is a fairly concise statement of the basic objective of the process, a step which is all too often neglected. If the fundamental goal of the ambulance service is to deliver the patient(s) to appropriate emergency facilities in optimal condition, the necessary ingredients can be more readily determined. The competence of the driver and attendant to achieve this purpose depends upon their training and on the equipment supplied with their vehicle. Having defined the type of patient and categories of injuries which will be served, the equipment requirements can be evaluated, and a realistic compromise achieved. The instruments developed for diagnosing and monitoring patients in emergency facilities may be redesigned or adapted for use in ambulances by utilizing telemetering systems as discussed in more detail in Chapter 6. Techniques for registering systemic arterial pressure repeatedly or automatically or in the presence of high noise levels (e.g., in helicopters) may be needed (see ultrasonic detection methods in Chapter 6). Equipment for resuscitation and support of blood volume need improvement. The equipment needed for each anticipated contingency should be explored by treating each component as a subsystem for analysis. For this purpose, the input–output relationship of each function to be served by the vehicle should be individually studied to arrive at optimized equipment and training. The same approach can be applied to components and organization of the emergency facilities. The anticipated goal or output for three types of facilities are distinguished in Fig. 3.7. If first aid be intended to prevent complications or minor injuries, standard emergency facilities may be developed to restore structure and function deranged by a wide range of environmental hazards. Trauma centers and shock-control stations can be dedicated to sustaining life in critically injured individuals even at the cost of extended or permanent disability (i.e., major surgery or amputation).

The most important point in this entire consideration is the fact that intensive efforts directed toward improving only one feature of such a system will be ineffectual. The entire system must be considered in terms of both its component parts and its relationship to the larger components of the health care system. For example, the finest emergency ambulance system in the world would fail in its mission without an appropriate selection of emergency facilities, strategically located and interconnected through organization and communication.

INSTRUMENTATION REQUIREMENTS FOR REMOTE AND INTENSIVE MONITORING

Evaluation of the condition of a seriously ill patient requires much more information than can be gleaned from the traditional "vital signs" (e.g., pulse rate, arterial blood pressure, electrocardiograms, respiratory rate, body tempera-

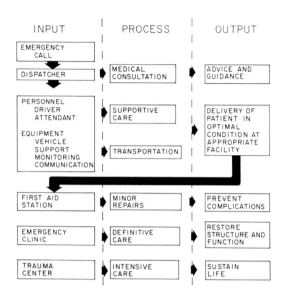

Fig. 3.7. The concept of systems analysis is illustrated by emergency services for which the input-process-output are indicated. Note that the objectives are different for the three levels of severity of injury, calling for different facilities (input) and care (process).

ture). A more scientific basis for management of patients suffering critical illness or traumatic shock could well depend upon data regarding blood volume, venous pressure, ventricular filling pressure or diastolic ventricular volume, ejection velocities, acceleration or time, and blood flow through critical arteries, including coronary, cerebral or key vascular beds (8). In addition it would be important to have reliable information regarding the oxygen content of venous blood, blood distribution in various vascular beds and changes in blood composition. New and noninvasive techniques for monitoring blood flow in various arteries, dimensions of heart chambers, oxygen saturation of blood, described in subsequent chapters, would be particularly applicable. Each new recording instrument shown to have value in a trauma center or shock station can be considered as a potential addition to the equipment in mobile surgical trucks, ambulances or helicopters, each of which may impose its own limitations and problems.

Despite notable progress, intensive care units have not fully utilized the fund of basic knowledge for assessment of cardiac function from an engineering point of view. Basic research on animals discloses that key variables needed to assess the functional performance of the heart ultimately include dynamic measurements. Variables which appear to be most sensitive and significant in assessing changes in cardiac performance include the rates of change of ventricular pressure *(dP/dt),* outflow velocity and acceleration from the left ventricle and power output in addition to absolute values of ventricular pressure, stroke volume or heart rate (9). Such observations encourage specialized instrumentation development designed to provide maximally useful information with minimal hazard and cost. Ultrasonic techniques for detecting and analyzing ejection velocities and acceleration of blood leaving the left ventricle are described in Chapter 6. These have not yet been fully evaluated on patients but hold promise of being safe and effective sources of data regarding the performance of the heart. Corresponding innovations are badly needed for monitoring the condition of patients with diseases affecting other organ systems by simple, safe and reliable instruments. Biomedical engineering is challenged to develop sensitive devices capable of detecting the changing function of the nervous system, respiratory tract, digestive system, genitourinary mechanisms, and metabolic activity as they might be affected by common diseases or pathological processes.

CONSOLIDATED FACILITIES AND SERVICES

Some of the most effective technological innovations are extremely expensive in terms of both capital outlay for facilities and ongoing costs of technical and professional staff. Typical examples include intensive care facilities, and special equipment such as therapeutic X-ray and radiation sources (cobalt bombs) or gamma cameras. Such facilities and services are installed in hospitals, large and small, for reasons of competition and prestige even when they cannot be utilized fully and efficiently. In some instances, insufficient utilization is not only financially wasteful but also hazardous to patients. For example, many urban communities contain several highly specialized facilities for open heart surgery; much more than enough to accommodate the patient load. The facilities are expensive and the techniques are complicated enough to demand the epitome of teamwork, which can be maintained at peak effectiveness only by regular and frequent scheduling (e.g., daily). Otherwise the safety of the patient is jeopardized. Mechanisms are needed to identify the regional needs for such sophisticated facilities and services and restrict the number to that which is optimal. In many parts of the country, this would involve reducing the number by as much as 50%. Precisely the same considerations may be applicable to other specialized functions such as cardiac catheterization, artificial kidney centers, and centers for intricate and demanding specialties, organ transplant facilities, rehabilitation and prosthetic services. One hazard in the process is the curtailment of individual initiative and innovative approaches that stem from diverse efforts even though they overlap. However, reducing the numbers of facilities and consolidating the patient loads can benefit patients, hospitals and members of the health team alike. Prime examples include obstetrics, large-scale automated laboratories, and multiphasic screening facilities.

CONSOLIDATED OBSTETRICAL SERVICES

An economic study of obstetrical units in the Unites States has provided a powerful economic incentive to reconsider the mechanisms by which pregnancy is handled in the United States (10). The underlying reason behind this is that many obstetrical services, particularly in small hospitals, are arranged to handle the peak loads, which means that during a large proportion of the time only part of the beds are occupied. It has been estimated that approximately 50% of the obstetrical beds in hospitals are vacant at any particular time. However, despite the fact that the beds are vacant, the services and the necessary personnel must be available to handle the peak loads. This is an extremely uneconomic mechanism for handling a problem of this sort. It has been estimated that a single hospital which is well organized and with sufficient bed capacity could handle all of the obstetrical cases in an urban district with populations half million to one million people and at the same time make a profit at lower costs than are currently charged. The economies which could be introduced by the increased bed capacity could also be furthered by an increasing reliance upon trained midwives to handle the recurring visits to the hospital by pregnant women and also to greatly encourage the use of home deliveries in uncomplicated cases. By a combination of these practices, a serious economic drain on all the small hospitals of a community could be eliminated and an efficient and well-managed obstetrical center in a metropolitan area could provide all the necessary services with improved maternal and infant mortality, and cost-benefit to all concerned.

AUTOMATED CLINICAL LABORATORIES

The advent of automation in clinical laboratories has greatly increased the volume and diversity of tests which can be carried out at greatly reduced cost (11). Chemical analysis of blood and body fluids is one of the most rapidly growing applications of quantitative methods in clinical medicine, increasing at about 15% per year as a sustained advance for as long as 20 years in large centers. The ultimate place of batteries of chemical tests in handling of patients is not certain but admission profiles have been advocated for routine testing of patients repeated at intervals for health screening. There are two major types of automated equipment.

Type A. Discrete Analysis Systems (with each reaction completed in a single tube)

1. Continuous (without intervention of an operator)

2. Discontinuous (operator performs certain steps such as transfering racks from one module to another)

Type B. Continuous Flow (reactions occurring in a continuous stream of reagents as a sequential analysis)

An example of the discrete analysis system is the Microlab by which batches of tubes (e.g., ref. 11) are transferred by an operator from one module to the next. Continuous discrete single systems (e.g., Robot Chemist, Clinomak) provide low cost analysis for small batches along single channels. Multichannel equipment for analysis of large numbers of samples include the Autochemist (see Chapter 4). The capacity of such sophisticated equipment is so great that maintaining a steady input of samples poses a major problem. They are particularly suited to health screening where enormous numbers of individuals are examined in prescribed numbers and predictable rates. They lack the flexibility which seems essential to routine clinical use unless they are located as central shared facilities serving large numbers of beds to attain sufficient volume.

Continuous flow analysis is accomplished by introducing samples into a stream of diluting fluid in small bore flexible tubing. Successive combinations of the sample with other flowing streams entering at different flow rates provide a variety of tests to be read out by photoelectric colorimetry, flame photometry, spectrophotometry and fluorometry. Bubbles are inserted between samples to prevent mixing or contamination. Multichannel types of continuous flow devices (i.e., Technicon) can produce up to 12 tests at present with ever expanding prospects.

In 14 years, United Medical Laboratories, Inc., in Portland, Oregon, has grown from a laboratory serving a handful of doctors to a scientific complex performing over 4,000,000 tests per month for over 20,000 physician-clients in every state and 20 foreign countries (12). They employ an autochemist which performs 24 tests on 107 samples an hour. They also utilize automated equipment for hematological examinations. Such commercial enterprises can be efficiently organized to provide prompt and accurate service at a profit with high incentive for maintaining minimal costs. The very large capital investment, technical competence, and capacity of automated laboratory equipment seems ideal for consolidation into services which can handle samples from a number of institutions in a locality. Although mass production laboratories may be subject to criticism for lack of precision or reliability, this is not an inherent defect. Indeed they are more likely to respond promptly to new techniques and standards as they become available. User satisfaction must be maintained or their survival is threatened (see also Elusive Normality, Chapter 5 p. 200).

AUTOMATED MULTIPHASIC HEALTH TESTING

During the past 20 years, multiphasic screening programs of various types have emerged, representing efforts to develop coordinated and integrated applications of many different test procedures in a highly organized and efficient manner. The automated clinical chemistry equipment described above was a central feature of most systems, supplemented by a variety of standard determinations (E.C.G., Blood pressure, etc.) and various other procedures. During the 1940's the programs failed to obtain the support of physicians, primarily because the quality or reproducibility of the tests were regarded as inadequate. In the 1960's improved equipment and techniques have restored a great deal of interest and support of the principle (13). An impression of the diversity of tests which are considered for various organ systems is indicated in the list prepared by an advisory committee to the National Center for Health Services Research and Development (Table 3.1, pages 94-95).

Although the list of tests in Table 3.1 appears to be quite inclusive, a very large number of additions can be expected through the applications of biomedical engineering. For example, a variety of the nondestructive or noninvasive techniques described in Chapter 6 can be considered for inclusion in automated or streamlined test batteries. The techniques for data analysis and display in Chapter 5 are clearly applicable to the problem of handling the masses of data collected by extensive testing.

A suggested organization chart for Automated Multiphasic Health Testing programs is presented in Fig. 3.8 (page 96) to illustrate the reliance placed on technicians and their relation to physicians of various appropriate specialties.

SHARED SERVICES

A highly organized series of tests conducted by technicians under medical supervision is a most appealing concept, but the original objective of detecting incipient illness before it is fully developed has obvious limitations. Comprehensive examinations of large numbers of apparently healthy individuals are extremely costly with a very small yield of treatable diseases. Further complication results from the large number of individuals with marginal, slightly abnormal, or equivocal test values, which can lead to neither diagnosis nor discharge without further costly examination. Experience to date suggests that multiphasic screening should be adapted to another, more critical role; namely the development of specific batteries of tests designed to distinguish the well and the worried well from the truly sick. This "filter function" of multiple testing may well become

TABLE 3.1

Automated Multiphasic Health-Screening Tests

---

1. Cardiovascular testing
   a. Systemic arterial blood pressure
   b. Electrocardiography
   c. Vectorcardiography
   d. Phonocardiography
   e. Peripheral vascular disease
      1. Plethysmography
      2. Oscillometry
      3. Pulse-wave velocity
      4. Thermography
      5. Roentgenography
      6. Surface temperatures

2. Central nervous system
   a. Electroencephalography
   b. Echoencephalography
   c. Rheoencephalography

3. Respiratory system
   a. Skin testing for T.B.
   b. Chest X-ray
   c. Sputum cytology
   d. Respiratory rate
   e. Throat and nose cultures

4. Gastrointestinal system
   a. Oral cytology (for Ca)
   b. Gastric acid determination
   c. Stool examination for blood
  *d. Liver chemistry "profile"
   e. Proctoscopy

5. Endocrine and metabolic disease
   *a. Diabetes
   *b. Thyroid
   *c. Parathyroid
   *d. Gout
   *e. Hyperlipemia
   *f. Protein and nutritional disorders
   *g. Liver conditions
   *h. Electrolyte disorders

6. Genitourinary system
   *a. Urinalysis
   *b. Blood tests
   *c. Cytology
   d. Venereal disease
   e. Breast cancer
   f. Cervical cancer

7. Eye
   a. Visual acuity
   b. Tonometry
   c. Retinal photography
   d. Visual fields

8. Ear
   a. Hearing acuity

9. Body as a whole
   a. Anthropometry
   b. Temperature

Clinical Laboratory Testing (see * above)

1. Urinalysis
   a. Glucose
   b. Protein
   c. Acetone
   d. pH, specific gravity
   e. Urine cultures
   f. Sediment examination

2. Hematology
   a. Hemoglobin, hematocrit, W.B.C., R.B.C.
   b. Differential WBC count
   c. Serology for syphilis

3. Blood chemistry
   a. Albumen
   b. Alkaline phosphatase
   c. Bilirubin
   d. Blood urea nitrogen
   e. Calcium
   f. Carbon dioxide
   g. Chloride
   h. Cholesterol
   i. Creatinine
   j. Glucose
   k. Glutamic oxaloacetic Transaminase (GOT)

TABLE 3.1 (Continued)
_____

3. Blood Chemistry (Continued)

l.	Inorganic phosphates	q.	Sodium
m.	Iron binding	r.	Total protein
n.	Lactic dehydrogenase	s.	Thyroxine
o.	Potassium	t.	Uric acid
p.	Protein-bound iodine		
_____

the single most important contribution of modern technology to effective health care delivery on a large scale (see Fig. 5.5). Both automated clinical laboratories and multitesting of patients in highly organized sequences are examples of an important trend to reduce the cost of sophisticated technology by establishing new forms of facilities (see also Fig. 3.5). Another prominent trend toward increased cost-effectiveness of medical care is represented by shared support services.

Hospitals have frequently been criticized for failure to keep pace with industry by adopting the automation and mechanization or assembly line techniques as a means of holding down hospital costs. This is an unrealistic criticism in view of the small size of the hospitals, since only those with over 500 beds have an average budget over ten million dollars. Industries of that range of cash flow cannot in general develop highly mechanized systems. Second, many functions of the hospital are direct services to the patient and are therefore not production type of activity. However, there are features of hospital functions which are clearly production oriented, including the food services, laundry, sterile supply service, radiology, and laboratory services. These are subject to large-scale and highly efficient procedures and can be accomplished by assembly-line mass production techniques if the size of the hospitals being served is sufficiently large.

The portions of a hospital which provide direct service to the patient are distinguished from the support services in the schematic drawing presented as Fig. 3.9. The various medical and surgical specialties dedicated to the care of patients with diseases of the various organ systems are directly evident to the patient. No doubt, the patient is generally aware that there are other activities going on behind the scene, but most individuals are quite unaware of the large and expensive infrastructure which is like the hidden part of an iceberg and contributes very materially to the size and complexity of the modern hospital. For example, the administrative offices, business offices, accounting offices, record rooms, all deal with paperwork which accumulates and accounts for some 25% of the total cost of health care (see Chapter 5). In addition, there are pharmacies, laundries, food services, purchasing, maintenance of the building and its contents, all of which contribute to the essential organism of the hospital but do not contribute directly to the treatment of the patient. Although the

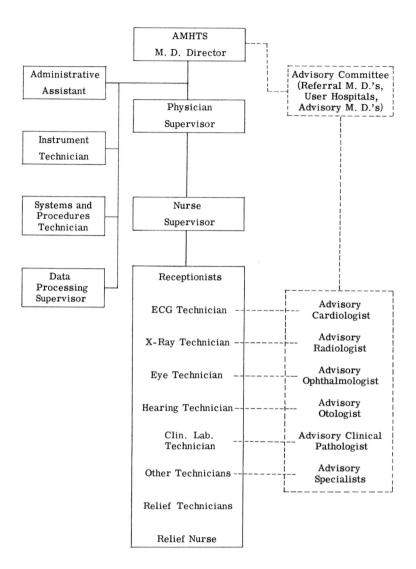

Fig. 3.8. A suggested organizational format for an Automated Multiphasic Health Testing and Services facility illustrates the reliance on technicians and their relations to various medical specialists (From the Report of the AMHTS Advisory Committee to the National Center for Health Services Research and Development USPHS, Volume 2, July 1970).

magnitude and diversity of services provided by small hospitals may be somewhat less than that in large hospitals, the relative size of the infrastructure is even larger.

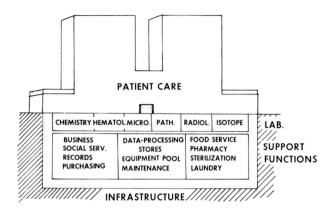

Fig. 3.9. Patient care in hospitals is supported by an infrastructure of laboratories and support functions which are susceptible to such industrial techniques as automation or mass production.

All the processes in the infrastructure can be considered as candidates for large scale automation or production techniques. Clinical laboratories have already been mentioned in connection with automated chemistry determinations above. Similarly, hematology, microbiology, pathology and radiology and isotope laboratories represent activities which can and have been automated to some degree. In addition, many of the administrative functions, computer activities, purchasing, maintenance are sufficiently nonspecific that they can be shared between hospitals with similar functions. The most obvious candidates for assembly-line mass production are the food service, pharmacy, sterilization and laundry. The significance of shared services are schematically represented in Fig. 3.10, with particular emphasis on the prospects of having common purchasing departments and equipment pools shared by several hospitals. This prospect is of extreme importance to the future of biomedical engineering because the increased effective size of combined hospitals would increase the prospect of using new forms of equipment of such great sophistication and capacity that they would be beyond the scope of the typical small American hospital. The main headings listed in Fig. 3.9 represent only a small fraction of the opportunities for options for sharing services among hospitals.

EXAMPLES IN UNITED STATES

Blumberg (14) elicited an astonishing array of functions which have been shared by hospitals. Representative examples of such functions, which he

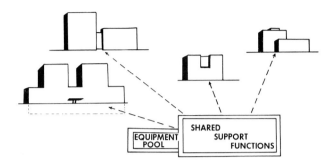

Fig. 3.10. One mechanism for reducing hospital costs may be the sharing of services and equipment pools among several institutions to take advantage of increased efficiency.

described in greater detail, are indicated in Table 3.2. Few of these examples have been tried out on large scale programs in the United States but progress in this direction is farther advanced in other countries.

EXAMPLES ABROAD

A variety of successful experiments in centralizing hospital services may be found in Europe. In Denmark and Sweden, where the hospitals are primarily owned by the counties, laundry and laboratory services at one hospital may be expanded to cover several adjacent units and increase the total number of beds being served. Scotland and England have a national health service, and a variety of schemes for sharing services. A central frozen meal system has been developed at Darence Park Hospital which is capable of utilizing a single continuous cooker for both boiling and frying, preserving foods by means of blast freezing. With a high volume of production it is possible to maintain a continuous operation which does not experience the peaks which are so common and inefficient in the smaller operating units. Central sterile supply services are also susceptible to operation on a shared basis among a group of hospitals. Weymes (15) has developed a concept for the implementation of central sterile supply services throughout Scotland in a system that would incorporate other support functions of hospitals and of service factories known as industrial zones. A description of the concept of various industrial zones has been presented by Knowland (16), Assistant Secretary for Planning to the Oxford Regional Hospital Board in England. The basic concept is to separate the nonclinical and supporting services in industrial sites. Evidence available thus far would indicate that such services are more economical when serving more than one hospital, particularly since this can greatly increase the total number of beds being served. It is generally

TABLE 3.2

Hospital Activities Subject to Sharing by Several Hospitals

Hospital administration
1. Combined administration
2. Industrial engineering efforts
3. Legal services
4. Public relations
5. Legislative representation
6. Accreditation
7. Liability insurance

Business services
1. Data processing-computer
2. Payroll preparation
3. Patient billing
4. Collection services
5. Accounting
6. Negotiating charges with third parties

Medical staff
1. Shared medical staff
   a. Medical director
   b. Chief of clinical service
   c. Pathology
   d. Radiologist
   e. Others
2. Concurrent staff meetings
3. Postgraduate education

Nursing services
1. Administration
2. Pool of temporary employees
3. Service manuals, procedures
4. Evaluation of procedures
5. Student nurse education
6. Graduate nurse education
7. Practical nurses
8. Nurses aides
9. OR technician training

Regional planning
1. Developing planning agencies
2. Implementing regional plans
3. Designated services to specific hospitals
4. Hospital utilization committees
5. Bed locator services
6. Planning public transportation
7. Parking
8. Home care services

Medical records
1. Management of medical records
2. Design of standard forms
3. Statistical analysis

Purchasing and storeroom
1. Group purchasing
2. Storage of supplies
3. Evaluation and testing
4. Central sterile supply service

Food services
1. Preparation of meals
2. Preparation of staff meals
3. Shared dietition

Other services
1. Joint laundry operation
2. Housekeeping services
3. Plant maintenance
4. Duplicating or printing services
5. Pharmacy
6. Blood bank
7. Coordination of ambulance services
8. Laboratory services

Personnel activities
1. Administration of personnel
2. Recruiting
3. Job descriptions
4. Negotiating wages
5. Job placement
6. Insurance coverage
7. Health career planning

conceded that efficient, automatic and mass production of laundry and central services can be attained in support of hospitals totaling some 5000 beds but numbers over 20,000 beds are under consideration. By joint use of transportation systems, mechanical equipment and maintenance personnel, the major advantages of industrialization can be achieved by such central services.

The applicability of shared services to the system of hospitals to be found in the United States, poses greater problems than have been encountered in places like England and Scandinavia where federal ownership of national health services is well established. Despite the competitive and diverse relationships between hospitals, means of establishing cooperative agreements between hospitals are now appearing with greater and greater frequency in various parts of the United States. The likelihood that such agreements would reach a scale necessary for full industrialization (4000–5000 beds) would seem somewhat remote. However, the economic pressures on the administrators of hospitals are becoming extremely intense and such efforts at shared services are acknowledged as reasonable approaches toward economy. One of the major problems is the need for standardization of supplies and equipment from all hospitals participating in a shared service whether the service related to linens or to surgical instruments. It will require rather careful negotiation to establish commonly accepted standards for two or more hospitals. But this is a process that has successfully been achieved in the countries with National Health Services such as Great Britain and Scandinavia.

The trends toward the construction of large medical centers and the affiliation of smaller hospitals into larger units, have important implications for biomedical engineering in the future. At the present time, the acceptance rate of new technologies in hospitals is relatively slow. However, as the hospital units become larger and more highly specialized, better organized and more receptive to technology it seems clear that there will be increasing opportunity for installing more sophisticated diagnostic techniques and new equipment with larger capital outlays in these units which at present would not be financially sound nor attractive to either the hospital administrator or the physicians on the staff.

A WORD OF WARNING

The prospects of cutting costs by sharing services is so appealing that a word of warning seems warranted. The vast majority of American hospitals are small, dispersed, autonomous, and organized in such a way that none is in a position to arrive readily at firm decisions (see the section on anomolous organizational relationships, page 64). The resulting negotiations tend to be long and complicated—more like international maneuvering than like merging of commercial enterprises. The incentives to engage in such prolonged discussions must be clear and convincing or they will break down. Unfortunately, the cost-accounting systems of hospitals are generally not definitive enough to determine accurately how much savings (if any) can be realized, even after sharing mechanisms have been tried. The fact that a wide variety of services have been shared in widely different settings indicates that with enough incentive, it can be achieved. More detailed information regarding the favorable and unfavorable consequences of sharing various services in an assortment of conditions needs to be assembled to serve as guidelines for organizations that may wish to entertain these prospects.

MANAGEMENT ENGINEERING FOR HOSPITALS

The concept of sharing support services by hospitals clearly stems from engineering and management concepts of improving cost benefits in areas which are thoroughly divorced from the services directly rendered to the patient (17). Such approaches are not strictly new although they have rarely been practiced until relatively recently. The principles of scientific management were conceived before the turn of the century and were promptly attempted in several hospital areas, including the surgical suite. The unsavory performances of self-styled "efficiency experts" made hospital personnel suspicious of methods improvement techniques until the last few years. The pressures for improved and expanded health services, rising costs, shortages of personnel, and increased confidence in management techniques have caused a resurgence of interest in appropriate methods of conserving resources without jeopardizing quality of health care. Management engineering programs in hospitals have developed during the last two decades. The activity can be defined as a "concern with the design, improvement and installation of integrated system of men, materials, and equipment." It draws upon specialized knowledge and skill in the mathematical, physical and social sciences together with the principles and methods of engineering analysis to specify, predict and evaluate the results to be obtained from such systems. This definition includes the principles and techniques of computer sciences, operations research, systems analysis, and management engineering.

The study of methods or procedures can be approached by a systematic sequence of analyses which provides a great deal of information with minimal interference with the ongoing work load. The kind of study selected depends upon the nature of the process and the objectives of the study but may include charting the sequence of procedures, diagramming the movements of people, studying organizational relationships or work distribution or random time sampling of activities (18). Since operations analysis and equipment development are integral parts of engineering, the application of these techniques to health care makes them legitimately a part of biomedical engineering. A growing number of hospitals and associations are obtaining the services of operations analysts or systems analysts to explore means of improving efficiency and cost-benefit.

Examples of acheivements attributed to management engineering include (a) reducing the average length of stay of Blue Cross surgical patients by one day through studies of the time elapsed between ordering patient services and the time the services were provided; (b) demonstration of a substantial savings if three Connecticut hospitals used one centralized maternity service; (c) 140 California hospitals achieved cost savings in excess of $18 million annually in three major service departments through participation in a cooperative, multi-hospital management engineering program.

A large and growing number of systems analysis programs have developed throughout the country, at least 18–20 are of substantial size. In general they are programs sponsored by research organizations or by hospital associations. They tend to have many features in common, but the approaches tend to be adapted to the particular situations encountered. For example, 15 hospitals in three states (Washington, Oregon, and Idaho) are participating in a coordinated program by a group of systems analysts from Battelle Northwest Laboratories. Their objectives are to assist hospitals in maintaining and improving quality of service, the use of resources, control of costs, identification of needs. The group methodologies are generally based on techniques originally developed at the University of Michigan or California's Commission for Administrative Services in Hospitals. Current programs entail studies of both personnel utilization (e.g., nurses or dietary employees) and quality assessment. Standardized forms are developed by the responsible individuals in hospitals working with the professional management engineers to facilitate data collection indicating the time devoted by individual members of the staff to specific tasks with provisions for comparison with standards. By quantifying these activities as much as possible, improvements in utilization or performance can be evaluated over periods of time so that the effects of improvement programs can be gauged. As an example, Nursing Standards Data can be collected on a form which indicates the length of patient stay, usual practices or procedures received, the procedures which may be performed by specialty nurses or other departments. The standard times required for a very wide selection of procedures are listed on the reverse side of the form so that the comparison can be readily attained between the actual time spent and the standard. In addition, the total time utilized for the various activities can be compared with the total predetermined standard to gauge individual performance. Obviously, a single chart of this sort has little value but a sequence regularly assembled over a period of time can provide an objective indication of how service and personnel effectiveness are affected by organizational changes or innovations introduced by responsible supervisors.

Assessment of quality of nursing care is being approached by a pilot program with random sampling checks of specific criteria individually tailored for different types of hospital wards and settings. The criteria include the kind of evidence of proper care that a nursing supervisor would look for, including the proper storage of medicines, discontinued drugs returned to pharmacy, state of order, and location of equipment on the ward, organization of nursing station and a large number of specific checks on the condition of the patients' charts, which indicate whether the laboratory work, orders, treatments and graphic sheets are being properly taken care of. In a systematic manner, these data are compiled at random intervals and evaluated to determine status and trends in the quality of nursing service. The ultimate value of such studies cannot be fully assessed but a body of data and experience is accumulating which may be of growing value as the health professions are required to face squarely the cost and benefit of what they do.

Systems engineering is most readily applied to procedures or processes in production operations where the value of the product has an established monetary value and the cost of producing it can be accurately assessed. In delivering medical care, the value of the product cannot be assessed by any existing standard and the accounting methods generally applied give no true picture of the costs. The great danger lies in the natural tendency to focus attention on those aspects which can be measured and expressed in numerical terms, neglecting intangibles which are absolutely essential but subjective in nature. The comfort, peace of mind, and satisfaction felt by a patient cannot be graded nor can the sympathy, warmth, or responsiveness of the health professionals or staff. Mechanisms for administering care must not disregard such factors.

References

1. Klarman, H. E. "The Economics of Health," p. 200. Columbia University, New York, 1965.
2. Hagedorn, H. J., and Dunlop, J. J. Health care delivery as a social system: inhibitions and constraints on change. *Proc. IEEE* **57**, 1894-1900 (1969).
3. Harris, S. E. "The Economics of American Medicine." MacMillan, New York, 1964.
4. Lee, P. R. Creative federalism and health programs for the poor. Pharos., **30**, 2-6 (1967).
5. Kamenetz, H. L. "The Wheelchair Book: Mobility for the Disabled." Thomas, Springfield, Illinois, 1969.
6. "Accidental Death and Disability: The Neglected Disease of Modern Society." Committee of Trauma and Committee on Shock, Division of Medicine Sciences, National Academy of Sciences, National Research Council, Washington, D. C., Sept. 1966.
7. Barton, G. M., and Barton, W. E. A comparison of medical and psychiatric emergency care in the U. S. and Russia. *Resident and Staff Physician*, **16**, 50-57, May 1970.

8. Rushmer, R. F., Van Citters, R. L., and Franklin, D. L. Shock: A semantic enigma. *Circulation* **26**, 445-459 (1962).
9. Rushmer, R. F., Baker, D. W., Harding, D., and Watson, N. Initial ventricular impulse: a potential key to cardiac evaluation. *Circulation* **29**, 268-283 (1964).
10. Guidelines and Recommendations for the Planning and Use of Obstetrical Facilities in Southern New York. Hospital Review and Planning Council of Southern New York, Inc. January 1966.
11. Moss, D. W. Automation in Clinical Laboratories. In "Advances in Biomedical Engineering and Medical Physics" (S. Levine, ed.), Vol. II. Wiley (Interscience) London, 1968.
12. Dexter, C. V., and Larsen, M. L. Centralized large-scale clinical testing in a commercial environment. *Proc. IEEE* **57**, 1988-1995 (1969).
13. Automated Multiphasic Health Testing and Services (provisional guidelines), Report of the AMHTS Advisory Committee to the National Center for Health Services Research and Development. U. S. Department of Health, Education and Welfare. Health Services and Mental Health Services, July 1970.
14. Blumberg, M. S. "Shared Services for Hospitals." American Hospital Association, Chicago, Illinois, 1966.
15. Weymes, C. "Planning a Regional Sterile Supply Service." Western Regional Hospital Board, Glasgow, 1968.
16. Knowland, Lt. Col. R. W. Area industrial zones. *Brit. Hosp. J. Soc. Serv. Rev.* **77**, 2224-2228 (1967).
17. Management Engineering for Hospitals. "A.H.A. Committee: Booklet on Management Engineering for Hospitals." American Hospital Association, Chicago, 1970.
18. Bennett, A. C. "Methods Improvement in Hospitals." Lippincott, Philadelphia and Toronto, 1964.

Chapter 4

HEALTH CARE DISTRIBUTION

REQUIREMENTS FOR REGIONAL PLANNING

A shortage of physicians is commonly blamed for the lack of available medical care to large segments of the population. The high level of physician's fees and the peculiarities of medical economics interfere with the common concept of supply and demand, as discussed below. Since there is a limitless demand for good health and medical care, the shortage of physicians will always be relative, perhaps even insoluble. The most important deficiency in modern medical manpower is the extreme distortion of its distribution, such that an oversupply exists in some situations (i.e., suburbia), and serious deficiencies exist in others (rural districts and central cities). As a result large segments of population in different geographical locations have little or no access to medical care as indicated in Table 4.1. Health care is no longer regarded as a privilege of those who can afford it but a right to which any citizen is entitled. At present no one person or no organization has accepted responsibility to see that health care is widely and fairly distributed to all who need it. Ribicoff (1) assailed the myth that the United States is the healthiest nation stressing that the excellent care available in the medical centers is not extended to some twenty million Americans who receive inadequate care or no care at all. A survey of 1500 cities and towns in the midwest showed that 1000 had no doctor at all and 200 had only one in 1965. Clearly, adequate incentives have not been provided to induce physicians to endure the stresses and inconveniences of practice in rural areas. They gravitate to the cities and particularly to medical centers because of obvious attractions.

DISTRIBUTION OF PHYSICIANS

The greatest differences in distribution of physicians are displayed by specialists with a ratio of 4 : 1 between greater metropolitan areas and isolated counties (Table 4.1). Although 45% of the population reside in communities with less than 5000, only about 19% of the physicians practice there. In rural areas the financially solvent can pay for health care but it is frequently available only after traveling considerable distances. The rural poor and migrant workers in some sections of the country never see a physician and do not participate in even public health measures which can so greatly reduce the likelihood of many infectious diseases (Fig. 4.1). The ghettos of central cities are also seriously deficient in health care. In many instances, provisions are made for the disadvantaged who do not take advantage of opportunities because of the need for baby-sitters, cost of transportation, long waits in clinics, suspicion of the system, and a wide variety of other deterrents.

The plight of physicians in isolated rural areas is difficult to resolve in view of the variety of problems they face. They generally constitute the only medical care available over a very wide area so they are subject to call at any time. Unlike the physicians in populated areas, they have little choice but to respond, since no alternatives are available in the vicinity. Both the physician and his wife may

TABLE 4.1

Distribution of Physicians (Physician–Population Ratios)[a]

			Active physicians—Nonfederal ratio per 100,000 population			
County group	1950 Population (thousands)	Number	Total	Specialist	G.P.	Hospital, etc.
United States	150,697	179,041	119	37	64	18
Metropolitan adjacent	109,272	148,498	137	45	67	25
Greater metropolitan	44,946	77,262	173	59	79	35
Lesser metropolitan	40,632	52,887	131	48	59	24
Adjacent	23,694	18,349	78	14	58	6
Isolated	41,425	30,543	74	14	55	5
Semirural	33,177	26,421	80	17	57	6
Rural	8,248	4,122	50	2	48	0.5

[a]From S. E. Harris, Reference 2.

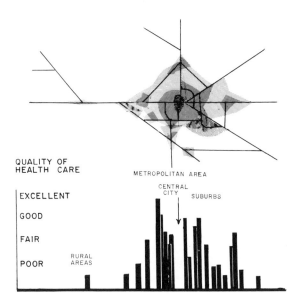

Fig. 4.1. Hospitals and other health care facilities tend to be concentrated in metropolitan and suburban areas while central cities and rural areas suffer serious deficiencies.

become frustrated for lack of forseeable relief. Isolated rural areas are frequently relatively undesirable places in which to live and there are ample opportunities to engage in practice under happier circumstances. Providing appropriate incentives to rural physicians will have to extend beyond merely financial return. Possibly some mechanisms might be considered, by which a substitute could be provided for a few months at regular intervals so that such physicians could engage in continuing education in some medical center.

Very few physicians are currently serving the ghetto areas of central cities. They function under conditions just as pressing and undesirable as the rural physician. The fact that help is nearer but unavailable to destitute individuals must be even more frustrating. The migration of the population to the cities and the rapid development of megalopolis will undoubtedly change the complexion of this problem during the next two decades (see below). The pressures on physicians who are attempting to provide medical care for the disadvantaged in both the central cities and rural areas can only mount. The trends in urban settlements as they may affect health care delivery will be discussed presently.

IMPROVING PHYSICIAN DISTRIBUTION

The main problem of providing adequate medical care in rural areas requires development of appropriate incentives drawing physicians away from the attractions of medical centers and metropolitan areas with their improved financial return, physical, social, and cultural surroundings, the intellectual stimulation of medical centers, facilities, colleagues, and consultants. With increasing specialization and migration to the cities, these incentives have clearly been inadequate in the past. Some rural communities have offered to build hospitals or clinics to order designed to attract a physician—some without success. Analysis of the problem has indicated that prior residence is a most important determinant of the selection for practice. This suggests special subsidies for medical students from rural areas might be awarded with a commitment that the student would practice in that rural area no less than five years after graduation. Another alternative is a form of public health service as a commitment to practice in either specified rural areas or in central cities in lieu of military service. If the military draft is discontinued, scholarships for medical students might entail corresponding commitments to those of the military for service in prescribed locations for three or four years after graduation. It might be desirable to encourage or require externships in remote rural areas or in central cities as part of medical curricula.

WAMI, a new concept of regionalized medical education for the states of Washington, Alaska, Montana, and Idaho is an example of an innovative approach to increasing the number of physicians and encouraging more uniform distribution in outlying areas. The curriculum is designed so that major portions of the first two quarters of course work (e.g., molecular and cellular biology, embryology, tissue structure, control systems and homeostasis, etc.) can be set up in existing educational institutions at various locations in the region with support and overall supervision from the University of Washington. A core curriculum from the second to the sixth quarter would be provided at the University of Washington for coverage of the major organic systems and introduction to clinical medicine. All the clinical training after the sixth quarter could be provided by private physicians in local community hospitals specifically selected, organized and reimbursed for training along pathways for family medicine or particular medical and surgical specialties. One of the major objectives of this approach is to increase the numbers of medical students coming from these less populated states, in hopes that larger numbers of physicians will elect to practice there.

DISTRIBUTION OF HOSPITALS AND HEALTH FACILITIES

Deficiencies of health care in rural and isolated areas stem from inadequate numbers and types of physicians and hospitals, both of which tend to be poorly distributed. Many rural and isolated areas are totally devoid of hospital facilities. Both the voluntary (nonprofit) hospitals and government-owned hospitals tend to be concentrated to an excessive extent in metropolitan areas. The voluntary hospitals are generally relatively small and have been intensely competitive, attempting to achieve full utilization. The best way to attract clients is to cater to the needs and wishes of the physicians who can be induced to use the facilities. As a result, hospitals have tended to develop where the doctors are in largest concentration; namely within larger metropolitan areas. For example, a single aerial photograph includes eight hospitals in the center of Seattle (Fig. 4.2). Harborview Medical Center is a county hospital and the remainder are all private hospitals, competing intensely to attract physicians and patients with a full spectrum of facilities and services. Locating many highly competitive hospitals in the center of a city may have been convenient for physicians in earlier times, but it is now functionally and economically disastrous. The access to hospitals by physicians, patients and the sources of supply is seriously complicated by the traffic and parking problems. Metropolitan locations with their noise, pollution, and congestion are not appropriate for the care of sick people. In addition, the exorbitant land values, logistics, and intense competition to supply extremely expensive medical and domiciliary services are contributing greatly to the large and rapidly expanding costs of health care.

ARE WE GETTING OUR MONEY'S WORTH?

The United States would be the world's healthiest nation if cost were any indication of quality (1). We spend more money on health and medical care than any other country in the world (i.e., $63 billion a year, $294 per person, and 6.7% of the gross national product). Comparisons based on available criteria (life expectancy, maternal mortality, infant mortality, distribution of care to the total population) reveal that a dozen or more countries score better than we do despite appreciably lower expenditures for health. For example, some of the countries of Western Europe spend considerably smaller amounts per person and a smaller fraction of their gross national product than we, and yet are superior to the United States judged by these critical criteria. This discrepancy will be considered in subsequent sections of this chapter.

BUYING MORE MEDICAL CARE

The traditional approach to financing medical care is the direct payment by the individual of physicians' fees, hospital charges and pharmaceutical prices to the vendor. This approach previously resulted in the best care being provided to the wealthy, the indigent obtained care through charity and all levels in between might suffer deficiencies or total lack of medical and dental services. The concept of paying fees for services directly by the patient to the physician

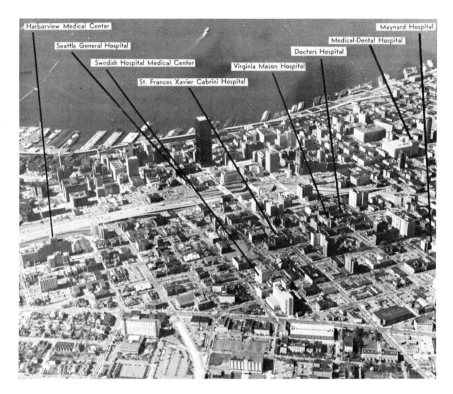

Fig. 4.2. Many small, competitive private hospitals tend to be grouped without co-ordinated planning in congested areas of cities. They are fundamentally uneconomic because of their size, composition, organization, and location as discussed in the text. (Photograph courtesy of the Seattle *Times*, May 10, 1970.)

has been the principle defended most vigorously by organized medicine. In place of this simplistic approach, a bewildering array of financial mechanisms have evolved. A significant move in this direction was the development of voluntary insurance.

VOLUNTARY HEALTH INSURANCE

The concept of obtaining protection against high cost catastrophy with low probability of occurance is implicit in insurance against losses from fire, accident or theft. A corresponding mechanism to defray costs of health care has become very widespread, such that three-fourths of the population have some kind of insurance for medical care (mainly for aspects of hospitalization). Something less than half have coverage for visits to physicians. Unlike the basic principle of insurance against devastating occurances, the probabilities that visits to physicians or hospitals are quite high and for this reason the insurance rates have progressively risen as they really serve more as a means of budgeting for expenditures. It is becoming increasingly difficult for many families to pay the progressively larger premiums. Expanding numbers of individuals are suffering from inadequate coverage by insurance. Many who need insurance protection most desperately are not covered, such as those with "high risk"—the aged and chronically ill who are left to fend for themselves. As the costs of premiums rose, more and more of the high risk patients were excluded from the programs. These individuals tended to turn to other insurance sources that would accept them increasing the concentration of individuals who would receive large payments and exposing such a program to a dilemma; either increasing premiums even more drastically or incurring indebtedness.

Neither of these courses are acceptable and the end result is a strong trend toward commercial insurance companies tending to deny those who are likely to need care and concentrating on those who are not.

The percentage of patients who have hospital or surgical insurance coverage is closely related to family income—ranging from 34% among those in families of less than $2000 income to almost 90% of persons in families of $7000 or more. The coverage of families of lower income groups with four or more children in almost negligible. This voluntary insurance is held predominately by those who need it least and is ineffectual among those who are most likely to find illness an economic catastrophy. The existence of voluntary insurance contributes to the rapidly rising costs of health care for several reasons. The insurance companies tend to merely pay the bills or charges and exert no effective pressure toward economy or efficiency. Indeed, both physicians and patients are indirectly encouraged to utilize expensive hospital services for ailments which could equally well be handled on an ambulatory basis or in less expensive facilities. It is becoming apparent that voluntary insurance alone is not an adequate answer. Group practice provides a mechanism for improving the efficiency with which quality health care can be provided.

GROUP PRACTICE

The inefficiency and waste of solo practice has been well documented and widely recognized. A group of physicians in sufficient numbers and diversity can provide more health care in an organized setting than could a corresponding number of practitioners functioning individually. Many advantages have been cited for group practice, with its opportunities for sharing facilities, personnel, and services.

The types of groups are quite varied, ranging in size from a few to several hundred and in organization from loose associations to highly structured clinics (i.e., the Mayo Clinic). Some provide comprehensive care while others are groups of specialists with common interest.

Independent plans have developed with a variety of sponsorship, such as employer-employee-union, medical society, fraternal, community, consumer cooperative, private physician, and commercial. Although the various types of independent health plans have grown rapidly in the past few years, only about 10% of physicians are involved in group practice, despite evidence that the net earnings of groups is greater, in addition to the remaining incentives (2). The arguments in favor of group practice seem more persuasive to this observer, particularly as a trend which is most likely to provide new emphasis on cost-benefit and a setting in which biomedical engineering may well provide valuable operations analysis and new technology. Prepaid, comprehensive health insurance programs provide an example of a mechanism which has built-in incentives for improved cost-benefit relationships.

PREPAYMENT COMPREHENSIVE GROUP INSURANCE PROGRAMS

The Kaiser Permanente plan originated in California during the depression years as a means of providing health care through prepayment to a group of physicians in an integrated clinic and hospital facilities. The evolution of this plan and its projections into the future were described by Garfield (3). It now has more than 2 million subscribers served by 51 outpatient clinics and 22 hospitals in California, Oregon, Washington, Hawaii, and in the cities of Cleveland and Denver. The program is organized as a nonprofit organization and is completely self-sustaining. The integration of the facilities and services has served to demonstrate that substantial savings can be achieved while providing widely acceptable medical care. In addition, three additional principles can be identified as operating in this system. The payments of the plans cover comprehensive care rather than specific services. Incentives are built into the system that individual members of professional staff stand to benefit from avoidance of wastage or overuse. The physicians on the staff are involved in the administrative and operational decisions that affect the quality of the care they provide.

Elimination of the fee for service through prepayment schedules has tended to encourage individuals to seek medical advice and care at earlier stages of illness and to participate in periodic examinations in an effort to benefit from preventive medicine. In contrast to individuals who directly pay fees for service, the subscribers to prepaid plans can take advantage of early treatment with prevention of complications. However, elimination of the fee for service also released a flood of subscribers that clogged the system. "Elimination of the fee was practically as great a barrier to early sick care as the fee itself" (3). The greatly increased appointments to enter the system on a first come-first served basis comprised a much larger proportion of individuals who were not really ill. They included individuals who were well, who had ill-defined symptoms, who were concerned about their health, who were slightly ill, moderately ill, and seriously ill. All of these people swamped the facilities and services of the group to the extent that the system groaned under the impact. It has become abundantly clear that restoring efficiency to the system requires a process by which the various categories of patients can be identified, separated, and handled in appropriate ways (see Fig. 5.5).

SEPARATION OF THE WELL, THE WORRIED WELL, THE EARLY SICK, AND THE VERY SICK

The establishment of entry priorities of individuals into a health delivery system required identification of the components of the entry mixture of patients in terms of their medical requirements rather than their ability to pay. Morris Collen, an electrical engineer and physician, has directed the development of a complex battery of equipment originally designed for health check-ups to detect illness for preventive care. This approach can pave the way toward developing efficient methods of screening large numbers of individuals rapidly, efficiently, and inexpensively for the purpose of channeling the well, the worried well, the early sick, and very sick into appropriate channels for proper handling in an integrated system. This concept of multiphasic screening or simply health screening is widely known and represents a focus of biomedical engineering on a problem of extreme importance for the future. Such multifactorial screening is regarded as an important method of dealing with a major influx of patients with a large proportion of healthy individuals (e.g., in a prepaid comprehensive group health plan). An uncontrolled deluge of patients into our already overextended and inefficient medical care channels resulting already from Medicare and Medicaid threatens collapse to the system unless we can develop some rational and effective mechanism for identification and diverting a mixed influx into appropriate channels in relation to the severity of their illnesses. Otherwise, our institutions designed to handle acutely ill patients will be clogged with worried well and early sick that could better be handled as ambulatory patients or in other types of facilities (see also Chapter 3).

All signs point to a rapidly expanding requirement for medical care for a rapidly growing clientele. The affluence of large segments of our present population has greatly increased the segment who can afford to pay directly for medical care or to share the risk of illness through insurance. The population growth continues to add new potential clients at a rapid rate. Migration of populations into the urban areas and immigration into the suburbs add to the overload on already congested facilities. The projected obligations and opportunities for biomedical engineering contributions must be predicted on a realistic impression of the nature and extent of stresses on the health care delivery capability of this country.

THE IMPACT OF FEDERALLY SUPPORTED HEALTH INSURANCE

Health care for all citizens is becoming widely recognized as a responsibility of society. One of the most prevalent approaches has been the development of insurance plans to cushion the blow of catastrophic expense and to spread the expenses over longer time and among larger segments of the population (4). It is claimed that 59 nations provide "state medical care," covering medical care services to certain portions or all of the population under a form of compulsory or subsidized insurance systems (5). The conviction that all modern prosperous and progressive nations should have some form of compulsory medical care is growing in many areas.

The United States is moving inexorably toward some form of general or compulsory health insurance scheme designed to extend the availability of medical care to all who need it. Federal support of medical care is not nearly as new as it might seem since the first Congress in 1798 passed legislation initiating compulsory health insurance program for merchant seamen who were required to make monthly contributions and received medical care through the antecedent of the United States Public Health Service. With the launching of the Social Security Act of 1935, a series of health insurance bills were introduced in Congress with remarkable regularity. Compulsory health insurance plans were proposed to Congress by President Truman in 1949 and 1950. During this period when almost endless debates were resulting in no decision, many of the countries in Western Europe were instituting large-scale programs to provide medical care to most or all of their populations by means of a variety of mechanisms. As a result these countries have adapted to stresses and strains which are just now being felt or threatened in the United States since the emergence of Medicare and Medicaid.

MEDICARE-MEDICAID

Medicare was a compromise consisting of a two-part extension of the Social Security Act (Title XVIII). Part A offered hospital, nursing home, and home health services financed by compulsory contributions of employers and employees through the social security system (with an earmarked payroll tax and trust fund). Part B created a supplementary voluntary medical insurance program with contributions from general federal revenues. Title XIX (Medicaid) extended the protection beyond the older age group (above 65) to include the needy recipients of other public assistance programs and ultimately to all "medically indigent." The enormous effort required to implement such legislation must be recognized by anyone considering modifications or changes in nation-wide programs. For Medicare alone, approximately 19 million people became eligible for benefits on July 1, 1966. (6). The complicated procedures required to register such an enormous group of widely scattered people, to develop and maintain precise records and to institute some sort of surveillance over utilization and quality of services was an enormous undertaking. The organizational effort for launching Medicare has been compared to that required for the Normandy invasion in World War II. It probably could not have been accomplished without the long previous experience of the Social Security Administration. The governmental machinery is impressive in its size and scope, but most of the operational work was and is accomplished by private organizations serving as intermediaries between the patient and the government, such as the Blue Cross. Association, and various insurance companies. Despite efforts to project the expenditures and restrain increasing expenditures, the costs have increased much faster than had been anticipated. For example, the rate at which increased costs of Medicaid and of Medicare rose above predictions are illustrated schematically in Fig. 4.3. This experience is common to all federally supported programs for health and is the most obvious complication in programs where the magnitude of the costs can be assessed.

In general, the Federal Government extended its support of medical care for the aged and indigent by supplying money to "buy" more services but made no provisions or proposed no mechanisms for adapting the system to the increased load.

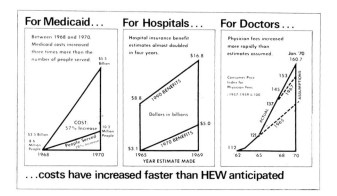

Fig. 4.3. The costs of Medicaid, hospitals, and doctors' fees have increased much more than anticipated. This is the common tendency in government-sponsored health care programs.

According to Bennett (7) both Medicare and Medicaid are political compromises that specifically exclude interference with the existing system of medical practice. For this reason, they have merely exposed the deficiencies of the current system without providing mechanisms by which it can meet the increased demands. If more funds are made available for either a product or a service without mechanisms for greater output, the inevitable result is an inflationary rise in costs. The increasing total costs of medical care have not produced a corresponding increase in the quantity or quality of health care. Indeed, this goal cannot be achieved without rather drastic changes in both the organizational relationships and tools, a process for which biomedical engineering is equipped to contribute. In most segments of industry, technological advances have generally been accompanied by fewer workers with higher salaries. In contrast, medical technology has increased both the number of workers and their wages. This is a sign that far too little attention has been directed toward proper uses of technology and to the elimination of excesses or waste in hospitals. There is a real need to reassess and restructure the health care capability of this country to develop a system which will meet the demand at lower costs. Improving the cost-benefit relationships will involve systems analysis, management engineering, and new technology in a more appropriately organized system.

NEXT STEP—COMPULSORY HEALTH INSURANCE?

Medicare and Medicaid have combined to greatly stress an overloaded and outdated mechanism for delivery of health care. Despite their predictable high and soaring costs, they represent only a first step in an inexorable sequence toward full medical coverage of the United States population. Already signs of political interest in this attractive plum are emerging in the form of tentative proposals for bills to be submitted to Congress in upcoming sessions. The political attraction of sponsoring and supporting this type of legislation seems utterly irresistible despite potentially disastrous effects that could result from premature imposition of such a stupendous load upon our belabored health care personnel and facilities.

HEALTH MAINTENANCE ORGANIZATIONS (HMO'S)

The latest in a series of proposed mechanisms for improving availability and distribution of health care delivery is the concept of the Health Maintenance Organization. The concept is not yet fully described or defined, but the general intent is to provide financial incentives to limit utilization of services and facilities beyond that actually needed by the patients. The contract mechanism is proposed as a new method of paying for medical care. The direct payment of fee-for-service has been extensively replaced by fixed payments of patients, families, or groups to third party (government or insurance carriers) who then payed the physicians for services. The basic change proposed in the HMO is to institute a fixed payment mechanism between the third party and the providers of health care. Thus a contract would be developed by which a designated organization of health providers could be paid in advance for complete medical care for a designated group of the population. This approach is clearly based on the successful use of financial incentives by prepaid comprehensive group health cooperative programs. It calls for the organization of health care personnel and facilities to provide comprehensive health services with standards of quality and utilization, with provision for peer review and the maintenance of central records for purposes of accounting and quality control. Appealing features of the tentative HMO proposals include better utilization of personnel, facilities, and equipment, services built around facilities, and the successful utilization of such approaches by prepaid comprehensive programs such as the Kaiser-Permanente program or Group Health. To date, incentives have not been provided which would encourage health care providers (physicians and health care facilities) to organize into Health Maintenance Organizations. If appropriate governmental inducements can be developed, the concept may be viable. Lacking such incentives, the concept has little prospects of realization.

Health care delivery in the United States is insufficient for current or projected demands. Included among the future sources of additional requirements and stresses are the unremitting changes in the population of the country, particularly the growth of the total population and its migration toward the cities.

URBANIZATION AND HEALTH CARE INSURANCE

The migration of people from rural areas and small towns to the major metropolitan areas is a phenomenon of the twentieth century. It stems in part from the movement of agricultural workers from mechanized farms toward the congestion, excitement, opportunity, and oppression of the cities; the scene of the action. The factors which impell this surge of humans in large numbers from the peace and tranquility of the countryside to the tension and clamor of the city are not fully understood, since it occurs in industrialized and in developing countries alike. The growth rate of modern cities portends disintegration and central decay unless the will, intelligence, and dedication of the people can be mobilized to develop and implement comprehensive plans including integration of essential components to provide the facilities and services for its functions. These requirements are equally applicable to the mechanisms for delivering medical care which have also grown haphazardly without plans or consistent guidance. The result is poor distribution of facilities and professionals, awkward location, and intense competition without mechanisms for integration. Such a system works poorly under current circumstances but threatens to break down under increasing load. As cities expand and merge, comprehensive plans will have to be developed and instituted promptly to provide the necessary services.

In this period of rapid change but widespread communication, we should make every effort to learn from the experiences and feasibility studies of others in total world laboratory. The most precious commodity we have is time and a failure to recognize the applicability of the experience of the others can waste time and resources which cannot be replaced.

Many of the countries of Western Europe have developed mechanisms for the delivery of high level health care to their entire populations by a variety of different mechanisms. The quality of medical care to be found in the best of the medical centers in Europe is fully competitive with the best that is offered in the United States. However, the average level of health care and its distribution to the population as a whole is unquestionably superior in several different countries. It would be the height of folly were we to fail to take into account the successes and problems encountered in these enormously expensive feasibility studies in Europe, selecting the most successful approaches which are applicable to our particular situation in the United States.

EUROPEAN EXPERIENCES WITH NATIONWIDE PROGRAMS

After World War II most of the countries of Western Europe initiated programs for extensive or comprehensive governmental support of medical care either through a National Health Service as in Britain or through subsidized insurance plans. In general these programs represented the assumption by government of major responsibility for the health, welfare, and economic opportunity of the entire population. The programs that emerged in each country exhibited major differences reflecting their previous history, traditions, and socioeconomic conditions. In some of these countries the results of a national health program has produced such high quality facilities and services that we in the United States can well be envious. (See the section on the Swedish prototype, page 122). Other countries have achieved wide distribution and availability of medical care to the entire population but with complications and less satisfactory results. A comparison of the various systems for delivery and support of health care in the various countries of Europe would not be possible and is not a primary objective in this manuscript. However, a recent period of study in Europe provided an opportunity to make some personal observations and arrive at some generalizations, some of which are pertinent to this discussion.

SOME BASIC CHARACTERISTICS

Scandinavia, the Benelux Countries, Germany, Switzerland, and Great Britain have progressed much further toward governmental responsibility for the health, welfare and economic opportunity of their citizens, than has the United States. Among those continental countries, there is full employment; even a labor shortage of such degree that foreign unskilled labor is actively recruited in many areas. These countries are small, compact, and unencumbered by many problems besetting the United States. In this environment of full employment, booming prosperity, and seasoned social welfare programs, the institution of nationwide health programs with governmental support was a natural sequence. In countries with well-established socialistic programs, the inevitable delay and red tape in many aspects of life produces much less reaction than would be encountered in North America. Furthermore, the medical profession is organized on a different basis in Europe; most hospitals are staffed by specialists and most general practitioners have no hospital privileges.

CENTRAL ROLES OF THE GENERAL PRACTITIONER

The general practitioners are the primary care physicians, making initial and repeated contact with their patients. When hospitalization is required, the general practitioner refers the patient to a hospital where the specialists on staff assume responsibility until the patient is discharged. Thus the general practitioner is the initial contact and the mainstay of the medical care systems of Europe with a distinct dicotomy between him and the hospital specialists. As a general rule, the specialists in hospitals are relatively contented, their pay is higher, they wield much power, the working conditions are advantageous and the patient load manageable. In contrast, most general practitioners in the same countries are harried, hurried, frustrated, and dissatisfied. In many instances, the patient load carried by the general practitioners is so great that their role of primary care is diluted and the quality of the patient care is necessarily reduced, often drastically (see below).

The reliance on general practitioners is generally so great that the medical care system could not function without them. In Great Britain, all adult individuals are expected to enroll with a general practitioner. Except in emergencies, patients must consult this general practitioner directly or for referral to hospital or outpatient facilities. The physician's income is often related to the number of patients on his list. According to Dr. Tostemark, the government of Denmark encourages the current balance of some 3000 general practitioners of a total of 7000 physicians in the country. Among the incentives are governmental support of the formation of group practices. The central role of general practitioners in providing the interface and integration of health care for virtually the entire population is very different from their role in American medicine where they are declining in numbers despite widespread desire for their services.

APPARENTLY FREE OR ARTIFICIALLY CHEAP

As a general rule, the public approves of services supplied by mechanisms which tend to reduce or obscure the true costs. Insurance tends to protect the individual from risk of high and unexpected expenditures in case of fire, theft, accident or injury. Many essential services and necessities are provided at reduced cost or apparently free by government. Under various circumstances, water, waste disposal, transportation, utilities, and other essential features of modern living are subsidized so that the consumer comes to regard them as essentially "free" or freely available. Removal of the economic restraints on utilization of these services inevitably leads to overuse and waste. Precisely the same result has occurred in the field of health. In countries where the government has assumed responsibility for providing hospitalization, physician's services, or drugs, the demand has swelled into a flood which tends to defeat the original purpose to some degree. The withdrawal of financial obstacles to medical care has frequently resulted in reduced access or inferior service through overuse, congestion, or excessive costs. Virtually all the health care programs in Europe are experiencing saturation. The escalating costs are approaching ceilings beyond which they encounter progressively increasing resistance from taxpayers.

In instances where the actual costs remain visible to the consumer, the tendency toward excessive use has apparently been significantly retarded. In the Swedish version, physicians originally charged the patients for their services on the basis of established fee schedules and the patients then applied for partial reimbursement from the government. In many of the programs, partial payments for drugs, glasses, or other products have been reimposed to stem the flood of demand. The fact that the patients retained the operational and financial responsibility for their own medical care seems like a sound approach to the ubiquitous problem of overuse and waste occurring when subsidized services are "apparently free." The concept that the consumer should have responsibility for the costs of such services (even if he is fully reimbursed) is attractive. When the responsibility is shifted to the government or to the physician to arrange for the payment of fees, a large proportion of the general public regards the services as their right and the constraints on overuse evaporate. Under these conditions the system tends to bog down through oversaturation depriving everyone of its full benefits. A look at freeways at rush hours is a case in point.

SIGNS OF SATURATION

The delivery of health care displays a characteristic found in communication systems that saturation of a channel produces deterioration in the quality of the output until the only product is noise. Experience has repeatedly demonstrated that segments of a population become very adept at milking a welfare system. In Holland, the average citizen pays about $160 a year for health insurance and then is not required to pay anything for a visit to a physician. The general practitioners' offices are crowded with a very diverse mix of clients, including a heavy concentration of individuals who would never dream of coming to a physician if they had to pay a small fee (even fully reimbursed). Such physicians have told me that all too often former patients made appointments merely to inform the doctor that they felt fine and had no complaints. A common attitude is that they paid for medical care and they are entitled to it whether they need it or not. A corresponding situation prevails in West Germany where 95% of the population have medical coverage. The daily patient loads carried by the general practitioners are so great that detailed histories or direct physical examinations are coming to be regarded as exceptional outside of hospitals.

SCREENING BY WAITING LISTS

Prompt access to hospitals in emergencies is almost always possible in any of these countries. Patients injured in accidents or with acute illness can be admitted without delay. In contrast, patients with serious, but not acute, need for hospitalization may be required to wait weeks for admission. Waiting periods of one or two months for entrance into hospitals is fairly common. In fact the magnitude of waiting lists and the duration of the delays are utilized as indicators of need to expand facilities or services. There is an unstated assumption that consistently prompt admission to a hospital for elective care is evidence of an oversupply and a waiting list of manageable proportions is an effective screening mechanism. Indeed, I was informed that a waiting period is one means of providing care to "those who really need it." This seems to neglect one important consideration; namely, that those individuals who have the time and inclination to wait around in physician's offices or on waiting lists of hospitals are not likely to be the most productive members of the community.

More objective screening techniques have potential value for identifying the general nature and extent of illness in ambulatory care and in assignment of priorities for hospitalization.

Among the criticisms leveled at a "National Health Service" are the waste of time, energy and money on care which any reasonable individual might be expected to provide for himself, the steady deterioration of standards due to unlimited demands on a "free" hospital service, lack of appreciation by the patients and the lack of satisfaction by the physicians. Criticism of the services, facilities, or functions of hospitals are relatively uncommon on the part of either the patients or the professional staff. The hospitals in many of the European countries (e.g., Scandinavia) are extremely impressive, particularly in terms of the very large size and high quality of the facilities and the functional interrelations between the larger and smaller hospital units. These characteristics are clearly the result of comprehensive planning on large scale, allocation of ample resources and a degree of organizational interaction among government, the profession, and the public unlike anything to be found in the United States.

Exploratory utilization of new techniques and technology in these hospitals is advanced beyond that in the United States even when the equipment comes from this country. For example, prototypes of computer applications are among the most extensive and imaginative in the world but the hardware is almost completely of American design. New construction of hospitals is very widespread and according to comprehensive plans derived from analysis of extensive information.

The traditional hospitals consisted of many different buildings housing the various medical and surgical specialties. The Karolinska hospital in Stockholm is a typical example (Fig. 4.4A). The new hospitals being added to such complexes tend to be large central blocks containing many components in well-integrated facilities and organizational relationships (Fig. 4.4B). The health care system of Sweden can serve to illustrate some of the benefits which can be derived from comprehensive regional planning, in a more favorable setting than exists in the United States. We can well afford to learn from their experience.

THE SWEDISH PROTOTYPE

Americans are reluctant to admit inferiority in any sphere of interest, least of all in such a central issue as the protection of health. The enormity of our current deficiencies and the drastic changes which would be required to correct them can best be appreciated by a rather odious comparison with certain highly advanced systems of health care developed under fortunate circumstances in the Scandinavian countries.

Sweden has been at peace since 1815 and has been spared the ravages and cost of two world wars. It is an extremely prosperous nation, having an increase in the gross national product of 400% between 1950 and 1967. It enjoys full employment, a reputation for high quality industrial goods, has little racial tension or foreign commitments and expends a very large proportion of its total taxable income on health and welfare. It is politically stable in the sense that the Social Democratic Party of Sweden has been in power since 1932 with but a slight interruption during World War II. This party's candidates have been repeatedly re-elected on the basis of a progressive sequence of social reform quite generally approved by the voting public despite the high levels of taxation required to support them. In such a setting it is not surprising to discover a highly organized, very effective system providing for health and welfare of its population of 8—9 million people in this small but prosperous country.

Sweden is a country about the size and geographical shape of California and has a population similar to that contained within California as of 1945 (Fig. 4.5A). The developments in health and welfare of Sweden should be considered in this context. A very large proportion of the taxes received by federal and local governments is expended for health care which is freely available to all citizens. Patients enter the hospitals by referral from their doctors and hospitalization is paid through indirect means. The costs of outpatient care are covered primarily by a combination of insurance and taxes but the individual patient pays a small token fee or charge for this service. A massive construction program of large and well-planned hospitals which are fully equipped according to the most modern concepts is now in evidence throughout Scandinavia. These magnificent structures with high degrees of integration and the most modern equipment and instrumentation are a tribute to long-range plans which are extensions of proposals originating prior to 1940 and continually updated to conform to changing times and ideas (8). Beginning in 1955 a compulsory health insurance scheme was introduced in Sweden which provided "free" hospital care. The patients chose their physician and paid him directly but received reimbursement of approximately three-fourths of their expenditures for medical care outside hospitals (including fees, laboratories, etc.). The reimbursement was dependent upon established charges for various types of outpatient care. Individuals who wished to consult private physicians outside of the health care system were free to do so but at higher costs without additional reimbursement. The cost of drugs and other medical aids were available free or at reduced cost. One favorable aspect of this scheme was the choice of physicians, payment directly to the doctor and subsequent reimbursement of most of the expenditure. This feature also provided an effective deterrent to the overuse or abuse of the system which has plagued other state supported systems (i.e., the National Health Service of Great Britain). Virtually all the population of Sweden can

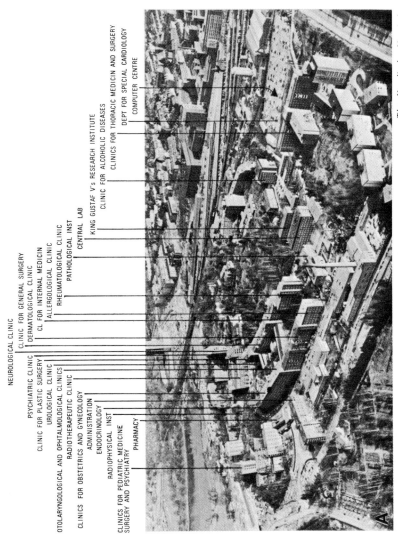

NEUROLOGICAL CLINIC
CLINIC FOR GENERAL SURGERY
DERMATOLOGICAL CLINIC
CL FOR INTERNAL MEDICIN
ALLERGOLOGICAL CLINIC
RHEUMATOLOGICAL CLINIC
PATHOLOGICAL INST
CENTRAL LAB
KING GUSTAF V's RESEARCH INSTITUTE
CLINIC FOR ALCOHOLIC DISEASES
CLINICS FOR THORACIC MEDICIN AND SURGERY
DEPT FOR SPECIAL CARDIOLOGY
COMPUTER CENTRE

PSYCHIATRIC CLINIC
CLINIC FOR PLASTIC SURGERY
UROLOGICAL CLINIC
OTOLARYNGOLOGICAL AND OPHTALMOLOGICAL CLINICS
RADIOTHERAPEUTIC CLINIC
CLINICS FOR OBSTETRICS AND GYNECOLOGY
ADMINISTRATION
ENDOCRINOLOGY
RADIOPHYSICAL INST
CLINICS FOR PEDIATRIC MEDICINE
SURGERY AND PSYCHIATRY
PHARMACY

The Karolinska Hospital.

Fig. 4.4. (A). The Karolinska is an Academic Hospital containing many specialty clinics and research institutes accommodated within many different buildings in a complex which is equal or superior in size and quality to any corresponding institution in the United States. (B) The Central Block recently constructed at the Rigshospital in Copenhagen is an example of the common trend toward the construction of very large, highly organized, and well-equipped hospitals in many large metropolitan centers of western Europe. 1, Central Block; 2, Administration; 3, Pediatrics; 4, Obstetrics; 5, Pathology; 6, Central Kitchen; 7, Psychiatry; 8, Auditoria; 9, Dermatology; 10, Doctor's residences.

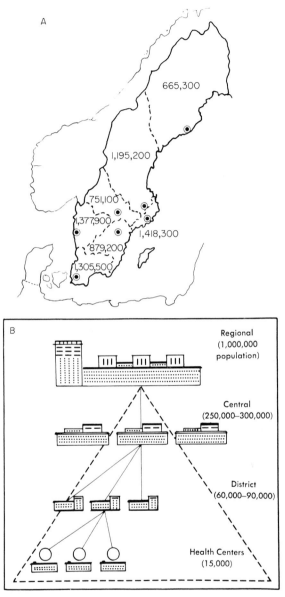

Fig. 4.5. (A) In accordance with long-range plans for health and welfare, Sweden was divided into seven regions, each with a population base of about one million persons. (B) Each region contains a full complement of regional, central, district hospital and health centers organized to provide comprehensive health care, largely under control of local county councils. (After A. G. Engle, Reference 14.)

afford to pay for medical care because of full employment. The Swedish government is pledged to provide work for anyone who wants it and unemployment is rarely a problem so that high-quality medical care has been effectively available to all who felt they needed it within the country.

Health insurance is remarkably comprehensive considering the fact that it covers the entire population with the following provisions.

a. All Swedish citizens and also foreigners listed with the census bureau (also Swedish citizens in other countries)

b. Free dental care for all children and pregnant women; reduced costs for other adults

c. Trips to the hospital or doctor's office

d. Physical therapy and convalescent care

e. Prescribed medicines free or reduced in cost

f. Sickness benefits with extras for each child—for unlimited times in general

g. Childbirth, mother receives care too plus daily payments and additions for multiple births. Additional payments are made for up to 180 days if the mother stays home to care for the child. This is important because a large proportion of Swedish wives are employed, contributing to family income. Such extensive care costs a great deal of money (i.e., 16.4% of the Swedish national income). The cost has also greatly increased as in all other comprehensive health service systems; multiplying three fold from 1960 to 1967.

Beginning in January 1970 a new law came into effect in Sweden which reduces even further the charges for outpatient care. The patient must now pay only 7 Swedish crowns (about $1.30) for an outpatient visit to a physician; this includes costs of laboratory tests. With this change in payment mechanism, the physicians in the system are paid directly by the government based on a rather complicated formula taking into account each physician's previous patient load, the hours he worked per week, and other factors. This further reduction in direct cost to the patient might be expected to stimulate a major increase in demand and lengthening of waiting lists but this has not apparently been excessive. One exceptional example was encountered where a waiting period of 10 months was projected for eye examinations. Emergencies are handled without delay. To the extent that a significant reduction in direct costs to the patient could be accomplished without a grossly apparent increase in utilization suggests that few, if any, patients had been unable to afford care because of cost under the previous three-fourths reimbursement plan. This may be considered a tribute to the widespread availability of health care throughout the country. The overall success of the health and welfare system is a sign of sound policies, long-range planning, and ample funding over a long period of time.

INTEGRATED HOSPITAL FACILITIES

The present organization of hospital care is built upon concepts derived by a government commission dating back to the early 1930's; namely, each county should develop and support a central specialized hospital containing a broad spectrum of medical and surgical specialties and the most modern and extensive equipment.

Beginning in the early 1950's, Dr. Arthur Engle became Director General of the National Board of Health in Sweden and provided sound leadership in evolving principles which are manifest in the current organization of hospitals and health care facilities (9). After analyzing the requirements for various clinical specialties it was concluded that certain activities call for strong centralization to provide the necessary facilities and support (i.e., neurosurgery, thoracic surgery, plastic surgery, radiotherapy). Certain other specialties should not be isolated from the department of medicine to preserve integration of medical care for the whole patient with particular problems involving cardiology, neurology, or endocrinology.

Studies of economic geography, population distribution, and migration were undertaken to discover the optimal size and configuration of geographical areas to provide the most desirable composition and distribution for designating seven different regions with a preferred site for at least one regional hospital in each (Fig. 4.5A,B). The results of this comprehensive study were embodied in a parliamentary act of 1960. The populations in the seven regions ranged between 700,000 and 1,500,000 people, a size believed to provide a necessary financial base for an effective health and welfare service. Each region contained a Regional Hospital, six of which are also called Academic Hospitals as they are closely affiliated with medical schools. These hospitals are very large, ranging in size between 1000–3000 beds with provision for all the medical and surgical specialties, comprehensive and sophisticated laboratories, extensive research, and service functions (Fig. 4.6).

Each region contains several counties (3–5) each of which supports one or more *central hospitals.* Many of the central hospitals in counties and municipalities are as large and well equipped as the academic hospitals. As a rule, the central hospitals range in size between 800 and 1000 beds and are equipped for large outpatient clinics typical of Swedish hospitals. The large outpatient departments are extremely important as a means of handling large ambulatory patient loads and avoiding unnecessary hospitalization. The central hospital also contains many additional services including rehabilitation centers, mother and child welfare centers, family planning, dental clinics, etc.

The local district hospitals are called *normal hospitals.* Prior to the regional planning these hospitals were found to be small, badly located and many were serving such small populations that they lacked provisions for round-the-clock emergency coverage or adequate surgical service. They were regarded as "a headache for the National Health Service." Many of these smaller hospital units have been discontinued or restructured so that they now serve populations of 60,000 to 90,000 people with limited services and few specialties such as internal medicine, surgery, anesthesia and x-ray. Diagnostic and therapeutic problems which are beyond the scope of these hospitals are usually referred to central or regional hospitals so that the patient may take advantage of their concentration of sophisticated facilities and highly trained consultants.

Small rural communities are commonly served by *health centers* or dispensaries staffed by one, two, or three general practitioners. In regions with concentrations of population up to 40,000 inhabitants, health centers may contain as many as 16 to 20 physicians including several specialists. In recent years there is a growing tendency to regard the normal hospitals and district health centers as integral parts of the central or regional hospitals to encourage effective utilization of their facilities. The echelon of facilities provides for patient access to care appropriate to the severity of his illness and integration of the professional personnel through increased communication and referral. On the basis of careful and extensive studies the National Board of Health has adopted a policy directed toward eliminating *all small hospitals or to expand them into normal hospitals of significant size, large enough to be economically sound (i.e., 400 beds or larger).* In very remote areas a one-doctor hospital or cottage hospital may be acceptable but this is an unusual circumstance. This trend toward the elimination of small hospitals because they are functionally and economically unmanageable is particularly striking when one realizes that 90% of the hospitals in the United States are smaller than 400 beds and would be in the category of size which would be considered for elimination by virtue of economic inefficiency in a country like Sweden.

The success of county council control over hospital organizations has led to the parliamentary act of 1961, which allows the county councils to take over supervision of district doctors and their stations. In addition, the mental hospitals are also being voluntarily incorporated into the system under county council control. The private physicians and small private hospitals which have functioned outside the National Health Service are disappearing in Sweden.

LOCAL GOVERNMENT CONTROL OF HEALTH CARE FACILITIES

In Sweden the overall responsibility for planning health and welfare services resides in the national government but the control and the primary responsibility resides within the local county councils. This local responsibility is comparable to the city and county support of schools in the United States. In a lower echelon are civil municipalities which are considered primary and represent the governing structure for cities and for rural municipalities. The size, distribution and population of the counties were originally designed to form an appropriate base for the health care system. Similarly the county councils, since their earliest organization, have assumed responsibility for management of medical care and the support of hospitals as a most important function. The county councils are composed of varying numbers of elected representatives many of whom have exceedingly long and stable terms of office providing a degree of continuity which is very important for the long-range planning required for health care. Under the medical care activities of county councils come the district nurse organization and the mother and child welfare programs. County councils are in charge of the public dental care programs, which are carried out at central and district dental clinics. Closely connected with the medical care responsibilities of the county councils are the county programs for nurses training in 21 county-operated schools. Infant nurseries, maternity homes, correctional institutions and special homes are also operated by the county councils. Since the administrative control and financial support of these institutions are centered at the local level, the success or deficiencies of health, welfare and educational processes are immediately apparent to the electorate of the county council members. This basic policy may have been a powerful factor in the development of medical centers of excellence in so many different sites in Sweden. There appears to be a healthy competitive attitude among the county councils in attempting to develop the finest possible health care facilities within their counties and in fact may even at times become overly enthusiastic in their efforts to outdo one another.

LARGE SCALE INNOVATIVE PROGRAMS

The progressive leadership of the county councils and the Swedish Health and Welfare Service have stimulated a substantial number of projects which are interesting as prototypes for Sweden or other industrialized countries. Examples of special importance to the issues in this monograph are briefly presented.

One of the most significant features of Swedish Health and Welfare is the consistent high quality of its long-range planning and the continuity required to carry out the plans while adapting to changing conditions. The development of an organization specifically empowered to initiate and to sponsor long range plans for medical care requirements, optimization, and function is characteristic of this country's effort (see Fig. 4.6).

SPRI—A MECHANISM FOR LONG-RANGE PLANNING

The Institute for Planning and Rationalization of Health and Welfare Services (SPRI) was formed in January 1968 and took over the staff functions of

Fig. 4.6. (A) Danderyd Hospital, one of six large medical centers in Stockholm (see Fig. 4.7), contains the Stockholm County Computer Center described in the text. (B) The Central Block in the Academic Hospital in Lund contains some of the most outstanding clinical facilities in the world, particularly the Department of Radiology. (Photographs presented through the courtesy of Dr. Gunnar Wennstrom, Swedish National Board of Health and Welfare.)

three separate groups—(a) the Central Board for Hospital Planning (CSB), (b) the Council for Hospital Operations Rationalization (SJURA), and (c) the Organizational Department of the Federation of Swedish County Councils. The purpose of SPRI was to encourage and coordinate the planning and studies of health and welfare services in Sweden, to gather and distribute information, to help with

the integration of health and welfare services and to approve standard specifications for hospital equipment. The institute is financed by a special fund into
which the government pays one-third and a foundation two-thirds. Thus SPRI is
not an integral part of the government and can serve an advisory role. It provides
the central guiding principles for use by the county councils and other health
agencies in Sweden through gathering data, initiating studies, and providing
sound advice and counsel.

The term rationalization as used in the name of SPRI contains an element
which would be familiarly known as systems analysis or optimization in the
United States. The program sponsors a series of widely distributed and carefully
conceived studies of the operations of various functions in health care services to
determine mechanisms by which the cost-benefits or cost-effectiveness could be
improved. This is an area of interest which is just now gaining a foothold in the
United States. For example, a team has conducted operations analysis in the
Karolinska Hospital to improve coordination between various forms of treatment and the functions of the medical departments. An objective scheme was
also worked out for the classification of patients with a view to suitable
treatment. The survey reported a mechanism by which outpatient, semiout-
patient and inpatient treatment could be carried out without interrupting the
patient–doctor relationship. At the same time, the waiting list could be reduced
by half while increasing the number of patients who could be taken in.
Efficiency was increased because a large number of patients could be cared for
wholly or partially outside of hospitals. Some other examples of the types of
studies and types of proposals stemming from SPRI are listed below.

DEVELOPMENT OF METHODS FOR ESTIMATING MEDICAL CARE
REQUIREMENTS (SPRI REPORT 3001)

This study deals with a central issue in the program of this organization.
The aim is to arrive at a better basis for estimating and evaluating the cost of
health services in outpatient and inpatient care and the relationship between
these types of care. In addition, the relationships of social welfare will be
elucidated. The first stage of the project was restricted to internal medicine and
the surgical specialties, but it will be expanded to include other elements of the
health care system in the future.

GUIDING PRINCIPLES FOR THE ORGANIZATION OF MEDICAL AND
NURSING CARE IN DIFFERENT SPECIALTIES

The object of this study is to clarify the different specialties and their
distribution in various types of facilities. Guiding principles have been drawn up

for two projects aimed at producing methodology and basic data for economic long-term planning in the medical services and for long-term planning for personnel requirements and personnel availability.

DIFFERENTIATION OF CARE IN THE OUTPATIENT TREATMENT OF A LARGE URBAN AREA

The frequency of visits in different age groups and types of illness will be charted and studied using different outpatient forms in Gothenburg, Sweden, to determine the reasons why the patient streams are steered toward different types of outpatient units.

ORGANIZATION AND ADMINISTRATIVE PLANNING

The two most important projects in this sphere are the outlining of a more efficient administrative organization for medical services and guiding principles as to how medical care blocks should be organized. The principles of distribution of responsibility are among the important problems to be studied. Proposals for organizational models for technical services will be made available to large- and medium-size hospitals.

PLANNING OF MEDICAL AND NURSING SERVICES

SPRI has begun to investigate the staff structure in the health and medical services. The first step was to map out the work of the laboratories. One of the subjects investigated was the training received by classes of staff and the distribution of work loads. A study of the profitability of automatic conveyor systems has been initiated. As the first step, the institute has investigated the possibilities of coordinating the information and has begun to collate the experience already obtained in this sphere, including the frequencies of traffic between the various units of the hospital. Investigations of central supply systems in hospitals and of centralized catering and prefabrication of food are included within their spheres of interest.

DEVELOPMENT OF TYPE PLANS FOR CONSTRUCTION PURPOSES

SPRI has been developing optimal plans for various types of hospital facilities and services. For example, a survey was made of the X-ray departments at 14 hospitals of different classes and sizes which will result in a publication illustrating the different floor plans and will be followed by a survey of the functional analyses of these facilities along with principles of design. Similar development work on intensive care departments and outpatient admissions departments are being made. Standards for laboratories and departments for the reception or admission of acute cases are also under study. The development of specific plans for rooms and room layouts found to be most effective in various institutions are thus translated and transmitted to planning groups all over the country.

EQUIPMENT STANDARDIZATION

SPRI took over the task of standardizing medical equipment from the Hospital Planning Board and is proceeding to provide standardization criteria for a wide variety of types of equipment and disposable items that are used in hospitals.

A private, nonprofit corporation sponsored by various public groups, including most of the county councils, and the Association of the Swedish County Councils has been set up under the name Landstigens Inkopcentral (L.I.C.). This new corporation purchases, stores, and delivers hospital supplies to any hospital ordering from the corporation. By this means, large-scale purchases can be made, enabling the purchasing group to hold prices from escalating and in some cases even reduce prices. This effort has also led to increased standardization of equipment in hospitals which improves the quantity-purchasing leverage and also may improve the overall quality. One of their efforts involved surgical instruments and was carried out by several men traveling throughout Sweden to discuss their needs with surgeons. Additional examples of SPRI reports are indicated in the list of titles in Table 4.2.

The existence of an organization with recognized responsibilities and resources to pursue studies, develop plans, and present proposals for optimization of facilities and services is a step of extreme importance in efforts toward the development and effective utilization of facilities, personnel, and equipment. Although criticism regarding productivity of this effort can be heard on occasion its very existence is an important feasibility study and the results of current operations appear to have value for many other groups and countries of the world.

THE COMPUTER CENTER FOR STOCKHOLM COUNTY

Stockholm county contains approximately 1.5 million inhabitants. It is served by an integrated complex of hospitals consisting of six very large central hospitals. Each of these has two normal hospitals as satellites and in turn four to six annex or supplemental hospitals as illustrated in Fig. 4.7. A general file handling system for the instantaneous storage and retrieval of different types of data is under implementation under a system which is called MIDAS (Medical Information System for Danderyd Stockholm). This MIDAS system is the initial effort in a long-range plan to serve the entire county of Stockholm. Under contract with the Univac division of Sperry Rand a major computer center has been located at the magnificent new Danderyd Hospital (see Fig. 4.6A), designed to ultimately provide comprehensive data processing and storage of very large quantities of information. For this purpose four large magnetic disc memories have been purchased, each containing storage for 125 million characters. Each citizen of Sweden has his individual number which is based upon the year, the month and day of his birth plus a 3-digit serial number for the identification of all individuals born on that particular day. This nine-digit figure is the basic identification number for each person in Sweden (Table 4.3). Each citizen also has a file containing a great deal of personal data for various purposes including taxes and census. The basic population data and files have been made available as the foundation for the storage of medical data. Thus the file is ultimately envisioned to include the ten-digit official identification number of each individual plus his name, home address, post office address, occupation, nationality, whether on pension or not, his vaccination for smallpox, his civil status, date of entry, preliminary income tax code, authority issuing drivers license, war registration, guardian's identification number, and present living conditions and military service (Table 4.3). To this information will be added critical medical information including his telephone number at home and at work, his blood group, any critical diseases which are tabulated in accordance with World Health numerical designation, any critical allergies or sensitivities and the type and number of vaccinations (Table 4.3B). Information will also be provided from previous inpatient care within hospitals. This will include the name of the hospital, the clinic, date of admission, terms of admission, date of discharge, terms of discharge, code of diagnosis, code of operation, code of anesthesia, if used. In a separate category will be information regarding previous X-ray examinations which will include the site at which the examination was carried out, the date and the classification code regarding the organ, the method of examination, and the final diagnosis.

As currently envisioned the data bank will include the information in Table 4.3 for all 1.5 million citizens of Stockholm. At a future time additional information regarding previous outpatient data will be added. This will include the name of the institution, the clinic, date of visit, and the codes for diagnosis, operation or anesthesia as they are applicable. By utilizing the enormous capacity of the disc memories a wide variety of combinations of these data can be obtained within seconds, merely by appropriate depression of keys on a terminal and observing the information as it flashes on a monitoring screen in directly interpretable verbal and numerical form. This project will have many

TABLE 4.2

SPRI Projects—Selected Titles

Planning department projects

3001 Planning of medical services and operational planning, stage I. Part 1 Development of methods of estimating nursing requirements.

3002 Planning of medical services and operational planning, stage I. Part 2 Drawing up guiding principles for organization of care in separate specialities.

3003 Review of medical care plan for a thinly populated area.

3004 Economic long-term planning in medical care. Advice and recommendations.

3005 Long-term planning for personnel requirements and availability in medical care. Advice and recommendations.

Organization department projects

4001 Studies of the conditions or organization for hospital blocks.

4003 Survey of the supervisory organization in hospitals.

4004 Development of and experiments with patients' cards/insurance certificates which can be read-out by computer.

4005 Evaluation of existing data-processing systems and recommendations of suitable component systems.

4006 Direction of and cooperation with experiment with integrated medical information systems in a small hospital.

4008 Structuring case records.

4011 Preliminary study of prerequisites for better utilization of available capacity by shift work and other organization of duties.

4015 Functional requirements relating teletechnical communications in wards.

4017 Study of effect of the design of the ward upon the work of the staff.

4020 Organization of technical services in hospitals.

4021 Profitability analysis of automatic conveyor systems (coordination of investigations).

4025 Work studies in the county council central laundries.

Construction department projects

5001 Specifications of requirements for teletechnical communications.

Part 1. Between staff themselves and between staff and patient in the nursing building.

Part 2. Between nursing building and surroundings—radio staff locators, telephones, etc.

Part 3. Radio and TV communication in nursing blocks, entertainment and supervisory work.

5002 Specifications for control and supervision of operation of installations.

5005 The BDC system (quantity recording, i.e., combining the conventional operational description and cost notes).

5006 Standards for X-ray departments (inventory, analysis, type plans, recommendations).

5007 Standards for operating theatres (inventory, analysis, type plans, recommendations).

5008 Standards for ward-based departments (inventory, analysis, type plans, recommendations).

5009 Standards for intensive care departments (inventory, analysis, type plans, recommendations).

5011 Standards for laboratories (inventory, analysis, type plans, recommendations).

5013 Standards for health centers (programme, analysis, type plans, recommendations).

Equipment department projects

6006 Lighting equipment: SPRI specifications for dental lighting and lighting in patients' rooms.

6007 Medical equipment: Reports on properties thereof and instructions for electromedical and intensive-care equipment and electronic systems for intensive-care departments.

6008 Radiological equipment: X-ray diagnostic apparatus (testing preparations).

6009 Diagnostic and medical equipment: SPRI specifications for oral thermometers and blood-pressure cuffs.

6012 Anaesthetics: SPRI specifications for anaesthetic apparatus, intensive-care rails, connections to anaesthetic apparatus, respiratory-blowers, catheters, respirators, and pulmonary ventilators.

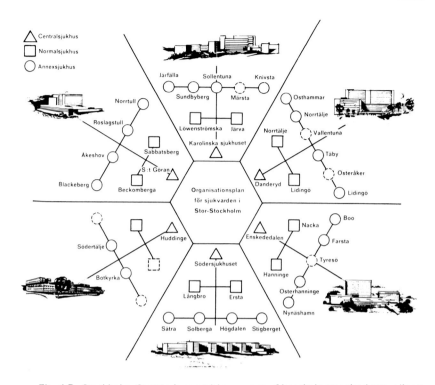

Fig. 4.7. Stockholm County is served by a group of hospitals organized according to effective long range plans to provide ample health care for the region. Patients are referred from the smaller hospitals to the Central hospitals if they require the most sophisticated care.

important implications for the future developments of computer applications because it will provide an opportunity to determine the usefulness of the data, and the problems encountered in compressing the data into the number of characters that can be accommodated in the system. This will require careful selection and establishment of priorities for the information having the greatest value for storage. The consequences of the system in terms of cost-effectiveness and cost-benefits will also be of great importance in the planning of major computer centers in other countries where such efforts are being developed as research projects. An experiment of this sort is difficult to envision occurring in a country like the United States where large segments of the population would be quite reluctant to provide large masses of data, freely accessible without assurance of security provisions. Furthermore, the superimposition of medical information on existing data banks established for tax purposes or for social security would appear to present almost unsurmountable difficulties. On the contrary, we can profit by the experience of such large-scale experiments in

TABLE 4.3

Developing Head-File on 1.5 Million Citizens—Stockholm County[a]

A. Vital statistics
1. Identification Number
2. Name
3. Home address
4. Post office address
5. Occupation
6. Nationality
7. Pension conditions
8. Smallpox registration
9. Civil status
10. Date of entry into No. 9
11. Preliminary income-tax code
12. Authority issuing driving license
13. Ward registration
14. Guardian's identification No.
15. Present living conditions (registration date, county, council, parish)
16. Military service conditions

B. Critical medical information
1. Telephone number (home and work)
2. Blood group
3. Critical diseases (See Table 4.2)
4. Critical over sensitivity
5. Vaccinations (year, month, type, No.)

C. Previous in patient care
1. Hospital (Institution)
2. Clinic
3. Date of admission
4. Terms of admission
5. Date of discharge
6. Terms of discharge
7. Code of diagnosis
8. Code of operation
9. Code of anaesthesia

} Arbitrary number of codes for each treatment

D. Previous X-ray examinations
1. Hospital
2. Clinic
3. Ward dept.
4. Date of examination
5. Classification code
   Organ (part of body)
   Method of examination
   Diagnosis

E. Previous outpatient care
1. Hospital (Institution)
2. Clinic
3. Date of visit
4. Code of diagnosis
5. Code of operation
6. Code of anaesthesia

} Arbitrary number of codes for each treatment

[a]From S. Abrahamsson, S. Bergstrom, K. Larsson, and S. Tillman, Dandaryd Hospital Computer System.

Sweden to estimate the cost-benefit and help decide if such medical information should and could be stored on the basis of local or state data banks in the United States.

COMPUTER SCHEDULING

At the Karolinska Hospital in Stockholm an effort is being exerted to expedite the acquisition of data about patients and to use computers for facilitating the initial stage of data acquisition. When a patient requests appointment at the hospital, he automatically receives in the mail a questionnaire which he fills out with answers to approximately 200 questions, constituting a rather comprehensive personal and medical history. The patient mails in the questionnaire to the hospital and a computer then determines what basic lab tests would be required to accompany this particular patient history. The patient then receives appointments for reporting to laboratories and the scheduling of the tests is designed to reduce the time spent by the patient to a minimum. When the laboratory tests have been conducted and the reports are in, the patient reports to a clinic and the physician is immediately presented with an extensive history and the results of basic laboratory examination.

HEALTH SCREENING IN VARMLAND

The population of the county of Varmland numbers nearly 300,000, fairly widely scattered over a comparatively large area around seven towns of which the largest (Karlstad) is about 40,000. The Swedish government authorized the health screening of 100,000 people during three years beginning in 1962 and was finished on schedule. The objective was to determine the feasibility of combined photofluorography with clinical laboratory determinations to assess the nature and scope of disability or suspected disease that could be detected in this way. To a photofluorography unit capable of 350 examinations a day was added a blood-sample collecting group. An automated blood analysis system was developed by the Jungner brothers (10,11) specifically for that purpose with a capacity of about 5000 analyses a day (400 patient samples per day). All inhabitants 10 years of age or older were offered photofluorography and all those 25 years of age were offered clinical chemical analysis, including hemoglobin, hematocrit, serum iron (for anemia), creatine (for kidney function), transaminases GOT and GPT (liver damage, etc.), γ-globulin content, β-lipoprotein, and cholesterol and protein bound hexoses and sialic acid for inflammatory changes. The results of the tests disclosed that 4 times as many diagnoses were found from combinations of tests than by a single test. This led to the concept (11) of "profiles" consisting of tests which may have diagnostic value in detecting abnormalities in the individual organ systems (see also Chapter 5). Probably the most significant result of the Varmland project was the stimulus given to the production of a sophisticated automated blood analysis system. The Autochemist is a complex machine designed and developed specifically for mass screening purposes lacking flexibility which is frequently required in clinical chemistry laboratories (see also Chapter 3). The current model, produced by AGA of Lindingo, is equipped with 24 channels capable of producing a wide array of determinations from blood serum (Fig. 4.8). High capacity data acquisition represented by the automated chemistry laboratory is in turn a powerful stimulus to the utilization of high-speed computers for data processing (see also Chapter 5).

PROJECTION OF FUTURE DEVELOPMENTS

Dr. Gunnar Wennstrom (12), National Board of Health and Welfare in Stockholm, anticipates a very large expansion of the total number of well-trained physicians and nurses during the period 1970–1980. The comparative figures for physicians, nurses, and total health care personnel are listed in the following tabulation.

	1950	1970	1980
M. D.'s	5,000	10,000	20,000
Nurses	19,000	50,000	75,000
Total health care personnel	81,000	150,000	(?)

The increase in the total number of physicians will result in progressive lowering even further of the ratio between doctors and patients. In 1970 there was one physician for every 750 people in Sweden. By 1980 it is anticipated that there will be one physician for every 450 persons in Sweden, a ratio beyond the fondest dreams for the United States. Because of the relatively small size and population of Sweden, there is concern at the national level regarding the total number of individuals who can effectively and appropriately be assigned to duties in the health care field. On this basis, the exact number of individuals in the paramedical fields of the future in unknown. It is anticipated that the rate of increase of the paramedical personnel will be smaller than in the categories of physicians and nurses and that more effective utilization of personnel and facilities will be a major goal.

The natural consequence of increasing so greatly the supply of physicians will be to produce a relative surplus, a situation which will also tend to follow the law of supply and demand and reduce the income of the average physician in the future. The educational process in Sweden is undergoing major reorganization at the present time with a substantial change in the structure in medical

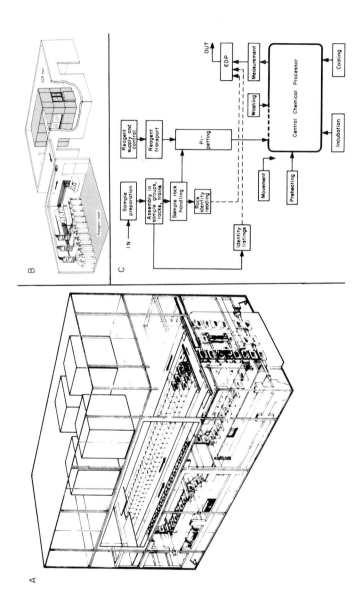

Fig. 4.8. (A) The Autochemist was designed and developed for a large scale project of health screening in Varmland and is capable of automatically performing a large number of chemical determinations simultaneously. (B) In a typical hospital installation, the analyzer is supplied by a very large group of chemical reagents from individual containers in an adjoining room. (C) The flow chart of the analyzer indicates the complexity of processing to provide such a large number of analytical processes. The capacity of the Autochemist is so great that continuous sources of samples must be supplied to provide efficient operation. (From Autochemist Handbook through the courtesy of Gunnar von Feilitzen, AGA Lindingo.)

schools as well as in the educational system in general. The major significance of these changes have been described in some detail (12).

Although the number of hospital beds for acute care is not expected to increase too greatly, significant efforts will be made to increase the number of facilities which require fewer personnel and cheaper mechanisms, particularly outpatient care. It is considered advisable to build widely dispersed health centers to cover the population groups in isolated or rural areas as described in the previous section. A great expansion is anticipated in the field of preventive medicine with such steps as general health controls, health measures at places of work and industrial health services.

The National Board of Health and Welfare publishes a report called RUPRO, which stands for Running Prognosis to provide authorities in organizations at all levels with statistical data on the trends of development in the services for health and sick care. The RUPRO-69 provides data regarding expected development over the period 1968–1975 with reference to outpatient care in hospitals and outside of hospitals, sick care in patients' homes, extended care facilities, hospital beds, personnel, education, cost assessment, and the population of local administrative areas as they change.

Waiting lists represent one major guide to the relationship between the demand for health care and the provision of health care. Reports indicate that a complete turnover of the waiting list was accomplished in two months. One new trend is the recognizition that it should be possible to adjust the size of the staff at any particular ward depending upon the circumstances. In 1969 approximately 1% of the beds were not occupied for lack of staff. For example, studies are being made to determine whether or not it is desirable and necessary to maintain hospital functions at full staff and activity over weekends, or whether substantial reductions in staff during weekends might be possible through proper organization and scheduling. Evidence has been obtained which supports the trends that the increase of salaries and of supplementary salary costs is growing faster than the other health and sick expenditures. This directs attention toward the organizational side of the service in continued planning activities. Economic studies are needed to evaluate alternative types of care such as hospitalization versus home nursing of the sick and specialized care in hospitals versus such care in ambulatory services (see also Chapters 2 and 3).

THE ROLE OF BIOMEDICAL ENGINEERING IN SWEDEN

The traditional forms of Biomedical Engineering activities have been clearly evident in only two academic institutions; namely, the Karolinska Hospital (Fig. 4.4A) and Chalmers Institute in Gothenburg. In Linköping, a new technical institute and medical school are being developed according to an integrated plan to encourage training and research in biomedical engineering. The applications of biomedical engineering are widespread throughout the health care system. For example, a large proportion of the SPRI reports contain important elements of biomedical engineering. Computers are being widely employed throughout the hospital systems in many different applications and configurations. In addition, the organizational structures of many hospitals contain provisions for engineering sections or departments for maintenance and operation of the complex equipment of the hospital and for data acquisition and analysis. The chiefs of such divisions have positions of responsibility and authority like the chiefs of the clinical laboratories. The commitment to biomedical engineering research and development is limited in scope compared with the program in the United States. They have concentrated attention on fewer selected areas, such as computers, prosthetics, and systems analysis, clinical laboratory methods, and some highly sophisticated techniques for data acquisition and analysis.

SUMMARY OF SWEDISH SUPERIORITY IN HEALTH CARE

The health care delivery mechanisms of Sweden excel in their widespread distribution of quality medical care to all citizens who need it. The staff and facilities of the large medical centers match the best of any country in the world. The planned integration of satellite, central and district hospitals with the large regional or academic hospitals provides a basis for cooperation and more effective flow of patients to facilities appropriate to the severity of their illnesses. Local control of health and welfare at the county level is a most valuable asset because it assures responsible management under rather direct surveillance of the voting constituents. The high overall quality of health care reflects expenditure of a large proportion of the gross national product of a prosperous nation. The long tradition of comprehensive planning and implementation of their plans reflects an unusually stable government with widespread popular support by the voting public. The net result of these factors is a national health service which is superior to any other in many respects. Other countries have much value to learn from the Swedish experience, but the lessons may not all be applicable in other conditions. The outstanding quality of health care facilities, coverage, and planning in Sweden result from a set of circumstances not likely to be encountered in any other country—certainly not in the United States. Sweden is a relatively small, extremely prosperous and homogeneous country with limited foreign commitments. Compulsory health insurance for the whole population was passed by the Swedish Parliament in 1946 to be effective in 1950 to provide a 4-year period of active preparation. We cannot expect to have such lead time in the United States. The cost of the health and welfare program is great, consuming some 37% of all government expenditures for social welfare benefits of which health insurance accounts for more than one-third.

The national taxation level is reported to have *tripled* in Sweden between 1945 and 1962, a fact which is not always mentioned in descriptions of the social benefits of a welfare state.

CONTRAINDICATIONS FOR THE UNITED STATES

The prospects of matching the relative quantity and quality of medical care of Sweden by upgrading the mechanisms in the United States seem very remote for many different reasons. We are a very large country with extremely grave internal problems and external commitments. It is difficult to imagine the American public submitting to total taxation levels up to more than 40% of total national output (Table 4.4) when our current level is less than 30%. It is even less probable that our national priorities will be restructured to provide the large relative expenditures for health and social welfare benefits.

TABLE 4.4

Taxes People Pay—In the United States and Other Major Nations[a, b, c]

	Direct taxes (on incomes, corporate profits, estates, gifts) ¢	Indirect taxes (sales and property taxes, excises) ¢	Social security taxes ¢	Total tax burden ¢
Sweden	20.2	13.9	8.2	42.3
Norway	13.8	15.2	9.2	38.2
Netherlands	13.2	11.2	13.4	37.8
France	6.5	15.9	14.5	36.9
Austria	12.0	16.5	8.2	36.8
West Germany	10.4	13.7	10.6	34.7
Denmark	16.3	16.5	1.9	34.7
Britain	13.1	16.2	5.1	34.4
Belgium	10.2	13.3	9.5	33.0
Canada	12.7	15.1	3.4	31.2
Italy	6.8	12.6	11.1	30.5
United States	15.5	9.1	5.3	29.9
Japan	7.9	7.5	3.5	19.0

[a]Total tax load, including national, state, and local levies, measured in terms of cents per dollar of total national output.

[b]Chart is based on 1968 data, latest available, from the Organization for Economic Cooperation and Development. Totals for Austria and Japan are rounded.

[c]"American Taxpayers' Bill—How Bad Is It, Really?" *U.S. News & World Report* September 7, 1970, p. 72.

The startling fact that hospitals smaller than 400 beds are being phased out in Sweden because they have proved economically inefficient is particularly disquieting in view of the predominantly small hospitals in the United States. Since we cannot phase them out for lack of substitutes, it will be necessary to develop mechanisms by which these small, autonomous, voluntary hospitals, poorly located and basically inefficient, can be welded into an effective, hopefully integrated system with greatly improved cost effectiveness. The contrast is represented by the differences between the indiscriminate collection of Seattle hospitals displayed in Fig. 4.2 and the integrated hospital system serving Stockholm (Fig. 4.7).

Local government at the county level is not geared to play the role of controlling health and welfare as proved so effective in Sweden. Our ability to develop and implement long-range plans is not comparable to that demonstrated by Sweden because of their long term governmental stability and support. The natural American tendency is to forge ahead without long-range plans, expecting to solve the unexpected complications as they arise. With the very rapid rate of change in our swiftly moving society, the need for accurate forecasting and comprehensive long range planning was never greater. The future promises even greater challenges than the past which means we must be prepared to make rapid and drastic adaptations which will tax our ingenuity and our incomes to the fullest.

Private citizens, potential patients, physicians, dentists, nurses, hospital administrators, economists, architects, and hospital industrial engineers can all contribute information regarding what is needed for health care, while engineers could provide input regarding what is possible. For such group efforts, the background and training of engineers for data collection and analysis, regarding complex systems, and development of complicated equipment and organizations and optimization of processes would be invaluable.

The government can play a key role in the process of implementing innovative plans for modernizing the health care facilities and services by providing mechanisms and incentives. The recent bill establishing a National Health Service Corps should improve coverage in rural areas and city slums. In addition, preferential schedules of charges can be offered through Medicare and Medicaid to those hospitals which develop approved plans and implement programs to increase the quality and effectiveness of their function through improved organization, shared services, distributed specialties and services, improved utilization of manpower, and other options described in this chapter. Although the financial incentives may stem from the federal government, the planning, control, and supervision should be on a local level, by multidisciplinary panels comprised of individuals interested and competent in the health field at regional, state, and county levels. These panels could be awarded the responsibility for review and approval of integrative and cooperative programs created from the many and diverse options on the basis of the peculiar requirements of the locale.

REGIONAL PLANNING: AN EKISTICS APPROACH

Many groups have been interested and involved in urban planning for many years and one of the most active developed within the Athens Technological Organization under the direction of Constantinos Doxiadis (13). Beginning in the 1950's extensive studies of the nature and characteristics of human settlements have been carried out by the Center for Ekistics (14,15). The term ekistics is derived from a Greek work meaning the foundation of a habitation, a city, or a colony. It is also related to an ancient Greek noun referring to a person who helps individuals to settle at a site.

> Human settlements are no longer satisfactory to their inhabitants. This is true everywhere in the world, in underdeveloped as well as developed countries (13).

The scale of the problem of the expansion of settlements can be seen by looking at a very old and a very new city at the same site. The old city is Athens which contained a population density over a period of 3000 years fluctuating between 64 and 72 inhabitants per acre. Despite the relative constancy of this density value over a long period of time, a remarkable change has occurred during the last 100 years and even more so during the last 50 years. The expanding city of Athens has spread over a much greater area and reached a population of two million people. The density within the limits of the municipality of Athens remains 68 persons per acre, which is very close to the average of the last 3000 years. However, the modern city of Athens has now spread over an area 40 times as great as ancient Athens. Many other major contemporary metropolitan city areas have exhibited a dynamic growth and show several common characteristics over the past 100 years.

1. They have covered from 10 to over 40 times their previously built up areas.

2. There has been no effective control or plan for their overall growth.

3. As a result the size and form of the present settlement is exceedingly confusing.

The situation in all urban areas which have grown rapidly without plan during the last century indicates that cities are becoming uncontrolled in size, in their structures and in their forms. Rapid technological change appears to have left mankind without the ability to achieve a simple and natural growth of cities in an orderly way. However, an examination of the growing city demonstrates that there are logical and predictable trends which underlie the growth and expansion of urban areas.

THE AGE OF MEGALOPOLIS

Papaioannou (15) projected impending development of Megalopolis through a process of expansion of metropolitan areas and growth along lines of communications. This new settlement is characterized by its large size in area and population, its high density of habitation, the inclusion of several large centers coalescing with strong interactions with each other and with the surrounding region. Analysis of extensive global data have led to projections that by the year 2000 A.D. the megalopolis will have attained such importance as to justify calling this period the era of the megalopolis. It is worth recalling that many current health facilities will be in active use 30 years from now. Population projections anticipate that the proportion of the earth's total population living in large cities will reach a maximum between 2000 A.D. and 2050 A.D. and it is assumed that somewhere between 36% and 53% of the population will be living in these large coalescing centers. Criteria for the definition of a megalopolis has been developed on the basis of the strength of the pull of large centers upon smaller centers in the vicinity. This formulation identifies several megalopolitan areas—(1) a region between Boston and Washington, D.C.; (2) the area encompassing Milwaukee, Chicago, Detroit, and Pittsburg; and (3) the area between Los Angeles and San Diego are currently regarded as megalopolitan areas by these definitions. By 1980 it is anticipated that Los Angeles and San Diego will be confluent with San Francisco. Many other examples in other regions of the world were predicted (15). Examination of these emerging megalopolitan areas suggests that the range of population will extend from 10 million people upward to 250 million inhabitants. Small megalopolitan areas will have populations between 10 and 35 million inhabitants, major megalopolises will have total populations over 250 million.

Although many individuals have regarded these prospects with horror and suggested that constraints be placed upon concentration of individuals and the migration into the cities, there is no successful precedent for such rigid control over people's movements except in regimented societies where freedom of choice is markedly restricted. An alternative approach to this problem is to analyze the past and project the future for the purpose of developing workable and accessible plans to facilitate evolution of effective urban systems. Recognizing the extreme mobility which has been made possible by combinations of mechanical power, rapid transit and freeways for cars, cities with daily urban systems for mobility of large segments of the population may be described by a circle with a 90-mile radius surrounding each population center. Within the

90-mile circles one can begin to see the merging of expanding communities which have become indispensible parts of the whole. The appearance of the United States as an area covered largely by daily urban systems with radii of 90 miles is illustrated in Figure 4.9. The thickness of the enclosing 90-mile circle is varied to indicate the relative population inside. In the eastern and northeastern parts of the United States the heavy concentration of population is already indicating the continuity and contiguity of urban centers tending to merge at their periphery. The 90-mile radii tend to cross governmental borders between states and counties so that their existence represents a growing need for regional planning without the inevitable obstacles encountered at artificial barriers or borders. One can also recognize the growing chains of population centers in Gulf states and along the West Coast of the United States.

THE EASTERN AND GREAT LAKES MEGALOPOLISES

Studies of the concentration of population and the trends toward growth can be represented best by two major belts of urbanization. One along the northeastern Atlantic coasts, extending from the region of Boston down to Washington and has been variously labeled Boswash or the Eastern Megalopolis. The second megalopolis is between Lake Michigan and Lake Erie. It requires little stretch of the imagination to envision the confluence of these two megalopolitan areas and their continued growth toward regions such as Ontario in Canada and a wide variety of other connections (Fig. 4.9B).

The projected time scale for these enormous changes in population is well within the lifetime of any single hospital which has been constructed within the last few years. Similarly, the highways which are serving the people at the present time will still be functioning when the growth of the populations in this country have reached the stage illustrated in Fig. 4.9B. Effective long range planning on a regional basis is of critical importance in view of the long time required to conduct studies, develop plans, to present them to public and political leaders with sufficient persuasive power to cause effective action. Delays in planning can only lead to major disaster in the future. The projections for the changes in the pattern of urbanization and the designs for living in the future inevitably will have enormous influence on the success of our planning for health care systems.

LONG-RANGE PLANNING—INTEGRATION OF HEALTH REQUIREMENTS IN URBAN PLANNING

The problems confronting the hospitals are merely an extension of stresses developing in cities of which they are an integral part. The cities are experiencing

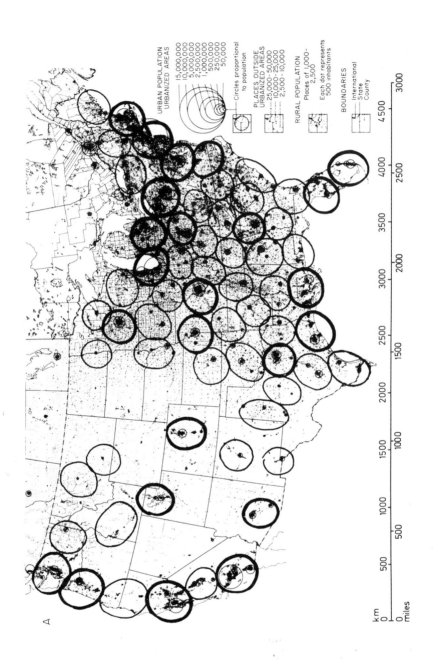

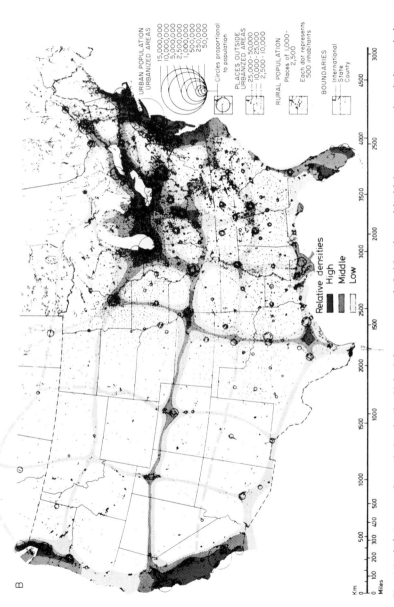

Fig. 4.9. (A) Centers of population in the United States are encircled by rings with thickness representing their populations within the daily urban travel systems of ∿ 90 miles. The confluence of these metropolitan areas is leading to the development of megalopolis. (B) Megalopolitan areas are well established from Boston to Washington, in the Great Lakes region, and in California. Future projections show development of megalopolises in other areas of this country through confluence of urban areas. (Reproduced through the courtesy of Dr. D. A. Doxiadis, Athens Institute of Ekistics.)

an expanding influx of people at sufficient rates to overload all their facilities and services. Such rapid growth stemmed directly from technological advances which made it possible to develop both hospitals and cities but at the same time outstripped the organizational ability to manage the resulting complexity.

Future prospects for overcoming these problems in the present period of rapid change depend heavily upon long-range, comprehensive planning for the future, taking into account the necessary interaction and integration of the components of the cities with specific provisions for a health care system. Current problems of urban centers and the future projections have successfully stimulated intensive efforts at forecasting the trends and developing comprehensive plans to contain and alleviate the stresses of the future. Although these plans take into account a very large number of geographical, social, economic and political factors and most of the essential services (e.g., communication, transportation, logistics of goods and services, utilities) provisions for health care mechanisms are not included among them. This crucial omission appears to result from an almost complete lack of input from the medical profession of data applicable to urban planning. The kinds of information which are required to tackle problems of providing adequate utilities, transportation, and other vital services are just not available from the medical community. Definitive data about the current status of medical care capability are difficult to obtain in the extreme because of the nonorganization of the medical nonsystem. The prospects of reliable projections for the future stemming from the organized bodies in the medical profession are a vain hope.

Imaginative planning on large scale for the future is even more remote. *The most important contribution of biomedical engineering may well be the development of innovative plans for the future.* The magnitude of the problem and the need for biomedical engineering involvement can best be illustrated by the most comprehensive regional planning effort undertaken within the United States to date.

URBAN DETROIT AREA STUDY—A PRIME EXAMPLE OF REGIONAL PLANNING

The Detroit Edison Company in cooperation with Wayne State University and Doxiadis Associates (16) have completed a 5-year regional planning study encompassing an area of 23,000 square miles for the purpose of projecting the nature and magnitude of the physical, social and economic growth of a large urban area for a period of 30 years and beyond. The first step consisted of analysis of the Detroit area in relation to its surrounding areas to place it in proper context. It included comparisons of the Great Lakes Megalopolis with the more familiar Eastern Megalopolis, including the forces at work in this large and complex area (Fig. 4.10A,B).

A very large number of variables were recorded.

Categories studied	Number of variables
Demographic structure	10
Demographic change	5
Economic level	7
Social characteristics	5
Housing conditions	7
Employment structure	6
Economic activity	10
Commuting to work	3
Local government finances	3
Farmland aspects	8
Transportation networks	4

The numerical data for each of these variables were subjected to a form of multivariate analysis to provide a priority list in decreasing order of importance. The first few such variables typically account for a very large proportion (e.g., 70–80%) of the variation between sample sources such as counties or different districts within a city. This analysis was applied to every county in an area slightly larger than the Great Lakes Megalopolis as defined by the initial studies. The first three principle components derived by these studies accounted for 68% of the variance. An index of urbanization (component I) accounted for 48%, an index of suburban character (component II) accounted for 13% of variance and an index of low socioeconomic status (7%). Enormous quantities of data were assembled and subjected to sequential analysis and comparisons to arrive at a comprehensive picture of current status and trends, with reference to population characteristics and distributions, economic characteristics, networks, utilities. The changes anticipated by the year 2000 were forecast on the basis of a very large number of alternatives based on different combinations of the variables of principle interest. By means of computer simulation an initial set of 49,000,000 possible patterns of development were reduced successively from 52,000 to 28 and finally to 7 alternatives in accordance with specified criteria. Included in these seven alternatives were the following major assumptions—(1) Detroit would remain the only high-order urban center in the area; (2) Detroit would remain the only high-order urban center but most of the expected growth be absorbed by smaller metropolitan centers; (3) Transportation networks (radial and grid) would provide two patterns of growth within the region; (4) two new metropolitan centers in the area would supplement Detroit; (5) one new urban center to supplement Detroit postulated either at the present site of Flint or of Toledo. After extensive comparative analysis of seven alternatives, the final recommendation and concept plan was formulated for a new urban center at the north end of Lake St. Claire of approximately the size of Detroit with integrated and interlocked networks and utilities for the year 2000, as illustrated in

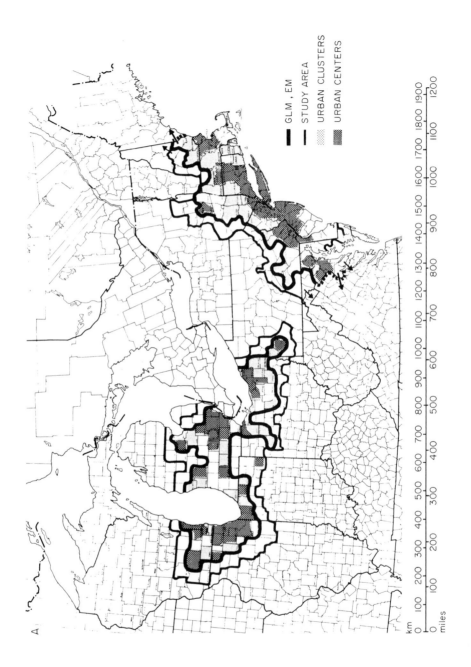

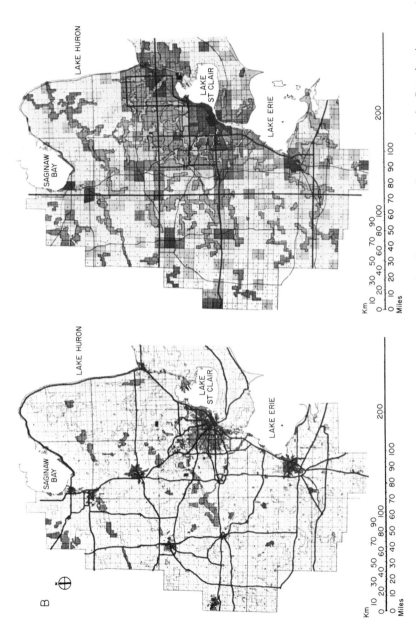

Fig. 4.10. (A) The urban centers in the Eastern and Great Lakes megalopolises are a part of a comprehensive Detroit urban area study. (B) The present Detroit urban area (left) is projected to expand and grow so much by the year 2000 that another city the size of present Detroit may be located on the north shore of Lake St. Claire (right). Comprehensive studies and urban plans should provide for health care systems but rarely include such considerations in any detail for lack of essential information from the medical profession.

Fig. 4.10. The results of this study are contained within three large volumes with documentation of the methodology, results and the bases for the alternative recommendations. The existence of such a plan certainly does not assure that any portion of it will be accomplished. This attitude may underestimate the impact of such detailed analysis and long-range forecasting because its very existence is very apt to initiate an upward spiral of property values in the area designated for the new urban center. Thus, such a plan may become a self-fulfilling prophesy.

The urban Detroit area study was completed on schedule in five years and it is anticipated that a corresponding analysis of an urban region around Cleveland or Pittsburgh could be accomplished in only three years using the methodology evolved in this initial study. In view of the rapid rate of change of modern societies and cities, the need for such long range planning will become increasingly apparent. The implications of such studies for new health care systems are obvious. However, I was able to find only one brief mention of medical facilities in the concept for future development (Volume III) and no evidence of any specific considerations of location, interaction or distribution of medical facilities. This omission would not have occurred in a country like Sweden (see above). If comprehensive planning is to play a significant role in providing for orderly growth and protection of our environment, the medical community has an obligation to undertake appropriate studies of future needs in a systematic manner. The experience and backgrounds of engineers with knowledge and understanding of medicine will be absolutely essential for the process of analyzing the interactions of so many factors of a complex system. This is the central theme of much of modern engineering training and practice. An example of how mathematical modeling can be applied to such an intricate problem is presented in Chapter 7.

INTEGRATING HEALTH REQUIREMENTS IN REGIONAL URBAN PLANNING

Despite the obvious need for including health care networks within the urban centers, Aldrich and Wedgewood (17) stated: "A health network is not provided among the many urban networks usually included in planning and design. Highways, water, gas, electricity, telephones, TV cables, fire stations, police facilities, parks, museums, railroads, schools, supermarkets, and bus lines get careful attention but not the needed network for health service delivery to children and youth. In looking at societal needs, health facility costs should probably not be considered as a separate item but rather as a component of services, housing, transportation, education, health, etc. If one were looking at urban plans for housing as a whole, which included a health network, the cost of housing, including health facilities, would be placed in better focus and the health costs would not appear so astronomic."

LOCATION OF HOSPITALS AND HEALTH CENTERS IN MEGALOPOLIS

In the past, hospitals have tended to function for very long periods so that most cities have a substantial concentration of health care facilities near the center where the noise, congestion, pollution and parking problems are most intense (see Fig. 4.2). The growing cities tend to expand along the major transportation channels which also provide great mobility for patients, doctors and logistics of supply. The predictable courses of urban growth permit planning of the optimal location of health care facilities of various sorts depending on evaluation of the advantages and disadvantages for each major group of individuals involved (patients, doctors, nurses, staff, services, etc.). It is possible to determine where to locate new hospitals for emergency care, chronic disease, obstetrics, rehabilitation, etc., for most effective function. Emergency hospitals or facilities probably should be near freeways and near regions of high accident rates, for example. Other hospitals might be variously located near city centers, in the periphery, in remote areas or along transportation channels in new towns, as needs dictate.

To evaluate health care facilities on a regional basis on the same scope and scale as the Detroit urban area would call for the development of concise information regarding such categories as location, interaction and environmental relationships of health manpower and facilities in large areas with sufficient detail that deficiencies or excesses can be recognized and accommodated. Studies of the nature of medical practice and hospital functions need to be carried out systematically and with sufficient precision that new additions or construction could be optimized in terms of the benefit they can provide the population in the region. Such studies are needed in the many distinctive regions of the country to the point that intelligent decisions and priorities can be established for future development to constructively meet the problems that can be projected into the future. The techniques for urban studies on large scale are just now being evolved. The health care requirements need to be incorporated directly into the regular processes of urban planning. If we accept a broad definition of the multidiscipline, such long-range planning becomes an integral part of biomedical engineering.

William Stewart (18) presented new dimensions of health planning, including the following characteristics (paraphrased from Max Ways).

1. An effort to improve planning by sharpening the definition of the objectives

2. A more systematic comparison of options by means of criteria based on goals

3. A more realistic assessment of consequences of various options

4. Assignment of responsibilities to those with greatest competence for decision making processes

5. Education and persuasion as the mechanism underlying implementation as a substitute for authoritarian power

6. Improved techniques for gauging the effectiveness and complications of various decisions or options

To these attributes Stewart (18) added two additional propositions. First, the planning process should have the broadest possible base of input from diverse but relevant sources (it not only improves planning but aids in implementation). Second, the planning should be accomplished and the decisions made as close to the site of the problems as possible. Although national planning is important and necessary, regional planning and local planning produce the highest probability of arriving at solutions which are relevant to the problems and consonant with the resources available.

References

1. Ribicoff, A., The "healthiest nation" myth. *Sat. Rev.* Aug. 22, 18-20 (1970).
2. Harris, S. E., "The Economics of American Medicine." MacMillan, New York, 1964.
3. Garfield, S. The delivery of medical care. Sci. Amer. **222**, 15-23 (1970).
4. Burney, L. E., Impact forces on rising costs of health care facilities. *In* "Costs of Health Care Facilities." National Academy of Sciences, Washington, D.C., 1968.
5. Schwartz, J. L., "Medical Plans and Health Care." Thomas, Springfield, Illinois, 1968.
6. Somers, H. M., and Somers, A. R., "Medicare and Hospitals: Issues and Prospects." The Brookings Institution, Washington, D.C., 1967.
7. Bennett, I. L., Conditions and problems of technological innovation in medicine, *Technol. Rev.* **72**, 43-48 (1970).
8. Engle, A. G. W., "Planning and Spontaneity in the Development of the Swedish Health System. The 1968 Michael Davis lecture." Center for Health Administration Studies Graduate School of Business, University of Chicago, 1968.
9. Tottie, M., and Janzon, B., *Regional Hospital Planning;* Current Trends in Health Services. The National Board of Health of Sweden, Grafisk Reproduction A.B. Stockholm, 1967.
10. Jungner, I., Calibration and standardization of the Autochemist with computer assist, 7th congress clin. chem. Geneva/Evian, 1969. *In* "Methods of Clinical Chemistry," Vol. I. Karger, Basel, 1970.
11. Jungner, G., Multichannel automated equipment (Autochemist), *Proc. Int. Symp. Early Disease Detection.* (In press).
12. Wennstrom, G., Training of health workers in the Swedish medical care system. *Ann. N.Y. Acad. Sci.* **166**, 985-1001 (1970).

13. Doxiadis, C. A., "Ekistics: An Introduction to the Science of Human Settlements." Oxford Univ. Press, New York, 1968.
14. Doxiadis, C. A., Ecumenopolis: The Settlement of the Future. A.C.E. Res. Rep. No. 1, Athens Technological Organization, Athens Center of Ekistics, 24 Strat. Syndesmov Street, Athens 136.
15. Papaioannou, J. G., *Megalopolises: A First Definition*, A.C.E. Res. Rep. No. 2, Athens Technological Organization, Athens Center of Ekistics, 24 Strat. Syndesmov Street, Athens 36, Greece.
16. Doxiadis, C. A., "Emergence and Growth of an Urban Region: The Developing Urban Detroit Area," Vols. I, II, III. Detroit Edison Company, Detroit, 1966-1970.
17. Aldrich, R. A., and Wedgwood, R. J., Changes in the United States which affect the health of children. *Ekistics* **28**, 227-230 (1969).
18. Stewart, W. H., New Dimensions of Health Planning, Prospectives in Biology and Medicine, Winter, 1968, pp. 195-202.

Chapter 5

DATA ACQUISITION, PROCESSING, AND DISPLAY

The quantity of information that can be obtained from patients for the identification and evaluation of illness is enormous and expanding rapidly compared with the situation only a few years ago. Thoroughness, the previous criterion of medical excellence (see Chapter 3), must now be tempered with judgment and discrimination to avoid obscuring the essential clues by masses of information not relevant to the current illness. Quantitative information which might have consisted of ten to fifteen determinations a few years ago, can now number in the hundreds, readily repeated on request. Effective selection and utilization of these rich sources of information demands the development of new attitudes and new methods of data processing and display.

The physician in solo practice is generally functioning at more traditional levels of information input, relying primarily on the medical history, physical examination, and a limited number of quantitative tests that can be performed in the office or by a readily accessible laboratory. If he is struggling to keep pace with a stream of patients awaiting his ministrations, the amount of time and attention he can devote to extracting information may be extremely limited. The large bulk of private practice is dependent upon local hospitals for quantitative testing in significant volume. In addition, commercial laboratories in growing numbers are beginning to provide chemical, hematological, microbiological, and clinical testing including examinations employing X rays and electrical recording. As hospitals grow, affiliate, and merge, the trends toward increasingly comprehensive laboratories will continue and gain momentum. The prospects of widespread multiphasic screening centers administering extended batteries of tests in the context of annual check-ups or for preventive medicine represents still another source of medical data that must be processed and handled efficiently to be of optimal use. The large hospitals and medical centers have been faced with the most drastic demands for data-handling techniques applicable to patients with various levels of illness. Laboratories represent only one channel through which information flows in health care facilities. Other major channels include routine admission information, routinely repeated measurements on the ward (e.g., heart rate, temperature, blood pressure, etc.), doctors' orders, treatment schedules, menus, nurses notes, discharge summaries, and the paper work required for payment from insurance companies or government agencies.

COST OF INFORMATION HANDLING IN HOSPITALS

It is commonly stated that paperwork represents about one-fourth to one-third the cost of taking care of patients in a hospital. One source of such figures was a study of three hospitals in Rochester, New York, disclosing that information handling costs were 22.6%, 24.2%, and 22.6% of the total operating costs, respectively (1). To illustrate the extent of personnel involvement, the percentages of available time spent by different categories of employees are listed in Table 5.1. A list of the most time consuming documents at Rochester General Hospital are indicated in Table 5.2; the 38 documents accounting for approximately half of all the information handling in the hospital. The total number of documents concerning each patient were tallied for each department of the hospital and the total was 316 different documents. Many additional documents were utilized for varied and sundry purposes by service divisions and the grand total of documents in the various departments was an astounding 860. Thus the 38 documents listed in Table 5.2 was only 5% of the total number in use but accounted for 50% of the cost of information handling. This type of study indicates the kind of initial insight required to assess the problems and alternative solutions for processing medical records.

The entire complex of health delivery mechanisms is now faced with the challenge to stem and channel appropriately the flood of data that is collected and stored during hospital care.

1. The quantity of data collected must be reduced by critical evaluation of the requirements and discrimination in the selection of information which is really appropriate to the patient and his management.

2. The information must be consolidated at its sources and presented to the physician in readily interpreted form.

3. The amount of information that is stored must be reduced to that which is truly meaningful by retaining the positive and significant data, and by separating out redundant or noncontributory values.

These objectives can be attained ultimately only with the full cooperation of the medical profession in identifying and specifying the essential information.

Direct studies of hospital paper work disclose that forms are developed for specific purposes and are typically retained long after the need has disappeared (2). The enormous number, diversity, and volume of documents in hospital operations represents one important opportunity for improved cost-effectiveness by developing procedures by which the type, quantity, and presentation of data is designed to be appropriate for the nature and severity of the patient's illness.

PATIENT RECORDS FOR VARIOUS LEVELS OF HEALTH CARE

Hospitals organized to accommodate patients with widely different degrees of disability have medical records with similar format and content whether the patient is acutely ill, convalescent, or is only under observation. Clearly the same admission data and routine tests are not equally relevant to all such patients. The future will probably see both categorization of patients within individual hospitals and more widespread use of different kinds of health care facilities as indicated schematically in Fig. 3.5. Under these conditions, the physician will necessarily indicate the level of health care required for the patient at the time of his assignment to the health care facility. This decision will correspondingly indicate the quantity and format of medical record appropriate to that particular level of illness. The great diversity of records which are rarely, if ever, subject to critical evaluation regarding their relevance or convenience represents one of the most vexing problems in modern medicine. A very small percentage of the data collected in a patient's hospital record is employed currently for his diagnosis, or treatment.

A critical evaluation of realistic requirements would undoubtedly reveal that the optimal quantity of information is far less than the currently available input. A large fraction of the forms in the medical record are filled out automatically. For example, routine admission to hospitals generally involves some laboratory work on blood and urine without consideration of its application to the specific patient (even if the same tests had been run elsewhere the same day). Much of the data obtained at regular intervals on the graphic record, or in the nurses' notes are "routine." Major elements of the history and physical

TABLE 5.1

Time Spent on Information Handling by Types of Employees at RGH

Department and type of employee	Percentage of employee's available useful time spent on info. handling [d]
Nursing	
One medical-surgical unit	
Head nurse and assistant head nurse	64
Registered nurses	36
Practical nurses	22
Nursing aides	9
Ward clerks	85
Total (except students)	34
All units—North- and Westside	28
Operating, labor, and delivery rooms	14
Central supply and I. V. service	6
Emergency department	28
Nursing offices	40
Total[a]	25
Administrative departments	
Accounting, etc.[b]	95
Other—administrative office, admitting, telephone, purchasing	
personnel, health office, etc.	66
Total	73
Radiology (diagnostic, therapeutic, isotopic)	
Radiologists[c]	35
Technicians	28
Others—secretaries, clerks, orderlies	70
Total	42
Laboratories	
Bacteriology (including prorated part of clerical)	17
Hematology	20
Chemistry	30
Pathology	37
Total (including photography)	28
Other	
Medical records	95
Interns and residents	30
Dietary (10 dieticians, 35%; 75 other, 3%)	7
Physical medicine (1 director, 50%; 3 clerks, 80%; 11 other, 5%)	22
Outpatient department	35
Pharmacy	26
EKG, BMR, EEG	36
Social service	45

TABLE 5.1 (Continued)

Department and type of employee	Percentage of employee's available useful time spent on info. handling[d]
Maintenance (chief and assistant, 36%; 42 others, including sec. 4%)	6
Housekeeping; laundry	<u><3</u>
Total hospital (958 equivalent full-time employees)	26

[a]Except students, whose information handling cost is zero by definition.
[b]Almost entirely information handling, see text for further explanation.
[c]Not on payroll.
[d]From R. A. Jydstrup and M. J. Gross (1).

examination are part of a ritual. Every 4 hours the oral temperature, heart rate and blood pressure may be taken and entered in the chart. The "obvious" objective of this procedure is to check the "vital signs" of the patient to be sure he does not unexpectedly experience a complication. The practice may be right or wrong—good or bad—but it clearly deserves to be critically evaluated. Physicians acknowledge other reasons for such a ritual. "It is a common and erroneous presumption that the nurse is required to feel the patient's wrists every four hours for a pulse count, when the purpose is only to assure regular nurse–patient contact (3)." The nurses' notes are first on the list of time-consuming items in Table 5.1 and yet their value to the patient is questionable, particularly if physicians do not consistently read them. An alternative use for nurses' notes is to provide protection for the hospital and its staff in case lawsuits might develop in the future. There may be more effective mechanisms to achieve protection against malpractice suits than the current practices. The information requirements for patient care are widely different from those needed in a court of law and this difference needs to be recognized and expressed by relevant procedures.

The information of importance in the medical record should be presented and displayed in forms that will attract appropriate attention. Recorded data should also initiate appropriate action, but this result is not always attained. In a study of 105 patients on a 31-bed ward over a 59-day period, 29 "drug reactions" were detected in 26 patients, but more significantly, 23 errors in drug administration to 20 patients were detected (4). Some 30% of the 104 patients had adverse reactions or were subject to errors in administration of drugs ordered by the physician. Whether such a study is representative of other wards or institutions could only be determined by additional critical evaluations, which are rare indeed.

TABLE 5.2

List of the Most Time-Consuming Documents at RGH[a]

Each taking more than 5 employee years
1. Bedside notes(nurses' notes)
2. Patient's chart as a whole (assembling
   and reassembling, filing and refiling,
   transmission, etc., not assignable to
   any specific part)
3. Doctors' notes (used for admitting,
   progress, and other notes)
4. Nursing care Kardex cards (actually
   a set of 2 different cards)

5. Radiological consultation form
6. Medicine ticket
7. Scratch-paper notes about patients
8. Patient's menu
9. Graphic chart
10. Miscellaneous requisition and charge forms
    for drugs, X-ray and many miscellaneous
    items (approximately 250,000 used
    per year)

Each taking 2 to 5 man years
11. Inpatient ledger card
12. Doctor's order sheet
13. X-ray film (all sizes)
14. Narcotics records (2 forms)
15. Emergency admitting form
16. Operative record
17. Necropsy report (front sheet and
    additional blank sheets)
18. Nursing team assignment sheets (2 forms)

19. Patient admission, discharge, and condition
    list (ward report)
20. Hematology requisition and charge form
21. Surgical tissue report
22. Bedside intake-output
23. Diet order sheet
24. Vendors' invoices

Each taking 1 to 2 man years
25. TPR work sheet
26. Purchase orders (2 forms West- and
    Northside)
27. Blood pressure and pulse graph
28. Medication record
29. Miscellaneous laboratory requisition
    and charge form
30. Chemistry requisition and charge form
31. Newborn record

32. Time card
33. Blackboards (23 or more)
34. Urinalysis requisition and charge form
35. Physical medicine history, examination, etc.
36. Nine-line addressograph plate
37. Daily audit and analysis of cash receipts,
    debits, and credits
38. Disease and operation code card

Each taking ½ to 1 man year
24 documents

Each taking less than ½ man year
More than 800 additional documents

[a] From R. A. Jydstrup and M. J. Gross, reference 1.

Considering the cost in time and effort devoted to laboratory testing, it is even more disillusioning to find that abnormalities may not be seen or may fail to elicit appropriate action by the medical staff. In one hospital, physician response to abnormalities in routine screening tests (urinalysis, fasting blood glucose, and hemoglobin) revealed "no response of any kind" to approximately two-thirds of the abnormal tests (5). As a result, attempts to rectify this deficiency included a workshop, specially organized to improve performance of the staff. This step was followed by repeated newsletter reminders. These efforts failed to improve the quality or quantity of responses. Finally, fluorescent tape was placed over the abnormal values which could not be seen without removing the tape. This approach produced a notable improvement in appropriate responses over a period of 6 months. The example may be an exaggerated exception which "could never occur in our hospital," but the multitude of factors that breed mistakes of omission or comission in medicine are too large to warrant complacency.

Physicians are generally accustomed to finding laboratory test values presented as numbers as they come from the laboratory. Each test result is expressed in a characteristic unit and must be interpreted in terms of the individual variability expected in that patient and in that particular laboratory. The process of assessing the normality of a particular value is actually a rather complicated statistical process carried out in the physician's head.

Monitors commonly utilized in surgery and intensive care facilities often display continuous records of fluctuating measurements. Both numerical notation and graphic presentations confront the physicians with the difficult problem of distinguishing abnormality or significant change by a process involving judgment rather than reliance on his amazing ability to recognize and interpret visual patterns. These observations indicate potential importance of improved display techniques which could attract the attention of busy health personnel and provide improved if not optimal utilization of the useful information which is contained within the records.

NUMBERS, GRAPHS, WORDS, AND PICTURES

Numbers are inherently nonspecific, since they can be used to count anything. In columns of figures, the individual values can be interpreted only by a rather complicated process of identifying the source or significance of the numbers, the units, and the relation to some kind of standard or reference presented or recalled from memory. Even with a great deal of experience, the process of appreciating and evaluating the significance of a column of numbers is laborious and fraught with potential error. For example, the data obtained from a patient in an intensive care unit (Fig. 5.1A) can be evaluated only by time-consuming analysis rather than by some form of pattern recognition (e.g., glaring inconsistencies can be detected only by long and detailed study).

Most of the current clinical laboratory tests provide values at relatively slow sampling rates as illustrated by the columns of numbers in Fig. 5.1A. In experimental animals it is now possible to record continuously a large number of

A

```
                    PATIENT MONITORING - PACIFIC MED. CENTER

  PATIENT LOG. UNIT  3  DATE  818                          TOWELL M
TIME SYS DIA MAP APID LAP  CVP  HR PRT RRT  MV    TV    CMP  NER WKIN VEQ DUP CO2  RQ     MIP INO ALO ECO  TI    T2 URINE
        817**2254OXIMTER AVD  7.0  HB 13.4  AO2 99  VO2 60
1117   0   0   0 0.00   0  0.0   0   0 14 19.53 1354 0.014  6.7 4.19-640 -30 575******  24  45  45  33  0.0  0.0  0.0
       TIDAL VOLUMES UNBALANCED   1619. -1354.
1118 116  74  89 1.48   0  0.0  65  66  0  0.00    0 0.000  0.0 0.00   0   0   0 0.000   0   0   0   0 38.7  0.0  0.0
1121   0   0   0 0.00   0  0.0   0   0 10 15.65 1444 0.000  0.0 0.00-676 -23 460******   0  45  45  35  0.0  0.0  0.0
       TIDAL VOLUMES UNBALANCED   1853. -1444.
1121 117  78  92 1.33   0  0.0  72  71  0  0.00    0 0.000  0.0 0.00   0   0   0 0.000   0   0   0   0 38.6  0.0  0.0
1130   0   0   0 0.00   0  0.0   0   0 10 13.09 1299 0.078  7.5 2.81-601 -21 408******  24  45  45  34  0.0  0.0  0.0
       TIDAL VOLUMES UNBALANCED   1616. -1299.
1130 108  60  76 1.64   0  0.0  66  66  0  0.00    0 0.000  0.0 0.00   0   0   0 0.000   0   0   0   0 38.4  0.0  0.0
1132   0   0   0 0.00   0  0.0   0   0 11 17.48 1543 0.000  0.0 0.00-650 -26 502******   0  45  45  33  0.0  0.0  0.0
       TIDAL VOLUMES UNBALANCED   1877. -1543.
1132 105  61  75 1.37   0  0.0  70  71  0  0.00    0 0.000  0.0 0.00   0   0   0 0.000   0   0   0   0 38.4  0.0  0.0
1140   0   0   0 0.00   0  0.0   0   0  4  7.76 1707 0.000  0.0 0.00-545 -14 238******   0  45  45  38  0.0  0.0  0.0
       TIDAL VOLUMES UNBALANCED   2005. -1707.
1140 102  62  76 1.12   0  0.0  63  64  0  0.00    0 0.000  0.0 0.00   0   0   0 0.000   0   0   0   0 38.3  0.0  0.0
1210 112  59  76 1.65   0  0.0  71  71 11 15.12 1343 0.064  5.5 2.55-583 -25 428******  24  45  45  32 37.9  0.0  0.0
1220   0   0   0 0.00   0  0.0  68   0  8 12.53 1470 0.018  9.0 2.35-518 -24 357******  26  45  45  33 37.9  0.0  0.0
1223   0   0   0 0.00   0  0.0   0   0 10 16.73 1629 0.000  0.0 0.00-638 -26 490******   0  45  45  33  0.0  0.0  0.0
       TIDAL VOLUMES UNBALANCED   2003. -1629.
1223   0   0   0 0.00   0  0.0  62   0  0  0.00    0 0.000  0.0 0.00   0   0   0 0.000   0   0   0   0 37.8  0.0  0.0
1230 102  66  81 0.79   0  0.0  66  53 13 19.81 1463 0.000  0.0 0.00-649 -30 510******   0  45  45  31 37.6  0.0  0.0
1240 104  57  71 1.35   0  0.0  60  57 10 19.82 1908 0.000  0.0 0.00-551 -35 560******   0  45  45  32 37.8  0.0  0.0
1241 108  64  77 1.17   0  0.0  62  64 10 17.73 1759 0.000  0.0 0.00-605 -29 468******   0  45  45  30 37.8  0.0  0.0
1250 139  75  95 2.20   0  0.0  93  93 12 20.74 1621-0.055  6.5 3.47-534 -38 567******  26  45  45  33 37.6  0.0  0.0
1251 128  69  87 2.18   0  0.0  84  85  7 12.36 1644 0.000  9.3 2.20-498 -24 375******  26  45  45  33 37.5  0.0  0.0
1300 112  64  79 1.47   0  0.0  65  63  7 14.16 1888 0.000  0.0 0.00-535 -26 448******   0  45  45  34 37.6  0.0  0.0
1310 122  70  88 1.49   0  0.0  84  84  4 10.76 2190 0.000  0.0 0.00-486 -22 306******   0  45  45  31 37.3  0.0  0.0
1312 110  69  84 1.21   0  0.0  78  78  9 17.35 1921 0.000  0.0 0.00-524 -33 498******   0  45  45  33 37.3  0.0  0.0
1315   0   0   0 0.00   0  0.0   0   0 14 26.44 1854 0.000  0.0 0.00-516 -51 690******   0  45  45  31  0.0  0.0  0.0
       TIDAL VOLUMES UNBALANCED   1555. -1854.
1315 117  74  91 1.21   0  0.0 104  84  0  0.00    0 0.000  0.0 0.00   0   0   0 0.000   0   0   0   0 37.3  0.0  0.0
1316   0   0   0 0.00   0  0.0   0   0 14 24.51 1693-0.019  6.9 3.62-485 -50 661******  26  45  45  32  0.0  0.0  0.0
       TIDAL VOLUMES UNBALANCED   1341. -1693.
1316 119  73  91 1.27   0  0.0  87  84  0  0.00    0 0.000  0.0 0.00   0   0   0 0.000   0   0   0   0 37.3  0.0  0.0
1321   0   0   0 0.00   0  0.0   0   0  7 11.20 1482 0.060  8.8 1.68-452 -24 324******  24  45  45  32  0.0  0.0  0.0
       TIDAL VOLUMES UNBALANCED   1263. -1482.
1321 119  74  92 1.09   0  0.0  75  75  0  0.00    0 0.000  0.0 0.00   0   0   0 0.000   0   0   0   0 37.3  0.0  0.0
1323   0   0   0 0.00   0  0.0   0   0  9 15.09 1606 0.066  8.4 2.39-500 -30 421******  25  45  45  31  0.0  0.0  0.0
       TIDAL VOLUMES UNBALANCED   1412. -1606.
1323 119  73  90 1.35   0  0.0  82  83  0  0.00    0 0.000  0.0 0.00   0   0   0 0.000   0   0   0   0 37.4  0.0  0.0
1330   0   0   0 0.00   0  0.0   0   0 11 19.15 1675 0.003  6.1 2.76-501 -38 531******  24  45  45  31  0.0  0.0  0.0
       TIDAL VOLUMES UNBALANCED   1463. -1675.
1330 138  73  93 2.23   0  0.0  89  90  0  0.00    0 0.000  0.0 0.00   0   0   0 0.000   0   0   0   0 37.4  0.0  0.0
1340 121  68  85 1.62   0  0.0  75  75  9 15.61 1732-0.052  8.7 2.65-496 -31 446******  25  45  45  32 37.4  0.0  0.0
1350   0   0   0 0.00   0  0.0   0   0  5 10.62 2048-0.008  7.2 1.18-327 -32 321-9.905  25  45  45  32  0.0  0.0  0.0
       TIDAL VOLUMES UNBALANCED   1251. -2048.
1350 108  66  79 0.95   0  0.0  69  70  0  0.00    0 0.000  0.0 0.00   0   0   0 0.000   0   0   0   0 37.5  0.0  0.0
1353   0   0   0 0.00   0  0.0   0   0  8 16.07 1886-0.063  5.2 2.16-477 -33 439******  24  45  45  32  0.0  0.0  0.0
       TIDAL VOLUMES UNBALANCED   1558. -1886.
1353 105  69  82 0.81   0  0.0  66  66  0  0.00    0 0.000  0.0 0.00   0   0   0 0.000   0   0   0   0 37.5  0.0  0.0
1400 127  65  85 2.19   0  0.0  69  68  9 14.97 1557 0.071 10.1 2.57-528 -28 425******  24  45  45  32 37.5  0.0  0.0
1410   0   0   0 0.00   0  0.0   0   0  7 15.32 1924-0.029  7.3 2.29-479 -31 424******  25  45  45  32  0.0  0.0  0.0
```

Fig. 5.1(A). Intensive care units frequently present many different measured quantities as columns of figures which represent much information but are extremely difficult to interpret.

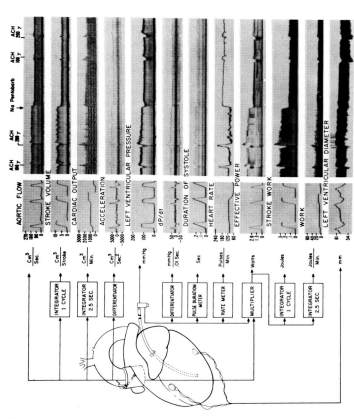

Fig. 5.1(B). Multiple simultaneous variables, directly registered or derived, can be presented as deflections on paper moving at various speeds.

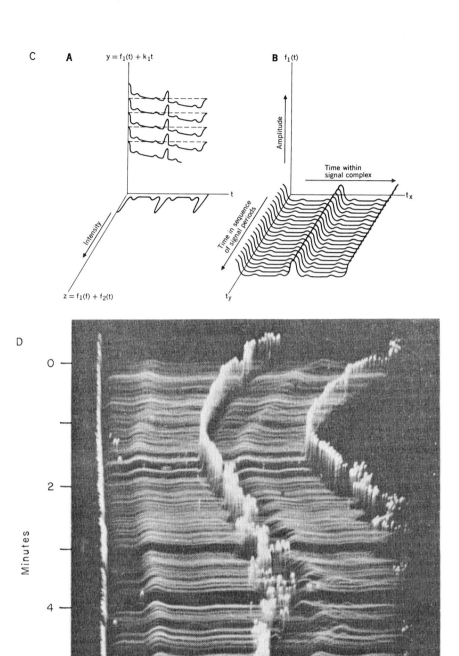

C

A

$y = f_1(t) + k_1 t$

t

Intensity

$z = f_1(f) + f_2(t)$

B $f_1(t)$

Amplitude

Time within
signal complex

t_x

Time in sequence
of signal periods

t_y

D

Minutes

0

2

4

6

significant variables derived by simple analog devices from chronically implanted flow and pressure transducers as illustrated in Fig. 5.1B. Recorded on
strip charts at relatively high paper speeds, it is possible to observe successive
waveforms of each variable, but they are stretched out over many feet of record
so that gradual changes cannot be readily observed, and the cost of paper is
exhorbitant. At slow paper speeds, the successive changes in these variables as
affected by changing conditions can be readily observed but the cyclic changes
in waveform are not visible. With the development of new and noninvasive
techniques similar information will be obtained safely and continuously from
human patients in the future as described in Chapter 6. Time compression of
cyclic phenomena by means of the "countourogram" (6) can be achieved by
downward displacement of successive cycles to pack data and yet permit
observation of changing waveforms, as illustrated in Fig. 5.1C,D. By increasing
the intensity of the trace with increasing amplitude of the signal, a kind of three-
dimensional portrayal is attained. These examples illustrate how various types of
presentations can be utilized to facilitate analysis and evaluation of various
information sources.

Written words are the most common vehicle for transfer and storage of
information during the diagnosis and management of patients. Written language
is so commonplace, we rarely have occasion to critically evaluate its relative
effectiveness. A moment's reflection indicates the enormous advantage of visual
images over word pictures in conveying information. For example, it requires a
great many words to convey the details of an impression gained by a trained
physician's brief glance at a patient. This brief exposure would permit the
physician to recognize this particular individual days later as distinguished from
an unlimited number of other individuals with the same kind of facial features.
The largest computer ever conceived is totally incapable of such a pattern-
recognition process.

ATTRIBUTES OF VISUAL IMAGING

The human visual apparatus not only registers the images on which it
focuses but can fill in information which is lacking by calling on previous
experience (Fig. 5.2A). The quantity of information gained is related to the
background and experience of the observer. For example, a seafaring man would
immediately sense that the sailboat suggested by two lines in Fig. 5.2 is a sloop
with jib furled and mainsail close-hauled on a port tack. The outlines of the
steamship can be discerned despite the clutter of many interlocking lines in
Fig. 5.2B, yet extracting the proper signal from such noise would challenge the
largest computer known. The human visual system can add information or
dimensions by recognizing familiar patterns, e.g., an impression of three

Fig. 5.1(C,D). Continuous waveforms can be packed into a small space while
preserving the waveforms by means of the contourograph. (Reproduced courtesy of
Webb and Rogers, reference 6, and IEEE Spectrum.)

dimensions can be obtained from lines on a plane surface (Fig. 5.2C). Very simple symbols can trigger a recognition process to provide a wealth of additional background material. The visual mechanism can recognize a familiar face unerringly even though the features have been intentionally distorted in all its details (Figs. 5.2D,E). Considering the remarkable capacity of the human visual system to consistently and accurately call up large blocks of information from very simple symbols, our continued reliance on words instead of symbols or graphic presentations to record and recall patient data seems inefficient and antiquated. The human capacity for interpreting meaningful patterns on the basis of past experience is incredibly effective but so commonplace that we give it little thought. Consider the process by which one recognizes a friend at great distance or the sound of a familiar voice badly disturbed by transmission over wires or radio waves. The incredible accomplishments of human perception became all too obvious when attempts were made to simulate or supplement them by the most sophisticated computers available. For example, the process of distinguishing obvious differences between white blood cells on a smear is accomplished easily by a technician and with extremely great difficulty by computer. Even more astonishing are the clues employed by physicians to recognize disease or estimate its severity. As a case in point, Maloney (3) presented photographs of two patients with comparable laboratory data in the early postoperative period (Fig. 5.3). At a glance, a physician could promptly predict which of these two patients will not survive—and have difficulty specifying the clues on which his opinion is based.

WORDS, THE FRAGILE WINGS OF THOUGHT

Transmission of information or ideas by means of words is an extremely tenuous process. Even the most common and familiar words may convey different meanings to different people, particularly when the context varies. As an example, Hyakawa (7) presented many different meanings for the very familiar word "frog" which are recognized by the context in which the word is employed, such as

a frog in the throat
a frog on the front of a uniform jacket
a frog holding a sword scabbard
a frog holding flowers in a bowl
a frog in the switch of a train track
a frog as a cluster of tobacco leaves
a frog on a horse's hoof
a frog in a pond

Ambiguity in the use of words is not confined to common writing or speech but is found in all aspects of biology and medicine. Consider the many meanings of words such as shock, fatigue, tone, angina, stroke, anemia, headache.

Fig. 5.2. Human visual perception has remarkable pattern recognition capacity. (A) Two lines suggest a sloop. (B) Outline of a ship can be distinguished from background "noise". (C) Three-dimensional figures are represented by two dimensions patterns. (D,E) Familiar figures can be instantly recognized even when intentionally distorted.

Active investigation of definitions, phrases, and codes by which such data can be most effectively elicited, recorded, and transmitted is certainly warranted by the growing crisis in handling and storing medical records (see section on medical nomenclature, page 190).

COMPONENTS OF THE PATIENT'S MEDICAL RECORDS

Physicians are extensively trained in the process of conducting a comprehensive routine for eliciting and recording information of value for clinical diagnosis during their medical school training. The main headings of the sequence are indicated in Table 5.3 consisting of the patient's history, physical

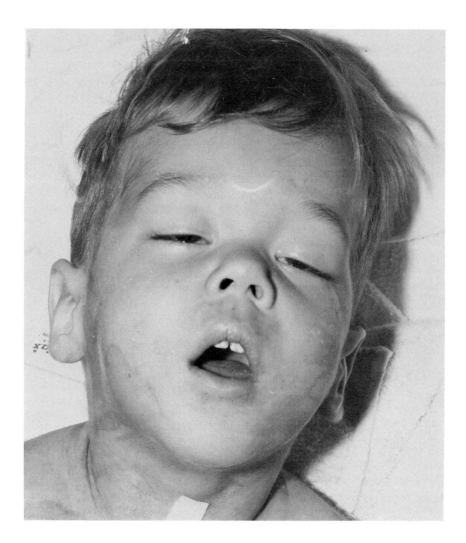

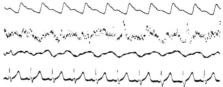

Fig. 5.3.

See facing page for legend

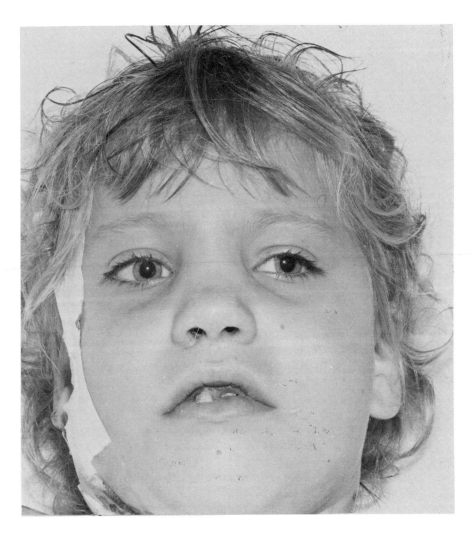

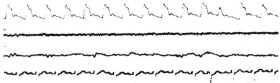

Fig. 5.3. Verbal description of these two children, no matter how long or detailed, would fail to permit recognition of the face on subsequent meeting. Even less would words convey the impression the patient on the right will survive surgery, the one on the left will not. (Photographs reproduced courtesy of Maloney, reference 3 and *Ann. Surgery.*)

examination, laboratory examination, differential diagnosis, and final diagnosis. An enormous quantity of additional detail is included under each of these main headings, producing a rather ideal objective which is rarely attainable in the "real world" of medical practice outside of academic institutions. Consequently physicians tend to tailor the nature and extent of their data gathering efforts in relation to their judgment of the patient's condition as the evidence unfolds. With increasing patient loads, strong economic and practical pressures tend to cause progressive curtailment in the quantity of information collected from histories or physical examinations and more reliance on laboratory testing (Fig. 5.4).

The shear bulk of medical records has multiplied with ever-expanding sources of diagnostic information and the increasing number of professionals (nurses, interns, consultants, laboratory reports, etc.) to the point that it is very difficult to locate vital information scattered among the mass of data. This process is further complicated by the trend toward using the medical record for diverse purposes, such as the basis for charges for materials and services, demographic and epidemiologic information, protection from malpractice suits, and many other ancillary functions. Only a small fraction of the typical hospital record is utilized for arriving at decisions regarding the illness of the patient. Each category of personnel which contributes to the medical record must also be able to find specific entries in the various scattered sections. The increasing tendency for health care to be delivered by a team rather than by a solo physician provides even greater incentive to facilitate rapid access and accurate interpretation of the information contained among the diverse kinds of data entered into the record. The traditional approach to the assembly of diagnostic information (see Table 5.3) appeared to be so deeply ingrained that basic changes appeared essentially impossible only a few years ago. In response to mounting pressures, there are many parallel efforts across the nation directed toward improving the efficiency of collecting, recording, displaying, and retrieving patient data which are collected so laboriously and at such great cost at the present time.

PATIENT HISTORIES

Physicians are heavily dependent upon the information gained from interview and interrogation of patients to obtain both background information and insight into the current complaints in terms of the time sequence of developing signs and symptoms. This process is time-consuming and costly, particularly when the physician conducts the initial interview himself.

A

SYSTEMATIC REVIEW: ✓ Check if no problem;   (Circle) if abnormal and describe in detail.

LAST   BY WHOM

PHYSICAL EXAM _Nov 1970____ ECG _?_____   CHEST XRAY_Nov & Dec 1970_ Harborview_

BLOOD, URINE TESTS _See Nov and Dec admissions to Harborview Med CTr._____

SKIN TESTS_____

IMMUNIZATIONS_____

GENERAL
Weight change   Fever-chills   Weakness ⟶ Sweats all the time - day and night and is visibly
Fatigue   (Sweating-night sweats) perspiring during the interview.
SKIN
Nail changes   (Itching)   Rash-eruptions ⟶ Has had dry skin and hair since age 9 yrs because
✓Headache   ✓Trauma   of "thyroid trouble"
HEAD
EYES
✓Vision-glasses Blurring   Photophobia   Wears reading glasses
Diplopia   Scotoma   Inflammation
EARS/NOSE/MOUTH
Pain   ✓Discharge   Vertigo ⟶ Decreased hearing in recent years -especially in the
(Deafness)_ _ _ (Tinnitus)_ ⟵   left ear with a ringing tinnitus.
Sinusitis   ✓Polyps   Postnasal drip
✓Epistaxis_ _ _ _Obstruction_ _ _
✓Teeth   Gums   Breath
Taste   Pain   Dentures
RESPIRATORY
Wheezing   (Dyspnea) ⟵   Hemoptysis ⟶ Dyspnea when climbing hills or at two blocks level-
Chest pain   ✓Cough   Sputum   no orthopnea or history of heart trouble.
BREASTS
Lumps   Pain   Discharge   Given water pills for high blood pressure in Spokane
CARDIOVASCULAR   However, these were stopped when high cholesterol level
Palpitation   ✓Pain   Dyspnea ⟵   found - now she thinks she has a normal blood pressure.
✓Orthopnea   Murmurs   (Blood pressure)
Cyanosis   Edema   Claudication
(GASTRO-INTESTINAL)
Appetite   Indigestion   Nausea-vomiting
Gas   Pain   Hematemesis
Jaundice   Hernia   Stool-shape, color   SEE PRESENT PROBLEMS
Constipation   Cathartics   Diarrhea
Melena   Hemorrhoids   Abdominal girth
GENITO-URINARY
✓Dysuria   (Nocturia) ⟵   ✓Hematuria ⟶ Nocturia 3-4 times/night for 2-3 months without dysuria.
✓Frequency   Urgency   Incontinence May have frequency during day - may be polydypsic.
SEXUAL HISTORY
Syphilis, Gonorrhea, Sores, Other
Epididymitis_ _Pain_ _ _ _ _ _Discharge
Gravida/Para/Abortions
Sterility   Impotence   Contraception
FEMALE - MENSES
Cycle/Duration/Amount/Menopause   None - since hysterectomy 1952.
Last Pelvic exam, PAP smear
Dysmenorrhea   Spotting   Irregularity
ENDOCRINE
(Goiter) ⟵   Glycosuria   Exophthalmos ⟶ Low thyroid since age 9 for which she takes 3 pills/day.
Treatment with hormones (thyroid) steroids) Dr. Clark HMC follows - he has done blood tests but no
ALLERGIC   RAI uptake.
✓Sens. to allergens, drugs, vaccines
Eczema ✓Asthma   Hay fever   Hives   Ganglion left wrist- removed
BONES, JOINTS & MUSCLES   Cyst of unknown type right thigh = removed
Trauma   (Swelling) ⟵   Pain-arthritis
BLOOD - LYMPHATIC
Anemia   Bleeding tendency
Pain   Lymph node enlargement
NEUROLOGIC
Syncope   Convulsions   Sensation
Gait   Coordination   Paralysis-strength Tongue got stuck out 3-4 yrs ago -- she was
PSYCHOLOGIC   given phenobarbital with good relief.
Memory   Mood   Sleep pattern
Anxiety   (Emotional disturbance) ⟵
Drug, alcohol problems

Fig. 5.4. Standardized and simplified notational systems facilitate recording of patient data which can be quickly scanned and interpreted. (A) Systems survey.

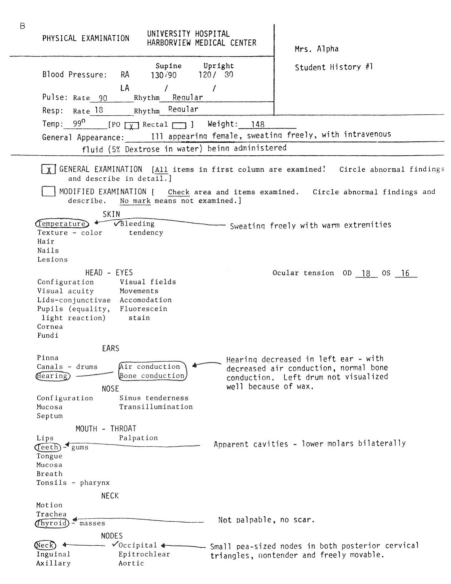

B

PHYSICAL EXAMINATION   UNIVERSITY HOSPITAL
                        HARBORVIEW MEDICAL CENTER   Mrs. Alpha

                          Supine   Upright   Student History #1
Blood Pressure: RA   130/90   120/ 80
                   LA   /   /
Pulse: Rate 90   Rhythm Regular
Resp: Rate 18   Rhythm Regular
Temp: 99°   [PO [x] Rectal [ ]]   Weight: 148
General Appearance: Ill appearing female, sweating freely, with intravenous
                fluid (5% Dextrose in water) being administered

[x] GENERAL EXAMINATION [All items in first column are examined!   Circle abnormal findings
       and describe in detail.]
[ ] MODIFIED EXAMINATION [Check area and items examined.   Circle abnormal findings and
       describe.   No mark means not examined.]

            SKIN
(Temperature) ← ✓Bleeding   Sweating freely with warm extremities
Texture - color   tendency
Hair
Nails
Lesions

        HEAD - EYES
Configuration   Visual fields   Ocular tension OD 18   OS 16
Visual acuity   Movements
Lids-conjunctivae Accomodation
Pupils (equality, Fluorescein
 light reaction)   stain
Cornea
Fundi

          EARS
Pinna
Canals - drums   (Air conduction)   Hearing decreased in left ear - with
(Hearing)   (Bone conduction) decreased air conduction, normal bone
          NOSE   conduction. Left drum not visualized
Configuration   Sinus tenderness   well because of wax.
Mucosa   Transillumination
Septum

      MOUTH - THROAT
Lips   Palpation
(Teeth) ← gums   Apparent cavities - lower molars bilaterally
Tongue
Mucosa
Breath
Tonsils - pharynx

          NECK
Motion
Trachea
(Thyroid) - masses   Not palpable, no scar.

          NODES
(Neck) ←   ✓Occipital ←   Small pea-sized nodes in both posterior cervical
Inguinal   Epitrochlear   triangles, nontender and freely movable.
Axillary   Aortic

Fig. 5.4(B). Physical examination. (From reference 15, reproduced courtesy of S. R. Yarnell.)

C

CHEST - HEART

Breasts	Masses
Chest - shape	Dullness
Symmetry	Fremitus
Resonance	Whisper
Breath sounds	transmission
	E $\longrightarrow$ A

Carotids	Murmur
Neck veins	transmission
Pulses - radial,	Bruits
femoral, (d.pedis)	Pulse character

— Absent bilaterally

Apex impulse - character, position
Cardiac sounds   Thrills
   Rate $S_1 S_2$ A_2 P_2
   Rhythm   Extra sounds
                $S_3 S_4$
                clicks
                murmurs
                rubs

ABDOMEN

Shape - scars   (Tenderness)   2+ tender, left CVA area
Organs - (liver)   character   and LUQ with some involuntary
  spleen, kidney   Peritonitis   guarding
Sounds   (Organomegaly)
Hernias   Masses - fluid   liver - down 4 fb on deep
               Sounds   inspiration

GENITALIA - RECTUM

Male	Female
Penis	Perineum - vagina
Scrotum - testes	Cervix - uterus
Prostate	Adnexa - ovaries

Rectum - sphincter tone, masses, guaiac

BONES - JOINTS

Spine	Deformity
Extremities	Joint mobility
Muscle bulk	Effusions
	Varicosities
	Edema

NERVOUS SYSTEM

Mental status
Speech
Gait   not tested
Cranial nerves   Fine tremor of fingers - both hands
(Motor system)   Sensory system
Coordination   Position -
Reflexes: R   L   vibratory sense
  biceps   ++ ++   Abnormal reflexes
  knee   +   +
  ankle   O   O

Fig. 5.4(C). Physical examination (From reference 15, reproduced courtesy of S. R. Yarnell.)

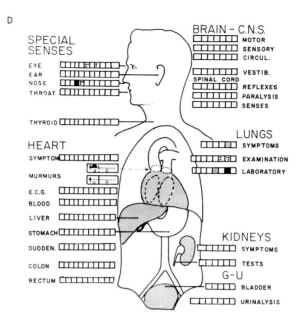

Fig. 5.4(D). Schematic example of a summarizing form.

COMPONENTS OF THE MEDICAL HISTORY

The information contained in the patient's history can be dissected into integral parts—(1) Personal history gives the name, address, age, sex, occupation, and other relevant information regarding the individual as a member of society. (2) the family medical history indicates the occurrence of conditions or states that might have genetic or environmental significance on the patient's condition. (3) The patient's past history lists the previous illnesses he may have experienced. All three of these categories are composed of information which might well be provided in advance by the patient using carefully prepared questionnaires without utilizing the time of the physician in transcribing it. This approach is being explored in several different types of settings. The initial screening history consists of a listing of chief complaints and a series of questions designed to elicit signs or symptoms stemming from the various organ systems. A relatively small number of rather standard questions can be utilized for this purpose. Thus the system survey is also readily adaptable to either a questionnaire or computer approach.

The organ system(s) that must be explored further on the basis of evidence of dysfunction deserve histories substantially longer and more detailed than the survey. Some of the more extensive forms of computer or pencil and paper approaches provide comprehensive information in this category, but it is difficult to render it as complete or meaningful as the physician's own interrogation. Efforts to use programmed techniques for eliciting the complete history, excluding the physician, are not likely to achieve ultimate success because they sacrifice the remarkable human capability to respond to subtle clues or recognize complex patterns. By recognizing subtle changes in expression, a trained physician may be able to identify topics that require more extensive exploration or even that the patient is intentionally giving misleading answers. Progressive experience will determine the relative proportions of the history gathering process which can be most effectively accomplished by computer, questionnaire, by physician's assistants, or by physicians in various settings, and with patients with various types and severity of illness.

COMPUTER APPLICATIONS FOR OBTAINING MEDICAL HISTORIES

Many different approaches have been proposed to facilitate the extraction of the desired information from the patient by interrogation using computers and by paper and pencil questionnaires. The most ambitious attempts have been complex branching programs utilizing high-capacity computers by which the sequence of questions is adapted to the sequence of answers. In some instances, the patients engage in a dialogue in which questions are displayed on a cathode ray screen and the responses are made on a computer keyboard. The choices of response may be (Yes), (No), (Don't Know), or (Don't Understand). Programmed instruction frames inform the patient how to proceed through a series of questions in which the response determines the next question in the sequence. At the end of such an interview, the computer may present a group of questions to determine the patient's reaction to the experience and to give their evaluation of this technique as contrasted with a physician's interrogation (8). The patient response to this process has been reported to be generally favorable. The computer was regarded as more thorough in general medical histories while the physician was believed to be more comprehensive in specialty interviews. The principal deterrent to this approach is the relatively high cost (i.e., $10 per history), even with large volumes of patients. One important extra dividend from exploring computers for obtaining historical data is the increased knowledge about the process as carried out by a physician. For example, Grossman et al. (9) reported that the physician generally records only a fraction of the information he elicits from the patient. "The items he chooses to record are those which he judges to be important or which defend his diagnostic hypothesis." Further, there is evidence that patients asked to answer exactly the same question on more than one occasion may give conflicting or even a reversal of their own responses. The content of automated medical histories tend to emphasize social history, medical history and review of systems. Acquisition of details about the patient's chief complaints is a more difficult proposition.

QUESTIONNAIRES FOR ELICITING MEDICAL HISTORIES

Questionnaires can be utilized for patient interrogation more cheaply than computers. The number of questions is often so great that the positive responses are lost or obscured and must be extracted and recorded in a more compressed form. Costs are necessarily increased if the data are reduced by the doctor's office staff or computer. In addition, valuable time is lost if they must be sent away for processing (editing) and return. One such system covers severity, duration, frequency and location of "all major symptoms" with sixteen pages of 1200 questions which can be sent away for optical scanning and print out within 24 hours (10). The report is essentially a compilation of positive information plus pertinent negative factors in the patient's history, allowing the physician to focus on significant features and avoiding the time-consuming task of reviewing large amounts of negative data. The obvious advantages of sparing the physician time from routine and presenting him with information of value in speeding up his diagnostic process seems most promising in the face of a widespread shortage of physicians. Fifty-two different systems for automated acquisition of patient history data have been compared with reference to design, performance, and cost in a "marketing survey" by Yarnell, Wakefield, and McGovern (11). Such critical evaluations and comparisons of various approaches will serve a useful function in aiding selections by individuals or groups interested in introducing these techniques in various clinical settings. It is beyond the scope of this particular monograph to consider the relative merits of various approaches in any detail.

The total data derived from the various components of the history are difficult to handle or reduce unless they are recorded in standard terminology. Mechanisms for compressing or compacting this information without sacrificing too much specificity or clarity are badly needed. If the physician has confidence that the patient has filled in all the positive answers and has not neglected to check any spaces, he knows by inference that the remainder of questions had negative answers so they may not need to be retained. The ability to eliminate from storage all or most of the negative answers is one of the main advantages of utilizing a system in which the same questions are always presented to the patient in precisely the same way. Since the physician rarely follows such a reproducable routine nor such an extensive interrogation as is now available in questionnaire or computer approaches, these techniques may prove to have significant advantages over the physician's version of a preliminary history, in terms of both completeness and data-packing.

SORTING AND SCHEDULING PATIENTS

Some potential pathways by which members of the population with real or imagined illness may approach health care delivery mechanisms are illustrated in Fig. 5.5. The preliminary sorting process to distinguish the well and the worried from the physically sick has been the province of the family physician or general practitioner, a situation still prevalent in Europe today. The depleted numbers of general practitioners in the United States has left an unfilled void in this sorting process that has not been adequately filled by the specialists in internal medicine and pediatrics. Substantial numbers of patients perform this process themselves by directly seeking care from specialists—a correct choice being dependent upon the patient's knowledge, judgment, or luck. Other patients apply for aid from diversified groups of physicians in clinics. The importance of this preliminary sorting process is bound to increase with the rapidly growing demand for health care stimulated by health insurance programs, population growth, and urbanization. Among many approaches, questionnaires can be effective for a preliminary sorting process to aid in appropriate scheduling of patients. For example, an on-line computer program developed at the Massachusetts General Hospital (9,10) was modified for scheduling new patients to be seen at the Lahey Clinic (12). The patient filled out the form in his own home and returned the completed form to the clinic. It was checked for obvious mistakes and the yes answers were key punched. A listing of the "positive" responses was prepared for review by the physician and the patient was simultaneously given an appointment to see a specialist identified by a point system applied to the questionnaire. The patient-prepared questionnaire has proved useful as a screening type history and for scheduling.

At the Karolinska Hospital in Stockholm, a questionnaire of approximately 1000 questions (13) is reportedly being sent to patients referred by their local physician to the hospital and awaiting an appointment. The patients fill out the questionnaire at home and return it to the hospital. A computer processes the information, and in addition, identifies appropriate laboratory tests suggested by the patterns of responses by the patient. The computer scheduling is designed to provide a sequence of tests with minimal delays. When the patient arrives at the hospital, the hospital physician has immediate access to an extensive history and a group of laboratory test results. The advantages and potential pitfalls in such systems will emerge with greater experience.

The single most serious mistake would be to attempt a universal system for acquiring data from patients. For example, a computer approach to a screening history would be disastrous if applied to patients who do not understand English, who are mentally retarded, or acutely ill. Similarly, presenting a patient with 1200 questions might be entirely inappropriate if his only complaint is a

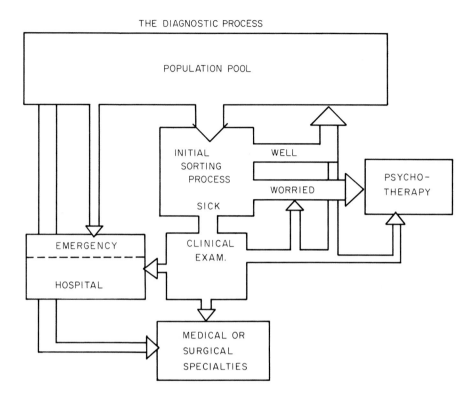

Fig. 5.5. Patients enter the health care system by many routes including a form of a filter system (i.e., family physicians) who separate out the sick from the well or worried well. Other patients go directly to specialists or to emergency clinics. Effective utilization of personnel and facilities require a more effective filter mechanism based in part on more effective utilization of information limited in the patient's history.

boil on the neck, a broken arm, or an obvious need for glasses. Medicine is inherently so diverse that an ideal or standard approach cannot succeed. The mounting problems of supplying health care calls for many different approaches, including education of the recipients.

Patients may provide incorrect information by misunderstanding the questions presented by computer or in paper-and-pencil responses. If increasing reliance is to be placed on patients preparing their own histories, they should be more actively educated. Patients are far more knowledgeable than they were twenty years ago because of frequent exposure to medical information presented by telecommunications and by popular magazines. Positive steps can and should be taken toward simplification and clarification of medical terms commonly employed. Authoritative information regarding the proper way of responding to standard questions in medical histories could be included in the presentations of mass communication to the general public. Progressive development and improvement of computer programs and medical questionnaires will undoubtedly lead to greater uniformity of terminology, forms and format as the most effective and concise wording of questions and responses become identified by scientific evaluation. In addition, the process of filling out such standardized history forms could be improved by developing carefully prepared instruction manuals, which could also be incorporated into the educational process in public schools—elementary and high schools. Such instruction could serve to greatly increase the information base of the general public so that they could play a more substantial role in their own medical care. It would also initiate the development of a health data bank by a growing proportion of the total population with information regarding baseline recordings and numerical values beginning at an early age. Unfortunately for some fraction of patients, awareness of the nature and significance and symptoms could induce increasing concern about their health and could even foster hypochondriacs. This process is clearly evident today and calls for improved methods of handling it, particularly with the imminence of apparently free or artificially cheap health care supported by federal funds (see Chapter 4). Restructuring and streamlining of basic mechanisms for medical management of patients with a wide variety of types and severity of illnesses is clearly needed. The "problem-oriented medical record" is such an approach which has attracted a great deal of intense interest in recent years.

PROBLEM-ORIENTED MEDICAL RECORDS

The traditional approach to the initial collection and transcription of medical data has changed but little over many decades, being deeply ingrained in medical students trained to conduct exhaustive histories and physical examinations. With the growing pressures from heavier patient loads, such thoroughness quickly becomes impractical. As a result, there is a remarkable spectrum of behavior from the compulsively elaborate to the sketchy and haphazard approach. Most of the medical records fail to convey the mental processes which relate the physician's interpretation of the patient's problems to the selection of diagnostic tests or therapeutic approaches. Recognizing this deficiency, Lawrence Weed (14) proposed a totally new approach that is adaptable to future applications of computers, technology, and screening techniques. The basic principle is a medical record which is organized around concisely stated problems presented by the patient. He conceived that each medical record should contain a complete list of all the patient's problems, including both clearly established diagnoses and other unexplained findings not sufficiently clear to indicate a specific conclusion. The list of problems was regarded as a sort of "table of contents" of the patient's records which would be supplemented and continuously updated by a "flow chart" during the subsequent course of diagnosis and therapy. Those signs and symptoms which combine into a diagnostic pattern can be represented by a single entity. The list of problems can also be separated into active and inactive problems in terms of relative urgency or importance. The problem list serves as a concise history, and the individual problems are numbered so that all successive laboratory tests, orders, or notations can be referenced to the specific problem to which they are related. By this means, the rationale of the physician's actions is much more evident to all who have responsibility to the patient during and after his period of care.

The success of this approach is highly dependent upon the formulation of the problems, careful analysis and appropriate response to each. If the patient clearly has a perforated ulcer, this is noted as the problem. If the diagnosis is not certain, the same problem would be listed as upper abdominal pain and a course of diagnostic procedures outlined to define the problem more explicitly. The temporal sequence followed in arriving at definitive diagnosis, and administration of therapy would be indicated by carefully organized flow sheets.

A "problem" is defined as any aspect of the patient which requires further attention for diagnosis, treatment, or observation. An example of a problem-oriented record sheet is presented in Table 5.3 to indicate the fundamental deviation from the traditional approach as presented to medical students in a syllabus for their course in physical diagnosis (15). This hypothetical patient exhibits several different consequences of chronic alcoholism and the problem list clearly indicates the physician's picture of their interdependence.

The strategy of confirming the diagnosis by further examinations, tests and therapy would be entered in an organized fashion under the appropriate problem number on subsequent pages of the chart. In this way any other member of the health team can obtain a clearer picture of the patient and the approach to his management which is being undertaken. The list of problems also constitutes a concise statement of the physician's diagnostic and therapeutic objectives. This approach makes possible an evaluation of the effectiveness or "benefit" of the patient's care.

COMPONENTS OF THE PROBLEM-ORIENTED RECORD

The personal information regarding the patient (name, age, sex, address, etc.) remains an essential ingredient of the record for which many physicians or hospitals have well-developed forms. This information may be placed on the inside cover of the chart for easy reference but permitting the problem sheet to occupy a position of prominence as the first sheet of the record. The problem list (see Table 5.3) would represent a continuous tally of the problems as they are resolved or accumulate with time. This list should be maintained current and can comprise not only diagnoses derived from all sources of information in the traditional approach: It may also include signs, symptoms, laboratory values, psychosocial problems, or other entries of interest or importance. Thus data which do not fit into a definitive diagnosis can be clearly registered with less chance of being forgotten or ignored. This approach also frees the physician from recording a "coded" diagnosis prematurely.

A treatment page can provide a running account of medication and treatments administered in accordance with specific problems and permits rapid and accurate review without searching through the chart. Such consolidation would be of great value to the house staff, nurses, pharmacists, and other therapists as a source of ready reference.

A data base consisting of the essential ingredients of the history and physical examination is supplemented by flow sheets which reference special laboratory results (i.e., X-ray, ECG, chemical determinations, etc.). These may be entered in chronological order under the appropriate problem numbers.

Many advantages to this approach can be recognized. The problem oriented record encourages the application of logic to the multiple problems that can be presented by a single patient. The significance of the cumulated data is more readily apparent to the physician, house officers, and nurses who have responsibility for care. The data presented and collected in such a manner are far more susceptible to computer processing. The effectiveness and soundness of the physician's course of action is rendered more clearly discernible. The data in such a record are far more useful for clinical or epidemiological research. Various approaches to diagnosis and therapy could be compared and evaluated through the more concise relationship between courses of action and their results.

These cause-and-effect relationships are of particular value for peer review, for cost accounting, for cost-benefit analysis and for scientific evaluation of the various approaches to diagnosis or modes of treatment. These advantages will be of prime importance in establishing quality standards and priorities for the delivery of health care in the trying days ahead. Serious consideration of major

TABLE 5.3

Components of Patient's Medical Records

Traditional record	Date	Problem no.	Problem	Date	Resolution
			Problem oriented		
I. History					
A. Personal information					
1. Name, address	12-6-70	1	Chronic alcoholism (X13 years)		
2. Age	12-6-70	2	Portal cirrhosis (2nd to No. 1)		
3. Sex					
4. Marital status	12-6-70	3	Impending hepatic coma (2nd to No.2)		
5. Occupation					
6. Place of birth	12-6-70	4	Esophageal varices (by X ray 5-22-69) (2nd to No. 2)		
B. Present illness					
1. Chief complaint(s)					
2. Course of present illness	12-6-70	5	Acute G.I. Bleeding (2nd to No. 2)		
3. Survey of organ systems					
a. Nervous system					
b. Cardiovascular system	12-6-70	6	Anemia (2nd to No. 5)		
c. Respiratory system					
d. Gastrointestinal system	12-6-70	7	Prostatic hypertrophy		
e. Genitourinary system					
f. Musculoskeletal system					
C. Past history					
1. Previous illnesses					
2. Surgical operations					

3. Allergies, drug reactions, etc.
4. Survey of organ systems
D. Family history
   1. State of health or
   2. Causes of death in family
   3. Hereditary illnesses

II. Physical examination
   A. General appearance
   B. Surface signs, vital signs
   C. Ears, eyes, nose, throat
   D. Thorax, (heart, lungs)
   E. Abdomen, (pain or masses)
   F. Genitalia
   G. Extremities
   H. Glands
   I. Etc.

III. Laboratory examination
   A. Clinical tests
      1. X ray, ECG, etc.
   B. Blood count
   C. Urinalysis
   D. Blood chemistry
   E. Microbiology

IV. Differential diagnosis
    (list of conditions which could be
    considered in arriving at diagnosis)

V. Definitive diagnosis

modifications of medical records is an extremely important step at this time because a critical evaluation of all the data processing and information transfer techniques is extremely timely. For example, the principles underlying the problem-oriented record may also be expressed in improved mechanisms for standardized and logical notation of the information elicited in the course of the data gathering process. Streamlined notation of positive findings in the systematic review of the patients past history and physical examination are some representative examples of such opportunities (15). As the Health Care mechanisms of this country prepare for the rising flood of prospective patients, great effort must be directed to improve the effectiveness with which patients can be provided care commensurate with their needs and to increase the effectiveness of each member of the team and component of the system.

MEDICAL NOMENCLATURE

The ultimate objective of clinical study is the accumulation of symptoms, signs, facts, findings, laboratory values and impressions so that a distinctive diagnosis can be intelligently selected from a list of possible solutions. In this process, medical terms are employed to represent various combinations of clues serving as a type of technical shorthand. Some diagnostic terms are quite specific, indicating both the nature of the abnormality and its locations (e.g., persistent patency of the ductus arteriosus or rheumatic pulmonary valvular stenosis). In contrast, lack of specificity is exemplified by a list of fourteen common diagnoses (Table 5.4), containing words from ten very different origins including physical signs (edema), symptoms (pain), pathologic lesions (inflammation), anatomic sites, laboratory determinations (anemia), clinical test (high blood pressure), functional states (epilepsy), traditional terms (shock). Consider the futility of attempting to study zoological specimens or organic chemicals if any particular specimen could be named in accordance with any one of ten different types of names! Medical nomenclature is an unstructured collection of terms which have evolved over the long history of medicine. Traditional terms (e.g., shock or stroke) have origins and meanings obscured by the haze of distant time and generally encompass not one but several different entities. Many diseases are identified by the name of some prominent physician, presumably but not always the man who first described it. Parkinson's disease or Paget's disease are not particularly informative names except perhaps to medical historians. The so-called classical descriptions of diseases are generally written very soon after discovery and at a time when least is known about its mechanisms or manifestations. Semantic confusion is fostered when several disease entities are covered by one term or when several terms may refer to the

TABLE 5.4

	1 Physical sign	2 Symptom	3 Pathologic lesion	4 Anatomic site	5 Laboratory sign	6 Clinical test	7 Traditional term	8 Etiologic agent	9 Eponym	10 Cause unknown
Rheumatic mitral stenosis			x	x				x		
Hodgkin's disease									x	
Sickle cell anemia					x					
Essential hypertension						x				x
Shock							x			
Stroke							x			
Pneumococcal meningitis			x					x		
Myxedema	x									
Idiopathic epilepsy	x									x
Bell's palsy	x						x		x	
Heart failure				x			x			
Tetralogy of Fallot							x		x	
Angina pectoris		x		x						
Adenocarcinoma of the breasts			x	x						

same abnormal function. For example, a patient with hyperactive thyroid function could be labeled hyperthyroidism, toxic goiter, exophthalmic goiter, Grave's disease, or Basedow's disease. Such situations can confuse physicians and confound computers. Current medical terminology is not computer-compatible and could not be converted by any means short of extensive or complete revision. It seems desirable to consider what could be gained by a new coding system even though a major restructuring seems very unlikely at present (16).

A new system of medical taxonomy could take many forms but it certainly should be designed to provide maximal information transfer. Some of the major specifications might well include the following:

1. Each letter, syllable, and sequence should have maximum intelligible information content.

2. The system must combine flexibility with consistency.

3. Computer compatibility should be assured (i.e., by an alphanumeric system.)

4. Additional data and new knowledge should be readily accommodated into the system with minimal reorganization.

5. Definitive diagnosis should be noted by a mechanism conveying comprehensive information about the specific patient and his illness. A new system of terms might be devised with as many of these favorable attributes as possible. The feasibility of achieving some of the stated objectives can be demonstrated by proposing a solution based on a series of propositions.

a. Individual letters may stand for words or phrases: In the years since the NRA, many government agencies have become familiarly designated by initials, particularly when the combinations of letters is pronounceable. For example, NASA very adequately conveys the meaning of the words National Aeronautics and Space Agency. It is not confused with NATO or with any other government agency. Words synthesized from initials sound at first like nonsense syllables. With familiarity and common usage they become adapted as meaningful units of speech serving as well or better than the original.

b. Enriching the meaning of individual letters by position senses: The quantity of information packed into a single word can be greatly increased by utilizing each successive letter to denote a particular category of meanings. This can be accomplished in the same way that an enormous variety of numerical values are conveyed by the ten symbols in our decimal system (i.e., by utilizing the principle of position sense). The number 7 has an entirely different significance if it is two places to the left of the decimal point instead of being two places to the right of this reference point. Its significance depends upon its location in the sequence. By the same token, a particular letter can be endowed with the meaning of an entire word or phrase depending upon its position in a sequence.

c. Alternation of consonants and vowels: A random selection of letters can produce pronounceable "mnemonic words" by alternating consonants and vowels. For example, if an assorted selection of vowels were placed in one box and another box contained consonants, one could select letters alternately from the two boxes and the resulting sequences would always be pronounceable. The resulting "words" would sound strange, just as any foreign language sounds like an unintelligible sequence of sounds until one learns to automatically assign meaning to the sound patterns.

d. Standardized spelling and pronunciation: Consistency and intelligibility of words can be greatly enhanced if every letter always has precisely the same pronunciation as in the Spanish language.

A DEFINITIVE DIAGNOSTIC CODE—AN ILLUSTRATIVE EXAMPLE

A physician arrives at a diagnosis by weighing evidence to determine what organ or organ system is involved, the pathological process involved, the nature of the functional disturbance, and its severity. One approach to developing an appropriate coding system would be to provide mechanisms for denoting each of these factors but packed into the most meaningful notation possible. Such a system must be capable of encompassing an enormous number of anatomic locations, pathological processes and functional disturbances with maximal discrimination and minimal ambiguity. One such mechanism is the arborization of a tree which develops thousands of branches from a single trunk. This basic approach has also been employed in automation of medical histories by computers, in MEDELA described in a subsequent section (see Fig. 5.8) and in many other logic systems. One could specify each individual twig of such a tree by a distinctive number or "word" if the branches at each successive fork are successively assigned a position in the word. For example, assume that the trunk divided into six main branches, each of these gave off six branches and this process continues for six branchings, the terminal twigs would number more than 40,000. One could discretely identify every branch by a distinctive six-digit number or six-letter word. Using this system one could begin at the trunk and follow the course unerringly to the specific twig one wished to identify. The same approach could be employed to logically arrive at a distinctive code word which contained a vast amount of information because each letter would represent unique information signified by its position in the sequence (16). To illustrate the principle, alternating consonants and vowels are assigned to designate the organ system, major components, minor components, pathology, functional disturbances, and severity as shown in   Fig. 5.6. The first three columns are capable of discretely identifying the location of some pathological process in the cardiovascular system. Each anatomical site may have its own set of pathological processes, each of which may produce a selection of functional disturbances. The state of current knowledge is undoubtedly inadequate to allow such definitive conclusions in many instances so a standard designation for "unknown" or "undecided" would be indicated. The same principles could be applied to the remaining organ systems. The advantages of such a scheme would include a high level of consistency and yet great flexibility. For example, a change in the patient's condition, additional laboratory data, or even new advances in medical knowledge might require a change in the designation which could generally be achieved by substituting one or two letters in the word. If multiple processes are present, multiple code words would be applied, one for each problem as in the problem oriented record (see above). Although the system looks complicated at first sight, so do the enormous number of Greek and Latin words that medical students must learn to apply. It seems quite possible that a system of this type could be developed as a kind of new technical language with pleasing combinations of sounds.

The words resulting from any such arbitrary system as this fail to convey useful information at the outset. With repeated use, the combinations of words would begin to conjure up mental pictures of patients as entities with certain physical and social attributes, pathologic changes and functional disturbance of specific components of organ systems. If medical students were required to repeat this process on every patient from the beginning of the training, certain combinations or patterns would recur with sufficient frequency that a group of letters would bring to mind a mental picture of a whole patient without conscious analysis of the letters or sounds. Ultimately more and more sequences of words would provide vivid pictures of whole patients, complete with their unique patterns of signs and symptoms. The scheme presented in Fig. 5.6 is intended to indicate the feasibility of a new diagnostic-medical terminology rather than a firm proposal. It is an expression of need rather than a proposed solution and additional details can be obtained by referring to the original article (16).

Efforts at simplifying and streamlining the medical record and the nomenclature should also be extended to improve the efficiency of utilization of the various health care personnel and facilities.

CLINICAL LABORATORY TESTING

Any discussion of objective types of clinical laboratory studies tends to focus on the electrocardiogram as a prime example. The sophistication of data acquisition and analysis of electrocardiographic signals is probably superior to any other type of medical signal analysis. For example, Caceres and his

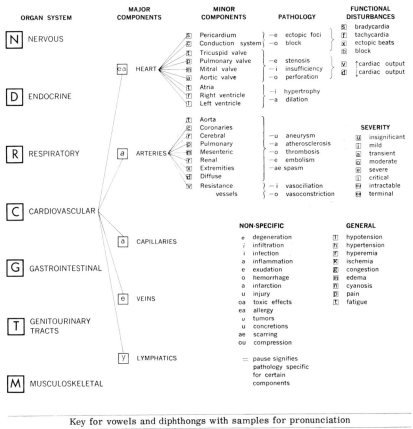

Key for vowels and diphthongs with samples for pronunciation

a		e		i		o		u		ou	oa	ae	io	ea
a	ah	e	at	i	eat	net	mite	it	no	mute	ouch	boa	haste	violet	create

Fig. 5.6. A consistent computer-compatible medical nomenclature could theoretically be constructed by means of a branching system producing more than 40,000 terms covering the various organ systems with six-letter words, each unique and pronounceable. (Photograph reproduced with permission, see reference 16.)

associates (17) developed a model center for processing electrocardiograms and spirograms at a rate of 4500 per month from over 40 use groups. The data were received at the Center by mail, by analog data transmission lines or by messenger (17) as illustrated in Fig. 5.7. Typical results are printed automatically and sent back to the user as exemplified by the data in Table 5.5. The large investment in research, development, and publications devoted to automatic electrocardiographic interpretation is widely known and does not warrant any detailed discussion here, except as it might be considered as an example of a process with principles widely applicable to many other information handling systems. It is not even clear how the experience gained in computer applications to the electrocardiographic interpretation can be effectively utilized for other forms of data than those which occur as bioelectric potentials such as the ECG,

TABLE 5.5

Medical systems development lab   Heart disease control program   Computer processed electrocardiogram   Screening demonstration[a]

Name
Number 401626   Tape 0022   Option   175   Date 06-19-68   Time 16--07--38
Height 71   Weight 175   Age 40   Male
BP   Unknown   Meds   Unknown

Diagnosis

Lead	I	II	III	AVR	AVL	AVF	V1	V2	V3	V4	V5	V6	
PR	0.00	0.12	0.08	0.12	0.00	0.12	0.13	0.12	0.12	0.09	0.12	0.09	PR
QRS	0.08	0.08	0.09	0.08	0.07	0.07	0.08	0.10	0.08	0.09	0.08	0.09	QRS
QT	0.35	0.36	0.36	0.37	0.33	0.36	0.42	0.36	0.37	0.38	0.37	0.37	QT
RATE	74	68	68	76	71	69	63	61	63	65	62	60	RATE
Code	3	3	2	3	2	2	3	3	2	3	2	3	
Cal	96	96	96	96	96	96	96	96	96	96	96	96	Cal

Axis in	P	QRS	T	C	R	S	STO	ST-T	QRS-T
Degrees	54	92	60	-78	84	179	210	150	32

D

4352   Short PR Interval
8381   QRS Axis range 85 TC 92

Exclude preexcitation
Vertical axis; normal if age below 30 or tall and thin

[a]From Caceres, C. A. and Barres, D. R., Computerized care—hospitals. *J. Amer. Health Ass.* **43**, 49–52 (1969).

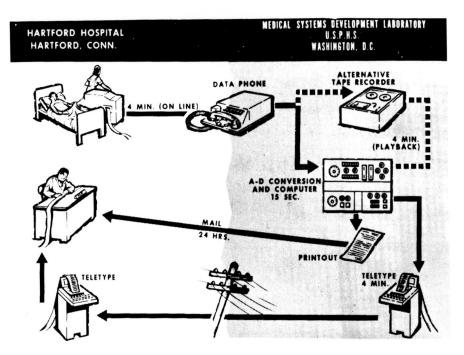

Fig. 5.7. Electrocardiograms can be recorded in Hartford Hospital and sent in real time to Washington, D.C. to be processed and returned. The process can be accomplished without the intervention of a physician but this is not essential. [Photograph reproduced by permission, courtesy of C.A. Caceres (17).]

EEG, and EMG. These signals are generated by the body in a directly recordable form without intervening transducers, and an enormous body of data has been collected over many decades with fairly well established diagnostic criteria based on both clinical comparisons and basic investigation on animals and man. Electrocardiographic analysis is undoubtedly the simplest and most direct of all the readily available clinical data sources including x-rays, spirometry, catheters, isotope application, etc. If correspondingly great effort will be required to achieve the same level of sophistication for the other kinds of data reduction, the total cost may well be prohibitive. Consider, for example, the problem of regularizing the interpretation of x-ray pictures.

INTERPRETATION OF X RAYS

The process of extracting information from X-ray plates is largely subjective, based on the experience and background of the examiner. The significant shadows or subtle differences in optical density must be identified and sorted out of complex patterns produced by overlying structures. The radiologist's impressions are generally conveyed to the physicians in charge of the case by rather long, unstructured descriptions assuming that the reader will attach the same meaning to the words as was originally intended. Most of the creative energy devoted to radiology has been directed to the improvement in the images to optimize the visual and perceived information as much as possible. The loss of information between the radiologist and the referring physician is substantial. The radiographic images can be stored in a computer and displayed on a TV screen but this process is extremely costly. Radiographic images are far more complex than electrocardiograms, and the diversity of data that can be extracted exceeds those presented in Table 5.5 by manyfold. To accommodate this enormous diversity, Brolin (18) developed the MEDELA system with a terminal consisting of a projector display of 16 mm film strips, each frame containing as many as 127 words or phrases standardized for use in regular radiological reporting. For example, radiological examination of just the nervous system included 129 frames totaling some 2500 words for use in such reports. To facilitate handling the enormous amount of data required for radiographic analysis of all the organ systems, information was organized like a branching tree (Fig. 5.8) and programmed so that a selection at one level decides which alternatives will be presented at the next level. In addition to words and phrases, alternatives could be presented in the form of sketches that proved most useful. The data from the terminal systems was stored in a computer (on tape or disc) and could be transmitted as hard copy on an output device at various stations in the hospital. This approach eliminated the need for typing of conventional reports since the radiologist learned to communicate directly with the computer. Due to the data compression, some 150,000 reports could be stored and easily retrieved from a single magnetic tape. Many different mechanisms resembling the MEDELA system are being explored in institutions throughout the western world. The promise of improving the rate and accuracy of information processing by such techniques is actually secondary to the increased knowledge and understanding of how the physician performs his functions. In addition, the process of preparing names and nomenclature in an organized fashion will undoubtedly prove useful in streamlining information processing regardless of the ultimate man–machine interface.

NUMERICAL DATA: Processing, Presentation, Interpretation

The clinical management of patients has been significantly altered where a large and growing selection of quantitative laboratory tests are readily available. In many institutions the dozen tests available in the clinical laboratories have

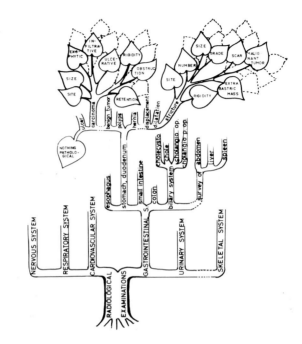

(A)

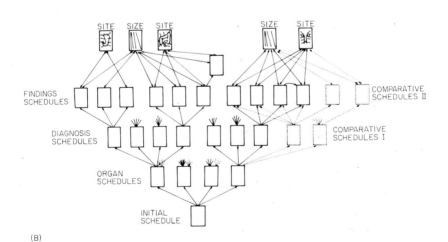

(B)

Fig. 5.8. (A) Interpretation of radiographs can be standardized by means of a branching system like that suggested for medical terminology in Fig. 5.6. (B) Reference frames in the Medella system are organized like a tree, each containing lists of terms of which provide standardized reporting of radiographs. (Reproduced with permission of I. Brolin, reference 18.)

grown in the past ten years to levels between 50 and 250 different types of analyses (19). The number of tests being conducted annually are expected to double in 5 years as it has every 5 years since 1946. The advantages provided by these new sources of numerical data have also generated serious problems related to their precision, standardization, reproduceability, processing and interpretation. The results of a test recorded as a specific number tend to convey a degree of confidence which is not always fully warranted. The advantages and problems resulting from the progressive availability of large numbers of quantitative tests have received so much widespread attention that only some of the pertinent highlights will be considered in this chapter.

ELUSIVE NORMALITY

The fundamental problem of distinguishing the healthy from the sick extends to the separation of "normal" from "abnormal." It comes as a shock to those outside the field of medicine that despite years of testing and millions of values, we still lack generally accepted "normal ranges" for virtually all the available types of numerical data. The sources of variability are many and may be considered in two categories—(a) analytical and (b) biological. The magnitude of measurement error varies greatly among the various testing procedures. The detection and correction of random and repeated errors will be more readily attained by more specific analytical methods and by the development of high-purity materials by the National Bureau of Standards to be reference substances for calibration of procedures in clinical laboratories. They currently include cholesterol, urea, uric acid, creatinine, and calcium carbonate more than 99% pure. Many additional substances are to be released in the future. Widespread use of such standardizing procedures would greatly improve the reproducibility of laboratory results from different institutions. The response to this important new option has been quite disappointing; hopefully acceptance will spread more rapidly in the future. No matter how accurate the testing procedures, a high degree of variability will persist owing to many factors affecting living organisms.

Complex living organisms display fluctuations of all recorded variabilities as an indication of responsiveness to internal and external factors. The frequency distribution of recorded variables about some mean is the most common representation of functional fluctuations found among a group of individuals or among measurements taken on one individual over a period of time. Since a single value or specified group of values cannot be designated as "normal" for groups or individuals, the problems of clearly recognizing abnormals are illustrated schematically by Fig. 5.9. If measurements are conducted either on two groups of "healthy" or "diseased" individuals or a single population containing a mixture, there almost always exists a borderline area in which values from normal and diseased individuals overlap (20). This borderline area does not always contain the same persons on repeated testing because each has his own range of individual variability as illustrated in Fig. 5.9. The most vexing part of this problem is the large number of potential sources of variation in any particular measured quantity exemplified by the following.

1. Age
2. Sex
3. Biological rhythms
   a. Daily
   b. Circadian
   c. Seasonal
4. Emotional status
5. Rest or exercise state
6. Environment conditions
7. Pregnancy
8. Genetic status
9. Menopause

Most of these are tacitly recognized but largely ignored in assessing normality and the list is far from complete.

PROBLEMS OF DATA PRESENTATION

Clinical chemistry data are most commonly presented to the physician as a group of numerical values with various units on some type of form (Fig. 5.10). The physician must assess the extent to which the particular value relates to a normal range for that laboratory and the particular individual as affected by factors like those above. When the "normal range" is not presented on the report, the physician is generally obliged to undertake this important evaluation as a mental exercise. If the normal ranges were presented in the form of bar graphs, the physician might be able to make a more direct comparison but the great divergence of absolute values produces such a great range of numbers that they can be included on the same scale with some almost disappearing from view (Fig. 5.10). Since the units for expressing the absolute value were arbitrary in the first place, some unit which was applicable to all would be much more useful. One such unit could be the standard deviation (SDU) which is a widely recognized expression of the deviation of a numerical value from the mean of some group of determinations (20,21). The standard deviation includes a statistically derived weighting determined by the variability encountered in the test group. The data from a patient could be expressed in terms of SDU's (or fractions thereof) merely by use of a computer readout or by use of a nomogram if technicians transcribe the data manually. Expressed in such terms, anyone could be readily instructed to identify which values were significantly different, requiring either retest or red flagging to call them to the attention of the referring physician. Carried one step further, the data could be manually or automatically plotted in the form indicated in Fig. 5.11, which illustrates how values deviating several SDU's from a mean can be rendered very obvious at a glance (21). The prime obstacle to such an appealing solution is the fact that the ranges of normal for group data from various laboratories have not been satisfactorily established as indicated above.

Another kind of graphical representation to bring out patterns of abnormality is an extension of a proposal by Wolff (22) designed to take greater advantage of the ability of the human visual system to identify subtle differences in patterns (Fig. 5.11B). On several intersection lines, scales could be arranged so that the mean value for each variable appears at a constant distance from the center to produce a circle. If laboratory values deviate from the mean value and are connected by a continuous line, the deviation of this new pattern from the reference circle can be easily recognized. Patterns of reproducible configuration would require that the same values were consistently recorded on the same scale. Perhaps more meaningful patterns would be obtained if the selection of the variables around the circle were based on tests having relations to common functions, organs or diseases. For example, the components of various profiles described below might form the basis for related test results for such a presentation.

Biochemical Individuality

The test values from the same individual fluctuate within a somewhat narrower range than those of a group as schematically illustrated in Fig. 5.9. On that basis individual normalization would make the laboratory determinations more meaningful since each individual could serve as his own control. Although the concept of biochemical individuality has a long history, the relatively narrow range of variability of the individual normal (see Figs. 5.9 and 5.10) has not been fully utilized in obtaining maximal value from quantitative analysis. For example, Williams (23) documented the limited ranges of "normal" values for some specific laboratory determinations (platelets, protein-bound iodine, serum cholesterol) as illustrated in Fig. 5.12. The relatively small deviation from a mean for individuals was observed in substantial numbers of tests over extended periods of time. Similar applications of patterns of test results have been employed by Casey (24) with growing acceptance in a clinical setting. For example, a metabolic profile consisting of 28 chemistries to evaluate the function of multiple organs is presented in a graphical presentation indicating degree of deviation from "normal" in a format similar to that in Fig. 5.11A.

The data that are currently collected and stored in hospitals or doctor's offices are not effectively utilized to provide a data base to establish the mean and range for the individual patient. Obviously the deviation of a test result obtained when an individual showed no signs of illness would greatly enhance

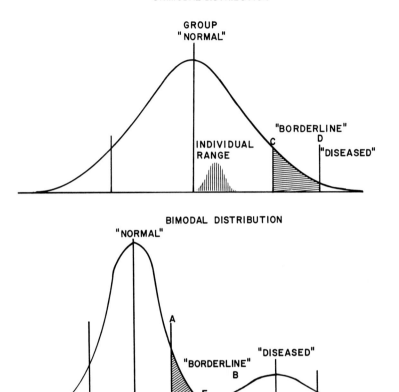

Fig. 5.9. The range of normal values commonly form a bell-shaped curve, extending above and below some mean value for the group. (A) Values from patients frequently overlap normal values at the ends of the curve. Individual variability (small bell curve) is much less than group ranges (see also Fig. 5.12). (B) If measurements are made on groups of patients and compared with groups of normals, the two distributions usually overlap providing a bimodal distribution. (After Jungner, reference 25.)

the significance of corresponding data when the same individual had symptoms or was suspected of relevant illness. For this purpose, a data bank would correspond to the acquisition and storage of individual medical data on the 1.5 million inhabitants of Stockholm county as described in Chapter 4 (Table 4.3). The American public is not likely to be enthusiastic to the concept of having the details of their medical records in some data storage facilities like the Social Security Administration particularly without rigid safeguards regarding its

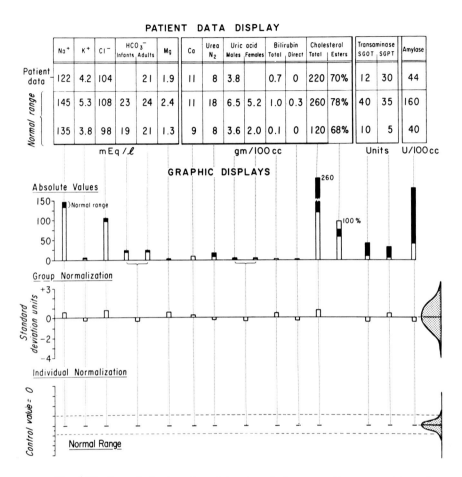

Fig. 5.10. Numerical values from laboratory tests are usually expressed in various units with widely different ranges of normal. By converting all these figures to a common reference such as standard deviation units, the extent to which values deviated from group normals can be more easily expressed. If the patients own normal range is known, the variability is greatly reduced.

release. An alternative possibility requires techniques for data selection and packing to the point that essential information might be carried by each citizen on his person after reaching a particular age level. It would be interesting to speculate how much clinical data could be compressed into the very small space represented by a durable card the size of a drivers license, or social security card, as suggested schematically in Fig. 5.12. Technology has been developed by which an entire page of information can be reproduced photographically as a dot the size of a period or colon. Even punch-card techniques provide storage for a

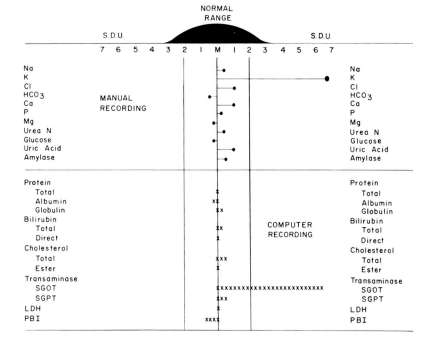

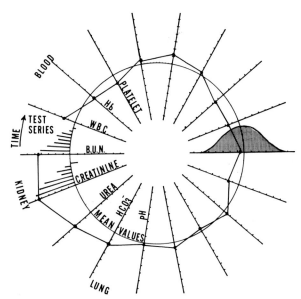

Fig. 5.11. (A) Test values that deviate significantly from the group range or individual range can be simply displayed by plotting individual values in terms of standard deviation units either manually from nomographs or by means of computer printout. (Reproduced courtesy of *J. Amer. Med. Ass.*) (B) Pattern recognition of abnormal test values may be facilitated by plotting patient data on radiating scales such that normal values describe a circle. Deviation from the circle could provide recognizable patterns [Modified from Wolff (22).]

large quantity of selected and edited information. Considering the rapid acceptance and utilization of credit cards, it seems quite possible that some kind of Medidata card could be developed and widely used if the data stored were widely accepted as *worth the time and trouble of storage for ready retrieval.* One essential ingredient required to establish the value of the data is the necessity for discrimination in the tests performed so that they represent reasonable priorities for utility and cost benefit. For this purpose it seems desirable to consider the concept of profiles.

Profiles. The growing number of available clinical tests demands more selectivity in their use than was applied when only a few could be obtained. Jungner (25) proposed test groups that might be particularly useful in eliciting evidence of diseases in various organ systems (Table 5.6). Within each test group are tests which may be specifically indicative of particular types of pathological processes. Such an approach could form the basis for structuring the test batteries in the clinical laboratories and the mechanisms for reporting the resultant patterns, reducing the number or percentage of tests which were not

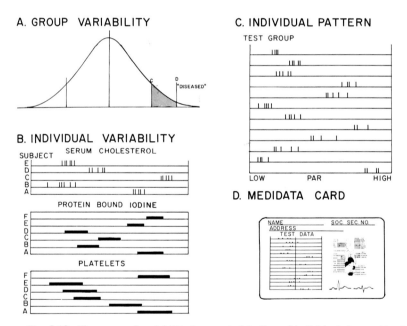

Fig. 5.12. The range of variability in repeated testing of individuals is considerably smaller than that for groups of individuals as illustrated by these plots based on data presented by Williams (23). One way to provide ready access to an individual data bank would utilize a "medidata" card like a driver's licence.

TABLE 5.6

Specific changes	Test groups
Heart/vessels	Cholesterol, total lipids, β-lipoproteins, BUN, creatinine, urinary protein, total protein, albumin, zinc sulfate test, sodium, potassium, chloride
Liver	Bilirubin, alk. phosphatase, thymol turbidity, serum iron, TIBC, GOT, GPT, LDH (amylase and/or others), total protein, albumin, zinc sulfate test
Lungs	ESR, Prot.-bound hexose, sialic acid, haptoglobin
Kidney	BUN, Creatinine, urinary protein (uric acid), hemoglobin, hematocrit, serum iron, TIBC, ESR, prot.-bound hexose, sialic acid, haptoglobin
Blood	Hemoglobin, hematocrit, serum iron, TIBC, total protein, albumin, zinc sulfate test, ESR, prot.-bound hexose, sialic acid, haptoglobin
Skeletal system	Calcium, phosphorus, alk. phosphatase, total protein, albumin, zinc sulfate test, ESR, prot.-bound hexose, sialic acid, haptoglobin
Endocrine Diabetes	Blood glucose (after glucose load), urinary glucose cholesterol, total lipids, β-lipoproteins, ESR, prot.-bound hexose, sialic acid, haptoglobin
Thyroid Parathyroid	PBI, T_3, cholesterol, total lipids, β-lipoproteins, Calcium, phosphorus, alk. phosphatase
General changes Inflammatory, neoplastic, etc.	ESR, Prot.-bound hexose, sialic acid, haptoglobin, hemoglobin, hematocrit, serum iron, TIBC, total protein, albumin, zinc sulfate test

[a] From reference 25.

relevant in automated chemical laboratories. Accumulated experience could also provide the basis by which the functional significance of the various components of the profiles could be weighted and ultimately reported in a highly streamlined form as a weighted average for the group. This sequence would tend to represent the ultimate in data packing and could foster storage for ready access by any one of several techniques (Figs. 5.11 and 5.12).

PATIENT MONITORING

The concepts of patient monitoring are so many and varied that the term can conjure many pictures (see Fig. 5.13) including arrhythmia monitors in coronary care units, displays of physiological function during major surgery, intensive care, or ward supervision by monitors at central nursing stations. To allay some of the semantic confusion, some of the different objectives for patient monitoring could be listed as follows.

1. Remote collection of data during normal activity (i.e., by telemetry)
2. To determine if a patient is alive or responsive
3. To inform the professional staff of a change in the patient's functional condition
4. To sound an alarm indicating a theoretical or medical surgical emergency
5. To provide retrospective data to help identify causes of such emergencies
6. To evaluate the effectiveness and course of intensive care
7. To protect the patient against serious events of low probability

Most of these objectives are being pursued in one or another context with varying degrees of success. For example, the shortage of nurses led to a great convergence of resources by many industrial concerns (more than 30) to provide monitoring equipment designed to permit a nurse sitting at a console to maintain surveillance over patients in various beds and rooms on a hospital floor. The results proved catastrophic for most companies because they all produced equipment to record various combination of heart rate, ECG, body temperature, blood pressure, and respiratory rate. These are the traditional "vital signs" but they fail to represent a reasonable coverage of the wide spectrum of disease states which can occur on any typical hospital ward, including the specialties. These commonplace physiological variables are the measurements that can safely be entrusted to those without a medical education. The health care system will not soon recover the interest or involvement of the large industrial companies which suffered major economic loses on this adventure in medical instrument development. A few of these companies have successfully marketed reliable equipment for monitoring the electrocardiogram to warn of excessive arrhythmias in patients who have suffered coronary occlusion and myocardial infarction. Others have produced equipment displaying these variables to aid in continuous evaluation of patients undergoing major surgery, within intensive care or trauma centers. In response to the use of instruments and allied health personnel, reducing direct contact between patients and medical and nursing professionals, the "surgeons invented the intensive care unit" to confine two or

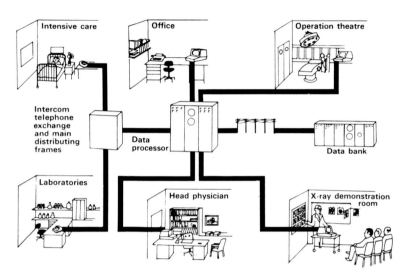

Fig. 5.13. The thoracic-SRT project.

three nurses in a small room with four or five patients. Maloney (3) urges restoration of the physician and the nurse to the bedside in such units by installing monitoring displays at the bedside with the computer in a more remote location. Of the seven reasons for monitoring listed above only the alarm function and the continuous evaluation of intensive care are prominently used.

COMPREHENSIVE REAL-TIME MONITORING FOR INTENSIVE CARE

One of the most sophisticated monitoring systems for intensive care has been developed at the Karolinska Hospital in Stockholm by Olaf Norlander (26). Utilizing rather standard computer hardware, a system has been developed with a number of clearly defined objectives such as—(a) the system has been built up around the patient from whom the data is collected; (b) it must be operated by existing ward personnel without special training; (c) the system must be able to gather information from existing sources (such as clinical laboratories, medical recourse, pharmacies, etc.) and deliver up such information to the sources if required; (d) the system must be compatible and supplementary to conventional chart systems and provide hard copy of information for continuous operation of the affected wards; (e) desired information must be available for rapid access from any one of many terminals strategically located in different locations convenient to both the bedside and the responsible staff. The primary emphasis throughout is to increase the safety of the patient and improve his care by more rapid and effective transfer of information, decisions, or signs of change (Fig. 5.13).

The system consisted of six graphic display sets (GRAFOSKOP), and 20 intercom devices. The displays can be connected to 13 different rooms by a cable-switch rack, including the office of three doctors, two surgical theatres, the head nurse's office and two rooms in intensive care. Provisions were made for simple insertion of many variables including six circulatory variables (five fluid-balance, six laboratory variables, six blood-gas determinations). Inserting the number 29 was acknowledged by a tone over the intercom terminal, the patient's name and census number was inserted from an addressograph stamp, a new tone indicated confirmation of the name and then the new variable could be fed in. If certain variables exceeded preset levels, another warning tone required recheck before it would be accepted by the data processor. Data from the computer storage could be instantly recalled on the viewing screen, including graphic summaries of previous determinations. This example serves to indicate the nature of current trends toward substituting or supplementing traditional charts with simple and effective computer functions.

COMPUTER DIAGNOSIS

Computers need to be viewed realistically as highly useful but often limited tools, which are capable of "elementary clerical work normally assigned to $100-a-week clerks and in numerous situations prove less efficient (27)." In standard operations computers perform three basic operations—(1) Add and subtract; (2) collate by matching such items as numbered checks against the same numbered account; (3) file, retrieve, and compare information and then furnish instant balance. All these functions are carried out so much faster than humanly possible, that one new computer can make more computations in a single minute than a mathematician could do by hand in 4000 years, but computers can also make mistakes on a prodigious scale because they cannot think or appreciate the fact that something has gone wrong. A realistic appraisal of the strengths and frailties of computers is badly needed by those who use them like the executive who said, "I'm vice president in charge of computer operations, but if I want to know what the computers are doing—or can do—I have to ask those kids in there."

Despite these apparent limitations computers have already carved a substantial role for processing huge quantities of medical data of the sort discussed above. In the analysis of the electrocardiographic patterns, computer programs have evolved which embody the combined experience and knowledge of experts. The interpretation is a more specific and concise result than those experienced cardiologists could have presented by expertise before. One example of the manner in which a computer can be programmed so that a physician can tap an enormous fund of cumulated knowledge was presented by Schwartz (28) in the form of a dialogue between the doctor and the computer using conversational algebraic language and a branching decision tree of the type illustrated in Fig. 5.8.

The criteria for ECG diagnosis have been greatly clarified by the process of developing effective computer programs. If the ultimate results of these efforts prove widely useful, the increasing confidence by medical practitioners will undoubtedly permit reporting of diagnoses much more concisely; interpretation by a few statements rather than the complicated report illustrated in Table 5.5. Electrocardiographic interpretation should not be confused with a diagnosis, but rather as one component of the more extensive evidences of disease or disability.

The diagnostic process is simple and straightforward when the pattern of signs and symptoms presented by a patient conforms closely with clear-cut criteria with which the physician has ample past experience. The approach to diagnosis in patients with uncommon, incomplete or confusing patterns of clues is often approached in two steps—(a) an initial assessment leading to a pattern suggestive of some number of diseases, and (b) a directed search for the defining characteristics which will help identify which item in that differential diagnostic list is the most probable cause of the illness (29). The prodigious memory available through computers has been employed to extend the list of possible disease states which could be responsible for various combinations of clues and to assign a kind of priority or weighting to each possible choice by a form of statistical analysis to elicit probabilities. Pilot studies to evaluate the use of computers in diagnosis have generally focused on limited categories of diseases for which substantial quantities of objective quantitative data may be available, such as congenital heart disease, thyroid disease, or blood disease, with a rapidly expanding application suggested by the lists in Table 5.7 (30). However, the relative proficiency of computer diagnosis compared with the performance of the physician remains obscure for the curious reason that such comparisons have rarely been attempted on any scientific objective basis such as percent of "errors." Statements of general approval are quite common but their justification remains to be established (31). In view of the analytical nature of such computer applications the scarcity of critical statistical analyses of results is astonishing.

TABLE 5.7

Principal Directions of Diagnostic Studies Made with Computer Aid in Different Clinical Fields[a]

Surgical clinic	Intrinsic valvular diseases
	Acquired valvular diseases
	Burn injuries
	Intestinal diseases
	Mechanical jaundices
	Chronic appendicitis
Oncology clinic	Cancer of the thyroid
	Cancer of the larynx
	Cancer of the lungs
	Stomach cancer
	Liver cancer
	Mammary cancer
	Tumors of the bone–joint system
Internal clinic	Stomach and liver diseases
	Rheumatism, infectious–allergic myocarditic
	Tonsillocordial syndrome
	Thyrotoxicosis
	Hematological diseases
	Acute poisonings
	Metabolic disorders
Neurological and psychiatric clinics	Brain tumors
	Brain–vessel diseases
	Brain trauma
	Some forms of mental diseases

[a] From Moiseeva and Usov (26).

References

1. Jydstrup, R. A., and Gross, M. J. Cost of information handling in hospitals. *Health Services Res.* (Winter) 235-255 (1966).
2. Anderson, H. C. A study of paper work performed by ward nursing service personnel in a Seattle hospital. Thesis for Master of Nursing, University of Washington, 1966.
3. Maloney, J. V. The trouble with patient monitoring. *Ann. Surgery* **168**, 605-614 (1968).
4. Hoddinott, B. C., Cowdry, C. W., Coulter, W. K., and Parker, J. M. Drug reactions and errors in administration on a medical ward. *Can. Med. Ass. J.* **97**, 1001-1006 (1967).
5. Williamson, J. W., Alexander, M., and Miller, G. E. Continuing education and patient care research. *J. Amer. Med. Ass.* **201**, 118-122 (1967).
6. Webb, G. N., and Rogers, R. E. The countourograph. *IEEE Spectrum*, June 1966.
7. Hayakawa, S. I. "Symbol, Status and Personality." Harcourt, New York, 1953.

8. Slack, W. V., and Van Cura, L. J. Patient reaction to computer-based medical interviewing. *Computers Biomed. Res.* **1**, 527-531 (1968).

9. Grossman, J. H., Barnett, G. O., McGuire, M. T., and Swedlow, D. B. Evaluation of Computer-acquired patient histories. *J. Amer. Med. Ass.* **215**, 1286-1291 (1971).

10. Irons, I. I., and Moldenhauer, J. G. Automated medical history system. *Med. Electron. Data* **1**, 65-67 (1970).

11. Yarnell, S. R., Wakefield, J. S., and McGovern, R. E. "A Review of Fifty-Two Different Systems for Automated Acquisition of Patient History Data: Comparisons of Design Approaches, Performance and Cost. Medical Computer Services Association, 1116 Summit, Seattle, Washington, 1971.

12. Rockart, J. F., Hershberg, P. I., Grossman, J., and Harrison, R. A symptom-scoring technique for scheduling patients in a group practice. *Proc. IEEE* **57**, 1926-1933 (1969).

13. Danielsson, H., Danielsson, T., Engkvist, O., Hall, P. L., and Mellner, C. H. "Questionnaires for Preventive and Internal Medicine." Department For Medical Information, Karolinska Sjukhuset, Stockholm, Sweden.

14. Weed, L. L. Medical records that guide and teach. *New Eng. J. Med.* **278**, 593-600 (1968).

15. Yarnell, S. R., and Libke, A. "An Introduction to the Medical History and the Medical Record; A Syllabus for Students Emphasizing the Problem-Oriented Approach," 2nd ed. University of Washington, School of Medicine, Seattle, Washington, 1971.

16. Rushmer, R. F. Numedcode: A cumulative, compact, computer-compatible medical nomenclature is both necessary and feasible. *Dis. Chest* **56**, 143-148 (1969).

17. Caceres, C. Computers in medicine; A model of the dimensions of biomedical engineering. *In* "Dimensions of Biomedical Engineering" (E. Salkovitz, L. Gerende, and L. Wingard, eds.) San Francisco Press, San Francisco, California, 1968.

18. Brolin, I. A computer-based terminal for radiological reporting. *Int. J. Man-Machine Studies* **1**, 211-235 (1969).

19. Brown, J. H. U., and Dickson, J. F. III. Instrumentation and the delivery of health services. *Science* **166**, 334-338 (1969).

20. Jungner, G., and Jungner, I. Interpretation of data obtained in laboratory screening programs. "Symposium on Multiple Diagnostic Screening Analysis." Dept. of Laboratory Medicine, University of Minnesota, May, 9-10, 1968.

21. Rushmer, R. F. Accentuate the positive; a display system for clinical laboratory data. *J. Amer. Med. Ass.* **206**, 836-838 (1968).

22. Wolff, H. S. "Biomedical Engineering." World University Library, McGraw-Hill, New York, 1970.

23. Williams, G. Z. Clinical pathology tomorrow. *Amer. J. Clin. Pathol.* **37**, 121-124 (1962).

24. Kerns, W. H. "Casey's Profiles" expand diagnostic role. *Mod. Hosp.* **108**, 122-127 (1967).

25. Jungner, G. Multi-channel automated equipment (autochemist). "International Symposium on Early Disease Detection." Elkhart, Indiana, Oct. 6. 1969.

26. Norlander, O. P., and William O. G. Real-time handling and display of data from intensive care and anesthesia. "Thoracic Data Project." Karolinska Hospital, Stockholm, Sweden.

27. Surface, B. What computers cannot do. *Satur. Rev.* July 13, 57 (1968).

28. Schwartz, W. B. Medicine and the computer; the promise and problems of change. *New Eng. J. Med.* **283**, 1257-1264 (1970).

29. Scadding, J. G. Diagnosis: The clinician and the computer. *Lancet* **2**, 877-882 (1967).

30. Moiseeva, N. I., and Usov, V. V. Some medical and mathematical aspects of computer diagnosis. *Proc. IEEE* **57**, 1919-1925 (1969).
31. Ledley, R. S. Practical problems in the use of computers in medical diagnosis. *Proc. IEEE* **57**, 1900-1918 (1969).

SUMMARY AND CONCLUSIONS FOR PART I
Health Care Crises
and the Technological Implications

The preceding chapters have outlined the nature and scope of the health care delivery mechanisms in this country, emphasizing some crucial factors that threaten accentuation of their problems and some approaches employed in European countries to deal with such stresses with varying degrees of success. Some of the major problems can be considered as complications of technology and many of the solutions suggest technological requirements, calling for active participation of engineering talent of different kinds. Engineers cannot be expected to function effectively or evolve realistic solutions to problems unless they are aware of the constraints imposed by the traditions, objectives, personnel, organization, and interactions which characterize the health care delivery mechanisms. Since many of the problems are multifaceted, solutions are inherently multidisciplinary and the potential role of engineering must be selectively identified in each instance.

THE NATURE OF CURRENT CRISES IN
HEALTH CARE DELIVERY

The fundamental problems facing modern medicine stem from an increasing demand for increased quantity and improved quality of health care for larger numbers of people presenting an expanding spectrum of illness with high expectations from a system which is currently incapable of responding to the needs of the consumers. Rising costs are the most visible aspects of the problem.

RISING COSTS

The uncontrolled upward spiral of rising medical costs can be ascribed to many different factors but essentially stem from demand increasing much faster than supply. Current efforts to "buy" more health care by supplying funds from a relatively fixed source of provisions for health providers is guaranteed to produce inflated costs. Engineering must play an increasing role in planning and achieving increased supply of these services.

INCREASED DEMAND

The basic sources of expanded demand include greater affluence by an increasing segment of the population, a growing population, urbanization, and a spreading attitude that access to good health care is a human right for every citizen. To meet these demands will require prompt and realistic responses by all participants in the health care system, including engineers.

HIGHER EXPECTATIONS

Widely publicized technical achievements have led the public to assume that the medical profession is capable of dealing with virtually any health problem, large or small. While it is true that many life threatening or potentially fatal conditions can now be handled by improved technology (e.g., artificial kidneys, open heart surgery, coronary care units, etc.), a large proportion of illnesses affecting masses of the population are not significantly affected by a visit to a physician or hospital. Many of the chronic illnesses have totally inadequate mechanisms and facilities. Technology must be developed to facilitate the identification and care of patients during mild or moderate illnesses, convalescences, chronic disability, aging, and mental illness.

DIVERSIFICATION OF FACILITIES

Medical care is oriented around the hospitals which generally aspire to provide relatively complete services for medical and surgical specialties even when they are relatively small in capacity. Thus, autonomous, nonprofit small general hospitals accomodate a majority of the admissions in the United States, (e.g., 90% of these hospitals are smaller than 400 beds). Extensive studies in Sweden have prompted phasing out of all such hospitals smaller than 400 beds as being economically unsound. A small city containing several small hospitals, actively competing for patients and staff by offering essentially the same services, would be better served with quality care at lower cost by a variety of mechanisms. They could affiliate and share support services (food service, laundry, laboratories). They could reduce unnecessary overlap in such sophisticated facilities as open heart surgery, radiation therapy, cardiac catheterization, laboratories, and specialized clinical laboratory equipment. They could establish equipment pools and common equipment maintenance facilities. Special facilities for care of mild, moderate illness, or convalescence and chronic illness could be established to provide care appropriate for the kind and severity of the disability. Certain hospitals could be designated for emphasis or specializing in obstetrics, or emergency service, or pediatrics or orthopedics, or physical medicine and rehabilitation, etc., relieving other nearby hospitals of the need to cover these services. Economy of scale could be achieved without such increased cost of sophistication that characterizes large modern medical centers. In addition, increased emphasis on facilities for ambulatory or outpatient care and better provision for home care is thoroughly warranted. There are recognizable engineering challenges implied by most of these possible trends. Most important is the need for screening mechanisms to identify, categorize, and channel patients to the type of facility or service appropriate to the particular forms and severity of disability.

THE FILTER FUNCTION

The family physician handled directly or referred to specialists any and all types of medical problems. He served as a kind of gatekeeper or guide for the health care system in earlier and less complicated times. Depletion of the numbers of general practitioners leave current health care systems without such a filter so that all the personnel and the facilities are congested with a complex mix of severely ill, moderately ill, chronically ill, consistently concerned, and healthy individuals. This assortment cannot be effectively handled as a conglomerate but will require extremely sensitive, simple, cheap and safe techniques for distinguishing the various functional categories of illness for channeling and scheduling purposes along with much more effective data collection, processing and display mechanisms. The requirements for new and improved bioinstrumentation and for data processing techniques will tax the creative ability of all the engineering talent that can be mustered to attack the many problems of differential screening.

OVERUTILIZATION

The current trends toward third-party payment of medical costs (insurance, Medicare, and Medicaid) foster overutilization of facilities and services by providing financial incentives to their use, by obscuring the costs and by relieving the patient of responsibility for meeting these costs. Experience with Federal support of health programs here and abroad has clearly shown that when services are artificially cheap or apparently free, they tend to always be overutilized. The patient should be aware of the cost and have some responsibility for payment even if he is ultimately fully reimbursed, judging by experience in both Sweden and Zurich, Switzerland. Finance mechanisms by which the patient can either pay directly or obtain a short term loan and actually pay the costs with partial or full reimbursement has substantial advantages in curtailing overuse. In addition it is a direct reminder of the value of the service and the magnitude of benefit derived by the program. When the doctor, the hospital, or an agency assumes responsibility for covering the costs of health care, the patient feels no constraints on utilization unless the costs are not completely covered. Indeed they commonly feel they are "entitled to whatever care they think they need" and may in effect demand more than is appropriate. Some of the pitfalls in providing health care on a national scale should be recognized and avoided by taking advantage of the applicable experience in foreign countries with extensive and long standing experience with various approaches.

REGIONAL PLANNING

The poor geographical distribution and lack of integration of health care facilities should be rectified on the basis of long range, comprehensive regional planning of a sort which is well developed in certain foreign countries and essentially undeveloped in the United States. Even major urban planning efforts generally fail to include provisions for appropriately located and integrated health facilities. This may reflect lack of knowledge and data for input into such planning efforts resulting from the relative independence and autonomy of the medical profession and the administration of most hospitals. Participation of medically oriented engineers is essential for development of regional plans to accommodate future requirements in a rapidly evolving society. For this and other purposes, there is a growing need for a wide variety of highly skilled and trained professionals and staff to meet the diverse requirements of the health care system.

MANPOWER SHORTAGES

The training of more physicians is the most common proposed solution to the health care crises. In view of the technological basis for many of the triumphs and travails of medicine, this is an extremely shortsighted position. Physicians may spend only very small fractions of their time in caring for patients and utilizing their unique training, judging by the few studies of practice which have been carried out. Much more information is needed regarding the time distribution of various professional and allied health personnel; an activity which calls for manpower, trained in systems analysis, methods improvement, and optimization methodology. These techniques of industrial engineering are destined to play important roles in analyzing and helping to correct sources of gross inefficiency in utilization of personnel and facilities. Most hospital administrators are poorly qualified to develop engineering specifications, evaluate maintenance and operations procedures, and oversee utilization of the complex equipment found in modern hospitals. There is a recognized need for greater numbers of technicians trained in the operation and maintenance of electronic equipment or test procedures of hospitals. Courses for such people are emerging in many community colleges. The increasing complexity of acquisition and processing of patient data in medical records is taxing the ingenuity of an expanding number of computer experts. They will be able to surmount such difficulties only with extensive cooperation of the medical professionals in the difficult process of editing, and organizing the large quantities of information currently being collected, both at the source and after recording. The future will surely see a growing demand for engineers with knowledge, experience and training in hospital functions, facilities, and operations. Training programs for such individuals are strangely lacking.

The potential contributions of engineering to medicine in the future will be considered in greater detail in Part II.

Part II

ENGINEERING APPROACHES TO HEALTH CARE REQUIREMENTS

INTRODUCTION

Many and diverse problems of modern medicine have been considered in Part I along with potential engineering contributions to some of the major crises. The full scope of engineering involvement in the total health care system is too extensive for comprehensive coverage in a book of this type, but representative examples may provide added insight into future technological impact. The development of equipment or tools for new research and clinical diagnosis represents one important bioengineering activity. Considering the large quantities of data currently being collected on patients as described in Chapter 5, the need for additional diagnostic instruments might be obscure. Large quantities of patient data are often assembled just because much of it is highly non-specific. Patterns, rather than one determination, are required to pin point definitive diagnoses. For this reason there is an ever-present need for more direct and unambiguous information hopefully obtained by simple, cheap, safe, and preferably non-invasive techniques. Examples of new and novel sources of information are described in Chapter 6, but these represent only the next immediate phase.

The use of mathematical models is a typical engineering approach to the analysis and understanding of complex problems. The application of models is extremely wide, extending from large-scale social organizations to quantitative descriptions of cells or components of cells. Representative examples of modeling are presented in Chapter 7 to indicate how such engineering techniques can be applied to problems of biology and medicine.

Biomechanics and biomaterials are aspects of *basic* bioengineering, which have clear application to specific clinical problems. The function of many tissues and organs (i.e., bones, muscles, etc.) can be quantitatively described in terms of changes in physical characteristics under various conditions of stress or load. This is the basic objective of Biomechanics. Diagnostic and therapeutic equipment often have stringent requirements to assure compatibility with tissues with which they come in contact (see Chapter 8). Artificial organs and prosthetic devices may interface with delicate tissues for protracted periods and must be made of materials that are consistent with physical and chemical properties of the tissues.

Modern medicine has entered an age of technology that will need to be accommodated by training of physicians in mathematics, physical sciences, and engineering. This type of training might be presented during premedical education, medical school, in postdoctoral training or continuing education (Chapter 9). Programs for the training of physician's assistants are emerging rapidly. Students of various engineering disciplines are being provided extensive training in biology and medicine to serve as independent or "hybrid" bioengineers or as engineering graduates with sufficient training and experience to serve effectively as members of multidisciplinary teams. The ultimate importance of diversified training in the field of bioengineering is revealed by projections of future engineering requirements of various clinical specialities as presented in the final chapter.

Most physicians find it difficult to identify unfulfilled technological requirements that would facilitate their care of patients. Similarly engineers are not prepared to propose medical applications of new techniques or technologies. Physicians do not know what is possible and engineers do not know what is needed. However, the value of training in both medicine and engineering becomes evident when the question of future requirements of medical and surgical specialities is posed. For example, a select group of medical faculty with interest and affiliation with the Center for Bioengineering were asked to consider the present state and future prospects of technology in their individual areas of medical interest. The responses included a remarkable array of innovative suggestions as revealed in Chapter 10.

Chapter 6

BIOMEDICAL INSTRUMENTATION

During the routine physical examination of a patient, a skilled physician gathers and interprets data regarding the status of certain organ systems with efficiency and speed which is not likely to be matched by any foreseeable combination of instruments and computers. However, the traditional approach to physical diagnosis is limited largely to those subjective sensations available to the physician. He cannot peer through the skin, he can hear only certain limited sources of audible vibrations, and he can feel only certain mechanical characteristics of internal organs. Despite these limitations, the precision with which experienced clinicians can arrive at correct diagnoses on the basis of sketchy and subjective impressions is quite remarkable.

During hospitalization and many other circumstances, the patient's condition is assessed most commonly in terms of a few "vital" signs such as blood pressures, heart rate, respiratory rate, body temperature, electrocardiograms, supplemented by laboratory tests which may be more directly related to the patient's illness. These have been chosen primarily because of availability, not because they are optimal solutions to established needs. In most hospitals, physicians have at their disposal additional and more sophisticated techniques for gathering data about internal organs by penetrating radiation (i.e., X rays and isotopes), by inserting tubes, catheters, or needles for measuring conditions internally or by comprehensive analysis of samples of blood and body fluids. Many of these tests are concentrated in major hospitals or medical centers and are not available for more routine care because of cost or hazard to the patient. For this reason, modern technology has made much less impact on the health care available to the general population through small hospitals or in physicians' offices. We can confidently expect future engineering to contribute to medicine by greatly expanding the number, reliability, and effectiveness of new instruments that are widely available and safe.

So long as new tests continue to be added to the traditional ones, progressively expanding the number and total cost of the procedures, the prospects of actually reducing costs of diagnosis through new technology are rather dim. Proper goals include the development of instrumentation which is so effective that a reduced number of tests can be used in place of a larger group of current procedures. This might significantly improve cost-effectiveness, a desirable objective which requires some reorientation of well established medical tradition; that the prime criterion of medical care is thoroughness based on maximal information (see Chapter 2).

DIAGNOSTIC PATTERNS VS DEFINITIVE TESTS

The current clinical descriptions of most disease states combine signs and symptoms many of which contain items bearing little obvious relationship to the basic disorder. For example, functionally significant disease of the heart produces most marked manifestations of the vascular system upstream so that diseases of the valves or myocardium of the left ventricle tend to produce disturbance of respiratory function (shortness of breath, cough, spitting up blood, etc.), while failure of the right ventricle is generally manifest by abnormal collections of blood or body fluids in the dependent regions. Hyperactive thyroid glands may produce tremor of the fingers, increased appetite, protruding eyeballs, sweating of the palms, and measurable increases in total metabolic rate. The signs of rheumatic fever include nodules under the skin, fever, skin rashes, swollen and painful joints, but the physician's concern is centered on inflammatory processes in the heart valves and heart muscle. Many local and generalized infections are accompanied by a diversity of signs and symptoms such as fever, weakness, changes in appetite, fatigue, muscle aches, etc. Diabetes not only changes the metabolism of sugar but also increases urine production, produces detectable odors, and affects appetite. As the causes and mechanisms of a disease state become sufficiently well known, a single specific test may establish the existence of the abnormality and its severity as a distinctive entity. A positive culture of a specific microorganism may definitively determine the cause of an infectious process. The diagnosis of hyperthyroidism may be established by the uptake of radioactive iodine and somewhat less directly by determining the protein-bound iodine in blood. A single blood smear may be sufficient to detect sickle cell anemia. Certain genetic deficiency diseases may be recognized by a single test. Immune reactions are quite specific for some types of systemic infections. Despite recent rapid medical progress, the number of disease states that can be diagnosed definitively by a single test remains very small and most of these are chemical, microbiological, or immunological in nature. Very few definitively diagnostic measurements of physical or mechanical dysfunctions are currently available. As a general rule, physicians have been forced to rely on sources of information which are available, instead of having access to more discrete sources of distinctive data. Herein lies a major opportunity to develop challenging long-range objectives for biomedical engineering, an expanding array of definitive diagnostic tests.

The total number and diversity of new and needed diagnostic instruments is so great that they cannot be considered in a presentation of this type. They are generally known and thoroughly described in other publications. Of greater significance is a systematic look at the various categories and options for instrumentation to facilitate identification of some opportunities for future approaches to unmet needs. It is convenient to consider several different categories of information as they might be applied to major organ systems as illustrated in Table 6.1.

NONDESTRUCTIVE TESTING: AN ENGINEERING CONCEPT

The concept of nondestructive testing was developed to test structural integrity and functional performance of complex and expensive devices. This approach corresponds rather closely to the processes employed by physicians attempting to accurately identify diseases with minimal harm or hazard to the patient. Many of the nondestructive testing techniques are common to both engineering and medicine (e.g., X rays, ultrasound, systems analysis) because the basic problems are so similar. A concentrated effort must be undertaken to assemble or develop specifically designed sensors which can elicit required information with sufficient accuracy, utility, and safety. We can extend wide availability by placing them in the hands of local physicians, physicians' assistants, or trained technicians in highly organized screening centers. The reasons for these requirements have been fully described in the preceding chapters. Many of the traditional diagnostic techniques have been by-products of commercial companies producing testing equipment for industry, modified for application on man. This process is continuing but many specialized types of

TABLE 6.1

Nondestructive Sources of Diagnostic Data

	External sources		
Organ systems	External signs or characteristics	Intrinsic energy	Spontaneous samples
Skin	Physical, mechanical biological, immun.	Temperature	Sweat odors
Nervous system	Coordination, reflex responses	E.E.G., effector responses input–output	
Respiratory system	Ventilation, speech	Movements Lung volumes Max. expirate	Exhaled air odors
Cardiovascular system	Impulses, pulses	E. C. G. Pulses Pressures Vibrations Sounds	
Gastrointestinal	Orifices	Sounds	Feces
Urinary system	Orifices	Contraction, stream	Urine
Genital system		Uterine, contraction	Secretions
Muscular system	Size, shape, texture	Contractile forces	
Skeletal system	Shape, symmetry		

diagnostic equipment need to be developed in accordance with the rigid and demanding specifications required for human applications.

EXTERNAL INFORMATION SOURCES

Diagnostic information elicited directly from the surface of the body is particularly valuable because its acquisition is easy and safe. Any characteristic which the physician can distinguish directly by his own senses could be recorded if the information were of sufficient value to warrant the effort. However, this type of information is so obvious and accessible that most such sources have been thoroughly studied and fully exploited. The familiar techniques need no mention here so only a few examples of innovations or prospects will be mentioned. Surprisingly, the most readily accessible organ in the body—the skin—is far from being exhausted as a potential source of valuable data regarding cutaneous and generalized disease (1,2) as indicated in Chapter 8.

INTERNAL ORGANS

The nervous system is a prime example of an inaccessible tissue, residing not only under the skin but for the most part within the bony cerebrospinal cavity (Table 6.1). The external signs of nervous system dysfunction take the form of disturbances of motor function such as muscular coordination, the walking gait, speech, and reflex response to stimuli. Opportunities for objective and quantitative additions to the neurological examination are suggested in a subsequent section.

The external evidences of lung functions are rather scanty, based on the shape and size of the thoracic cavity, the movements of chest and abdomen during breathing, and the control of respiration during speech. Serious deficiency of lung function may produce a bluish discoloration of the skin due to inadequate oxygenation, but this occurs at advanced stages of pulmonary disease and has limited diagnostic significance. Photoelectric oximeters provide more objective evidence of changes in arterial oxygen saturation (i.e., from the ear lobe), but they have not been adopted for routine clinical applications. Greater reliability and significance is provided by use of oximetry of the blood (see subsequent section).

The heart beat is indicated externally by a sudden outward displacement of a portion of the anterior chest wall which can be observed, felt, or recorded by a wide variety of simple instruments. Similarly the veins are directly visible in dependent regions and the pulsation of the arteries and veins have long served as sources of clues regarding function of the cardiovascular system. These external changes are manifestations of various forms of intrinsic energy within the cardiovascular system (see Table 6.1) as indicated in the next section.

The characteristics of organs within the abdominal cavity are not readily accessible and provide external evidence of disorder primarily at the inlet or outlet orifices (Table 6.1).

The muscular and skeletal systems are readily accessible under the skin, and loss or distortion of their form or function are readily appreciated by external observation supplemented by relatively simple sources of additional information.

The traditional practice of medicine has always depended very heavily upon three sources of information for clinical diagnosis—(a) past history and present illness, (b) physical examination by the physicians, and (c) objective tests performed in clinical laboratories. Technological progress has centered primarily in clinical laboratory testing procedures for which the physicians have progressively relinquished responsibility. He no longer insists on performing urinalysis, blood counts, blood pressure, or electrocardiographic recordings himself. In addition, X rays of the chest have largely supplanted the art of percussion for estimating the size of the heart or for identifying consolidated disease areas in the lungs. However, the patient's history (see Chapter 5) and the physical examination have remained in his diagnostic role and have been jealously guarded from the incursions of technology.

THE PHYSICAL EXAMINATION—DOCTOR'S DOMAIN

Physicians gain much information of great value by apparently casual appraisal of the appearance of the patient and by physical examination of the external surface of the body. The conformation, texture, color and many other aspects of the patient are assimilated rather automatically in his overall evaluation. In point of fact, physicians have difficulty specifying the multiple sources of the information on which they base many clinical diagnoses. Although it is theoretically possible to make objective or quantitative measurements of anything a physician can observe through his own senses of vision, hearing, touch, temperature sense, etc., the greatest challenge to biomedical engineering is to develop techniques which are sufficiently quick, reliable, safe, and useful that they can effectively supplement the exquisite sensitivity of human senses coupled with analytical capabilities of the human brain. There is little prospect of providing diagnostic batteries of instruments which can compete successfully with physicians in terms of merely detecting the presence of abnormal signs during routine physical examination. However, the examining physician records his subjective sensations in impressive-sounding but vague terms in the patient's medical record. He depends upon his uncertain memory to conjure up the previous condition of a patient many days, weeks, or months later. The uncertainty is greatly reduced when a physician can refer back to numerical values which reliably indicate the status of some structure, function, or control. The deployment of diagnostic equipment which is accurate, reliable, and safe in the hands of technicians can present to the physician valuable information which could greatly facilitate his decision-making processes. Viewed in this light we can anticipate that biomedical engineering should contribute to future physical diagnosis in at least four different ways.

a. New technologies for extending the human senses by means of new transducers and energy probes (i.e., ultrasonics, thermography), amplification (i.e., electron scan microscopes, television, etc.), and new clinical laboratory tests or studies: These innovations will tend to improve the quality of health care but may also increase the cost unless they are designed and utilized with optimal cost-benefit relationships in mind.

b. New technologies which will enable physicians' assistants and trained technicians to gather objective or quantitative information which can be presented to the physician for his interpretation: Such approaches should conserve the time of physicians for the role for which he is uniquely trained; namely interpretation of data, diagnostic decisions, a choice of therapy and evaluation of results.

c. New techniques must be developed to reduce the cost of medicine while maintaining high quality. This approach undoubtedly will involve centralizing groups and batteries of diagnostic instrumentation in efficient organizational frameworks so that the data can be collected, edited, analyzed, and displayed to the physician to provide maximum relevant information at minimal cost.

d. Improved methods of handling and displaying information as described in Chapter 5 must be developed.

SENSORY LIMITATIONS OF THE EXAMINER

The sensory nerves which convey information about the external environment in man have exceedingly high degrees of sensitivity. The range of amplitudes of both sound and light which can be perceived is exceedingly large. However, the time scale within which changes can be perceived is much more discretely limited.

For example, phenomena can occur too rapidly to be observed by the unaided human senses. The beating wings of a humming bird appear as a blur to the unaided eye, but can be converted into discernible motion by means of high-speed photography. Similarly action can be too slow to be perceived. It is impossible to observe the opening of a flower or the movement of a glacier with the unaided senses, and yet time lapse photography can render these movements visible. In the same way, the range of both sound and light which can be perceived by the senses is quite limited, in view of the overall range of frequencies which exist in nature. The senses are responsive to the narrow range of visible light and are not sensitive to the very wide range of electromagnetic frequencies and ultrasonic frequencies. Whole new batteries of instruments can be developed which will be important supplements to the human senses in the identification of the location, placement, and function of internal organs (see also Fig. 5.1).

SENSORY SUPPLEMENTS FOR THE PHYSICAL EXAMINATION

PALPATION

The doctor may place his hand on a patient's forehead to detect fever but he generally relies on a thermometer to provide more objective evidence. Electronic thermometers can now reduce the time required for this measurement to a fraction of that required with a mercury thermometer. The surface temperature of the skin is not a reliable indicator of core temperature because it is the main effector organ in temperature control. The distribution of surface temperature can be directly displayed by both infrared recording techniques (thermography) or by direct applications of liquid crystals which change in color over small and predictable temperature ranges.

Of these only thermography has elicited much interest from the diagnostic point of view (Fig. 6.1). The changes in distribution of surface temperature of the skin may indicate underlying pathological lesions since warm areas may represent inflammatory or cancerous processes and cool areas may signify inadequate blood supply. Commercially available thermographic equipment displays images on television screens showing temperature differences as slight as $0.2°C$. It has been employed in examinations for carcinoma (i.e., breast tumors), burns, frostbite, and rheumatoid arthritis (3).

Palpating pulses provides valuable information concerning the level of the arterial pressure, heart rate and some idea of rate of ventricular ejection. Venous pulsations (i.e., in the neck) have well established diagnostic significance. Countless efforts have been expended on the objective recording of arterial and venous pulsations from the skin surface. These methods have never become well accepted, presumably because the time devoted to their recording is excessive in relation to the information content. Most of the recording transducers are placed directly on the skin, tending to load the vascular walls and to distort the signals. A number of techniques are now available to record skin displacements without actually touching the skin (e.g., based on electrical capacitance, ultrasonics, or fiberoptic techniques). Similarly, physicians frequently place their palms over the chest wall to feel the impulse of the heart. A large body of data has accumulated through objective recordings of these chest wall displacements often called apex cardiograms. The loading of the chest wall by contact transducers is known to distort the signals just as in the case of pulse recording. The ultimate application of such objective recordings remains to be determined.

PERCUSSION

In earlier days, physicians took great pride in their ability to elicit information about heart size and lung aeration by sharply rapping a finger in

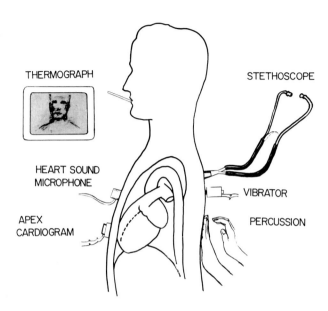

Fig. 6.1. Physical examinations are the special province of physicians depending largely on their subjective senses. Supplemental technology is available in the form of thermography for surface temperature distribution, phonocardiography, and apex cardiography for heart sounds and apical impulse and the prospect of vibrators to elicit information about resonance to supplement percussion.

firm contact with the chest wall with the fingers of the other hand (Fig. 6.1). This tapping induces vibrations in the chest which were interpreted both by the vibration sense in the fingers and the elicited sounds interpreted by the ear. This procedure has been largely replaced by x rays, but there may be information of value regarding the status of the lungs in the technique if it were rendered more scientific. A transducer which induced vibrations of fluctuating frequencies could be employed to determine frequencies at which the chest contents resonate as affected by changes in the relative distribution of air and fluid in the lungs. These resonance frequencies could be documented and correlated with various types of disease. Percussion is also used over the abdomen to detect and locate collections of gas in the bowel. Such a process could also be rendered more accurate and specific if desirable or necessary.

The mechanical energy involved in lung function is clearly manifest in respiratory gas exchange. The movements of the diaphragm and chest wall producing the exchanges of air in and out of the lungs are readily observable and subject to measurement. However, it is much more convenient to monitor the movements of the air directly in terms of tidal volume, pressures, vital capacity, and the maximum rates of expiratory gas flows as evidence of adequate function and reserve. Instruments to measure lung function in these terms are now commercially available but are not seeing adequate utilization in routine diagnosis outside of specialized facilities.

AUSCULTATION

The movement of gases in and out of the airways of the lungs produce audible vibrations which can be detected through a stethoscope applied to the skin of the thorax (Fig. 6.1). The physical characteristics of these sounds which lead to conclusions by the physician have not been identified on any scientific basis. The process of analyzing these characteristics could automatically produce instrumentation that might supplement the physician's senses on a more objective basis.

Audible sounds and murmurs are produced during contraction of the heart, and these could be recorded routinely, but rarely are. The advent of sound spectrum analyzers in industry led to extensive explorations of heart murmurs in hopes of finding patterns of frequencies or harmonics which might identify discretely the sources and causes of heart murmurs. For example, it was hoped that the sound frequencies produced by a pulmonary valve with congenital deformity could be distinguished from sounds from a similar type of deformity in the aortic valve— much like one recognizes a familiar voice over the telephone. Extensive tests have established that the common heart murmurs from deformed heart valves or holes in the partitions of the heart resemble jet noise lacking harmonics or consistent patterns (4) which might have allowed differentiation on the basis of sound patterns (Fig. 6.1). The anatomical location, time sequence and intensity patterns provide important clues for interpretation which can be aided by electronic devices (5).

SELECTIVE ATTENTION FOR HEART MURMUR EVALUATION

The heart sounds and murmurs emitted by the heart occur in regularly repeated sequences, necessitating that attention of the physician be directed to each component alone for several beats, intentionally ignoring the remainder of the noise in the cycle. Thus he listens for a few heart beats to the first heart sound, then directs his attention solely to the interval between the first and second heart sounds, identifies the characteristics of the second heart sound, and then directs his attention to the interval immediately following this sound to pick up abnormal sounds during the filling time. This process requires extensive training and some physicians never become particularly skilled in the process. By means of simple gating circuits, the process of selective attention to the succession of sounds can be consistently duplicated for a physician as illustrated in Fig. 6.2. A technician can be easily trained to operate such simple equipment during routine recording of electrocardiograms and in addition provide objective numerical values for the intensity of sounds appearing during the two or three critical intervals when murmurs of significance are to be found. By this means, those patients whose hearts have no evidence of abnormal sounds in the systolic or diastolic intervals could be identified. Those patients with sounds of intensity above some critical level at the selected periods of the cycle could be categorized as requiring particular attention of the physician who would subsequently examine them. In this way, physicians could be spared the need for spending time repeatedly and unnecessarily listening for abnormal sounds when they can be demonstrated to be absent by devices more sensitive than the human ear. In general, the pressures, pulses, vibrations, and sounds from the cardiovascular system have been widely exploited but there remain important opportunities to improve data acquisition, particularly with equipment which can be placed in the hands of trained technicians to collect information to be presented to physicians for evaluation.

INTRINSIC ENERGY DETECTION

A major contribution of technology to clinical diagnosis is represented by the recording and analysis of the various forms of energy emanating from the various internal organs and detected at the body surface. These techniques have proved extremely valuable as nondestructive testing techniques and have been explored and exploited very intensively and are generally familiar. The prospects for major theoretical or technical break throughs in these areas are so remote that only a few points will be discussed.

The cardiovascular system is the source of both electrical energy (ECG) and a variety of mechanical energies. Arterial pressure recorded indirectly by means of a cuff is probably the quantitative clinical measurement with the longest history and widest application. The pressures in the heart chambers, arteries and veins have provided valuable information regarding the functional conditions in these sites, particularly since the advent of pressure gauges with high fidelity and adequate frequency response. Direct recordings of arterial and ventricular pressures are generally obtained through catheters (see the section on inserted probes, page 253). The external signs of cardiovascular function are manifest by pressure pulses which can be sensed by touch but are rarely recorded clinically.

Electrical potentials constitute the most familiar example of intrinsic energy from the central nervous system as it is recorded by electroencephalography (EEG). An enormous body of information and experience on

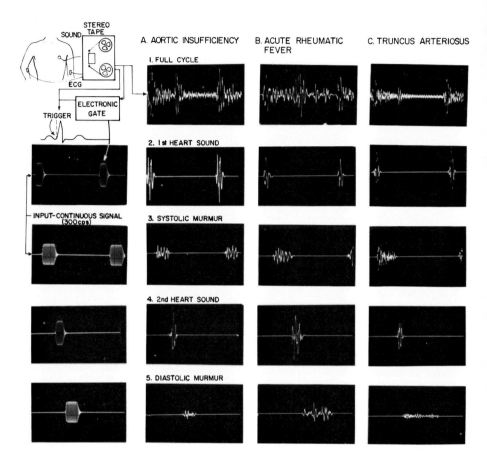

Fig. 6.2. Evaluation of heart sounds and murmurs can be facilitated by simple electronic gating circuits which permit the physician or a technician to direct attention to selected short intervals in each heart cycle in sequence, e.g., the first heart sound, systolic interval, second heart sound, and diastolic interval. Interpretation of these sounds and murmurs can be improved by eliminating the distraction of the many and varied sound occuring during each cycle. (Reprinted with permission of Medical Research Engineering from Reference 4.)

electroencephalography has accumulated in the literature, including techniques of transmission and computer applications in the analysis of the waveforms. There is no need to say more here. Central nervous system function is manifest externally primarily by activity in effector organs such as the skeletal muscles, smooth muscles or glands. In addition, neurologists rely heavily on information presented on X-ray films of the head and on the applications of ultrasonic energy probes (see also Fig. 6.8).

THE NEUROLOGICAL EXAMINATION

The functional status of the nervous system is most commonly evaluated by eliciting an extensive series of reflex phenomena, the particular domain of the neurologist. The neurologist applies a stimulus to a particular site on the patient and observes an anticipated reflex response. This process might be rendered more consistent and scientific by adopting the approach of the electrical engineer evaluating the function of a complex electronic circuit.

STANDARD STIMULI AND OBJECTIVE OBSERVATION

The functional integrity of the central nervous system in man is assessed by a neurological examination; by applying stimuli which are not standardized and by observing responses subjectively. Familiar examples include shining a light into the eye and observing the contraction of the iris or tapping the knee and observing the resulting kicking movement of the foot. A remarkable amount of useful information is obtained very rapidly by these techniques but great reliance must be placed on the physician's discrimination and memory. An engineer faced with a complex electronic circuit commonly introduces a known perturbation at some critical point in the circuit and measures objectively the magnitude of the response. By this means, objective and numerical responses can be more accurately compared and recorded for future reference. The same kind of circuit analysis could be applied to the central nervous system which is already equipped with test points and output responses. The functional state of various segments of the nervous system could be tested by applying standard stimuli to selected sites with provisions for recording objectively the magnitude and dynamic response of the output. Successive determinations over periods of time would permit far more definitive information regarding progression of disease processes. By obtaining numerical instead of descriptive data, the recording, storage and retrieval of vital information could be far more effectively handled (i.e., by computer—see Chapter 5). Critical evaluation of the central nervous system presents a challenging application of engineering techniques and technology.

The concept of *standardized stimulus-quantitated response* can be indicated by some possible applications. For example, the state of balance between the two divisions of the autonomic nervous system (sympathetic and parasympathetic) is of great importance in the control of many internal organs. This balance could be quite readily tested by introducing a standard light stimulus into the eye and registering changes in pupillary diameter. This is not done clinically but should represent little technological problem in a society capable of discrete measurements from the moon and distant planets. The response to standard stimuli should distinguish changes in the reflex despite variability induced by other sources of input to the reflex. Reflex control of arterial blood pressure could be explored by applying standard external pressure variations on pressure receptors in the carotid artery in the neck and observation made of the magnitude of the change in blood pressure or heart rate. Such measurements would be of particular value in assessing the reflex relationships in patients developing high blood pressure as an expression of aberrant controls. During surgical anesthesia, regulatory mechanisms are usually depressed or distorted and the changing sensitivity of the carotid sinus reflex during surgical operations could be of substantial value to the anesthetist. Sudden controlled lung inflation can reflexly induce altered pressures in the respiratory tract. Also a sudden depression of the jaw induces a contraction of jaw muscles (e.g., the jaw jerk). These two reflexes have potential value in assessing the state of reflexes during anesthesia. Neural circuits involving the auditory and vestibular functions of the ear are also subject to input–output studies, as are a wide variety of other complex functions of the body.

The availability of equipment for objectively recording reflex responses could have value in setting standards, in training of physicians and physicians' assistants and in the detailed examination of patients with neurological diseases. Experience gained in these roles would help identify the tests which have the largest information content in relation to time and effort expended.

SAMPLES AS SOURCES OF INFORMATION

An important source of diagnostic information has long stemmed from inspection or analysis of the excretions or other spontaneous samples exuded from the body. Some are extruded from the body as byproducts of organ function, others are removed from the body by various sampling methods. A large array of analytical techniques have developed for extracting information from such samples and the number and diversity is increasing rapidly.

SPONTANEOUS SAMPLES

The urine and feces are traditionally recognized sources of information regarding the function of the kidneys and gastrointestinal tract, respectively. The secretions from the skin (sensible and insensible perspiration) salivary glands, or from genital tracts are not generally analyzed for their diagnostic information content. The partial pressure of oxygen and carbon dioxide in exhaled air from the lungs, coupled with the lung volumes provides the necessary information for evaluating pulmonary function. A new and interesting approach is the analysis of characteristic odors from the body (skin and lungs) at the Illinois Institute of

Technology. By gas chromatography and stepwise discriminant analysis it has been possible to distinguish vapors of human origin from cooking odors (6). Olfactronics is the name applied to identification of characteristic odors of the type which occur with particular diseases such as infections (diphtheria, typhus, typhoid, etc.), are related to respiratory or gastrointestinal tract (pulmonary empyema, bronchiectasis, stomatitis, Vincent's angina) or other conditions such as pemphigus, diabetes, uremia, gout. The ultimate usefulness of this approach is far from clear. Its importance lies in the fact that a new avenue of exploration is opened in the never-ending search for additional information by totally nondestructive techniques.

SAMPLING METHODS

By far the most common sample obtained for examination is blood plasma or serum removed through a needle from a convenient vein. The interpretation of variations in composition of blood poses problems because this complex fluid is in an unstable equilibrium between input and output of its constituents from many component organs of the body. For example, the quantity of adrenal hormones is very different in venous blood draining directly from that gland as compared to a sample collected from another vein in the body. Similarly blood withdrawn from the coronary veins or sinus, from the hepatic vein or from a renal vein contains information of much more specific value than can be obtained from a mixed sample. In contrast, mixed venous blood samples from the pulmonary artery are needed for computing cardiac output in accordance with the Fick Principle. The anatomical specificity of blood samples is one fundamental reason for using catheters rather than needles for obtaining blood samples. Needles are frequently employed to obtain samples of body fluids (e.g., cerebrospinal fluid) or samples of tissue from various organs (bone marrow, kidney, liver, muscle, etc.). Specialized instruments have been developed for obtaining samples of tissues from hollow organs such as the stomach or bladder under direct visualization. Techniques for obtaining biopsy specimens can be greatly improved.

The engineering contributions to techniques for obtaining samples of blood, body fluids and tissues are relatively insignificant. The many new and useful techniques for extracting data from such samples are major advances through biomedical engineering.

ANALYSIS OF SAMPLES

An indication of the scope of electronic instrumentation in laboratory measurements made on patient specimens (7) is presented in Table 6.2. The colorimeters employed with automatic chemical analyzers now produce a flood of quantitative information in large medical centers and testing laboratories. Spectrophotometry and chromatography provide a diversity and specificity of chemical determinations undreamed of relatively recently. The physical and chemical properties of samples can be analyzed in terms of the many different variables listed in Table 6.2. From such vast amounts of information new insight into metabolic, endocrine and other biochemical abnormalities can be recognized. They are useful in evaluation during therapy of fluid-electrolyte disturbances and problems of nutrition, respiratory, renal, hepatic, and endocrine function. Measurements of metabolic intermediaries, end products and the exchanges of tracer substances are most useful in analyzing for biochemical errors or malfunctions.

AUTOMATED CHEMICAL ANALYSIS

The impact of automation in clinical laboratories is generally known. The large numbers of chemical determinations of wide diversity at low cost provides a wealth of information previously unavailable. The output also contains a large number of determinations that the physician would not ordinarily order and some controversy persists whether the additional data is beneficial or whether the growing lists of numbers of laboratory reports are tending to obscure the significant values. The various types of automated chemistry equipment have been mentioned briefly in Chapter 4 and are presented in much more detail by Moss (8). Two categories of discrete analysis systems are recognized—(a) Single channel analyzers capable of dealing rapidly with relatively small batches of specimens and retaining a good flexibility in changing from one type of analysis to another, and (b) elaborate systems capable of handling large numbers of samples on which a set range of analyses are performed on each one with minimal intervention by the operator and relatively straightforward patient identification and data processing. The full requirements for clinical chemistry determinations have not been met. For example, serum sodium and potassium by flame photometry were not included in the first commercial instruments. These two ions represent 20–25% of the work load in many hospital laboratories so they need to be included or considered in the total picture. The trend toward growing numbers of clinical tests of increasing diversity and complexity is really only beginning. Dickson (9) indicated the direction of current research by citing some projects being carried out in various laboratories.

1. A high resolution analysis system that has identified 130 components that regularly appear in urine was reported from Oak Ridge National Laboratory. Also under development is a fast clinical analyzer system to automatically load, centrifuge, and read spectrophotometer output on an oscilloscope with a computer output automatically printing results.

TABLE 6.2

Applications of Electronic Instruments in Medicine; Laboratory Measurements Made on Patient Specimens[a]

Analysis of biochemical constituents
  Colorimeters, automatic chemical analyzers
  Spectrophotometers for visible ultraviolet, infrared spectroscopy, and fluorescence
  Emission spectrometers, flame photometer, emission spectrograph
  Chromatographs (paper, gel, starch)
  Gas chromatography, amino acid analyzers
  Gas analyzers for oxygen, carbon dioxide, nitrogen, carbon monoxide, helium, and
    radioactive rare gases (krypton, argon)
Analysis of physical and chemical properties
  pH meters
  Electrophoretic apparatus
  Nuclear magnetic resonance detector
  Electric conductivity meters
  Viscosimeters
  Osmometers
  Mass spectrometer
  Thermal conductivity meters
  Paramagnetic oxygen analyzers
  Ionization detectors
  Oxygen electrodes, blood-gas analyzers
Analysis of physical properties
  Densitometers
  Light amplifiers
  Electron microscopes
  Ultraviolet microscopes
  Electronic micrometry
  Electronic scanners, cell counting, cytology
  Isotope detectors, scalers, counters, integrators, plotters
Miscellaneous laboratory apparatus
  Fraction collectors
  Deionization apparatus
  Centrifuges, ultracentrifuges
  Titration apparatus
  Ultrasonic cleaning equipment
  Automatic equipment for staining of pathologic specimens

---

[a]From W. A. Spencer *et al.* (7).

2. An automated system for analyzing steroids in urine by paper chromatography is under study at the Medical College of Virginia. Such data is intended for use in detecting endocrine disturbances and cancer patients whose steroid patterns may be distorted.

3. Faster, more accurate determinations of vitamin B_{12}, important in pernicious anemia, are being developed at Marquette University and automated isozyme separation and identification for diagnosing heart attacks is a goal of the University of Wisconsin.

4. X-ray emission spectroscopy and related approaches are being employed at M.D. Anderson Hospital in Houston to perform multiple determinations on a single drop of blood.

The influence of available automated hardware on the clinical laboratory is exemplified by Technicon whose "Autoanalyzer" is claimed to have 90% of the market for such devices, having sold or leased 13,000–15,000 blood serum analyzers over the world. The 12-channel unit with a two-man crew can perform about 90% of the clinical laboratories "wet" chemistry determinations at a rate of some 800,000 procedures yearly. Other manufacturers are producing automated analyzers with increasing sophistication and/or volume. For example the "Autochemist" produced by AGA in Sweden (see Fig. 4.8). Beckman Instruments, American Optical, and Hycel, Inc., have also entered the field. DuPont has designed an automatic analyzer based on a prepacked kit or packs of reagents which has a basic capability of performing 30 tests but can be expanded to 60 different tests. The results are expected to be delivered at rates of 50–100 per hour.

These impressive developments are not without problems, the greatest of which is quality control and standards. The increased acceptance of automated equipment by many different laboratories indicates that significant progress has been made in this direction. With such greatly increased economic incentives and competition, the stringent requirements are bound to be attained and maintained. An important requirement is the development of sources of standard reference materials for calibration of instruments. Nowhere in all of modern medicine is the role of engineering more visible than in the development of automated testing with its attendant problems of data handling.

CELL STUDIES

More rapid and precise counting of blood cells is now possible as an important adjunct to hematology. The Coulter Counter is fed blood samples and automatically performs multiple dilutions, electronically measures red and white blood cell counts, determines mean cell volume electronically and the hemoglobin content of blood by a colorimetric method at theoretical rates of 180 measurements an hour. Allowing for calibration and standardization a more realistic rate is around 100 patient samples per hour. Apparatus for differentiating the various types of white blood cells automatically is still in the research stage. The traditional differential counting of white blood cells is an expensive process currently performed by technicians with a microscope. The process is difficult to automate because it involves sophisticated pattern recognition (i.e., detecting the shape of cell nuclei). One approach to distinguishing nuclear shapes of white blood cells requires sequential measurement of the curvature of segments of the nucleus (10), a difficult problem even for a digital computer. Photoelectric scanning systems for converting various color intensities into digital signals show some promise of performing differential white blood cell counts.

Immature red blood cells appear in abnormally large numbers when new cells are being generated rapidly (i.e., certain forms of anemia). They contain abnormally large amounts of RNA at first, diminishing during the first 24 hours. Accurate estimation of the blood marrow response is rendered possible by a flourescent staining technique which was described by Thaer (11). Techniques of this sort are subject to automation and represent reasonable approaches to automation in hematology, in general, and other specialized forms of cytology.

AUTOMATED CYTOLOGY

Automation has proved extremely successful in clinical chemical determinations but has made little impact on many other categories of procedures such as microbiology, serology, cancer cell detection, or chromosome counting. The general problems of automation for cellular samples were considered by Kamentsky and Melamed (12). The techniques for cell collection, preparation, and transport are still in developmental phases and include slide scanning methods, transparent tape transport and fluid transport systems. The slide scanning is merely mechanization of the standard microscopic viewing sequence. Cells can be passed beneath sensors as a continuous stream either on transparent tape or in flowing liquid. In any case identification and classification of cells by the kinds of pattern recognition quickly accomplished by technicians are extremely difficult to achieve by automated devices coupled into computers. A clinically useful device must identify many cells in a short period. Using high resolution optical techniques, the amount of data required by the computer for identifying cells at rates of 10,000 to 1,000,000 per second would require binary data acquisition at 10^6 or 10^7 bits per second and even higher rates with multiple levels of gray or color. The most promising method is based on zero resolution microscopy and fluid transport, similar in concept to the fluorescent microscopy described for identifying immature red blood cells. The scanning spot is increased to a size larger than the cells so that no cellular detail is represented in the image. The light transmission or scattering of different cells is monitored after specific differences are recognized or induced (i.e., by selective fluorescence or staining). It is reasonable to anticipate the development of a family of instruments from current prototypes with a range of specifications suitable for mechanizing classification and enumeration of cells for many different laboratory tests.

ELECTRON MICROSCOPY, SCAN, AND PROBE

Dr. John Luft

In the history of science, progress had depended upon both brains and instruments, with each dependent upon the other for the greatest productivity. Although many instruments are analytical and easily quantitated because their end products are meter readings, there is another class of scientific instruments which give an image, such as telescopes, cameras, and microscopes. These images are easily appreciated qualitatively, but the information content of a single picture is so large that it is often tedious to extract the significant quantitative data from it. Despite these difficulties, the advantages which have come from microscopes which permit visualization of smaller and smaller details of the world around us, have encouraged continued development of these instruments. The electron microscope permits one to see details 100 times smaller than was possible with the light microscope.

The electron microscope became a reality in 1945. Even then it was considered in both of its forms: the transmission microscope in which a broad electron beam penetrates through extremely thin specimens (Fig. 6.3A,B), and the scanning microscope in which a narrow beam of electrons sweeps over the specimen in a raster and an image is built up on a TV monitor tube in proportion to the number of electrons which are scattered from the beam by the specimen. Although each form of the electron microscope has advantages and disadvantages, it was clear from the first that the scanning electron microscope (SEM) would never attain the resolution which the transmission microscope could reach. Thus, the scanning electron microscope was put on the shelf because of the desire to see the smallest details of the physical and biological world.

Fig. 6.3. (A) This transmission electron micrograph was taken from a thin (50 nm) section of mouse tissue at the interface between muscle and tendon. Two or three banded collagen fibrils extend diagonally across the micrograph in a narrow slit between a muscle cell (lower left) and tendon (upper right). One collagen fibril continues without interruption from corner-to-corner, a distance of 2.8 μm. The collagen fibrils appear to be enveloped in dark material. This is a consequence of using a special stain (ruthenium red) during the preparation of the tissue in order to identify certain cement substances at cell surfaces. (Mouse diaphragm, acrolein/rutherium red/osmium tetroxide fixation; otherwise unstained. $\times$100,000.) (B) Electron micrograph of a capillary with a red blood cell occupying the lymen. The fine structure of capillary endothelium can be visualized, including many small vesicles. Striated muscle fibers and mitochondria can be seen in the corner of the figure, illustrating the very close proximity of the blood to the energy conversion units in muscle.

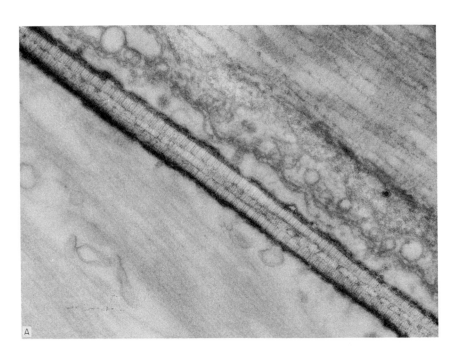

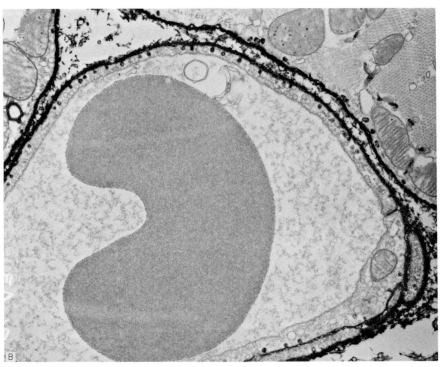

245

The expectations of the transmission electron microscope (EM) have been realized: it can routinely resolve specimens of thickness of 3-4 Å (0.3–0.4 nm) of high density (high contrast) and 10–20 Å (0.1–0.2 nm) with biological material. The penalty is that of thinness. To attain these limits, biological material must be put into solution or be separated into molecules and dried onto a thin support film (stretched over holes in 200 mesh screen), or embedded in plastic and cut into slices 500Å thick (50nm) and stretched over similar screen. The results are represented by pictures of collagen filaments which reveal a striated banding pattern in detail to 1.5 nm. The biochemists are very pleased with the bands, but the engineer desperately needs to know how long the fibrils are and the nature and location of the terminations. No one has ever seen the *end* of a collagen fibril, or at least an end which could be identified with confidence (see Fig. 6.3A). In the crucial case of wound healing, where cut collagen fibrils in the skin somehow reunite to form a mechanically strong fabric, the details are missing. A better method of getting the data is necessary.

Now that high resolution is easily available, one can look again at the scanning electron microscope (SEM). Its singular virtue is the fact that it looks *at surfaces* rather than *through slices,* and it can look deep into fissures as well. For this property, limited resolution of about 20 nm is an acceptable price to pay. One need look no longer between the wires in the support screen at a thin section, but can visualize a continuous and extended surface at any magnification desired. The SEM still does not permit the tracing of a single collagen fibril deep into a large volume of tissue—it is restricted to a surface, but the surface can be very irregular. It is probable that special tissue preparations, following natural cleavage planes in connective tissue, will permit an analysis of the texture and pattern of collagen fibrils over long distances. The preparation of the tissue for the SEM is much easier than cutting sections, and the image which one sees in the SEM is dramatically interpreted by the eye in three dimensions, so that the work is very much faster than anything possible in the transmission EM. Obviously questions of the type posed regarding the nature, distribution, and relationships of collagen fibers are typical examples of the kind of information which may be sought by means of combined approaches using the electron microscope and the electronscan microscope.

The potential applications of the SEM for investigating the *texture* of tissues are so numerous as to preclude discussion here. Preliminary work from University of Strathclyde, however, has proven the feasibility of the instrument on various biological materials. Connective tissues, from loose "areolar" tissue to tendon, need examination. Cartilage, particularly articular cartilage, is a natural target since the important element is the surface itself. The internal surfaces of blood vessels, where atherosclerosis begins, must be examined (Fig. 6.4A,B). Where the disease is spotty or focal, the SEM is enormously useful. Large areas of blood vessel surface must be examined, most of which is normal, to find the early isolated lesion. One could perhaps accomplish here in 15 minutes what would require six months with the transmission EM.

It is a safe generalization that almost nothing is known *in detail* of the strengthening mechanisms employed in the vertebrate, and the SEM together with the transmission microscope, offer the best combination for the foreseeable future to attack this problem. These two instruments are complementary rather than competitive. It is important to recognize that this field of tissue mechanics is virtually unexplored, so that most of the early work will remain at the research level. However, as phenomena are discovered which are sensitive indicators of disease processes, the diagnostic potentialities will become obvious.

The electron microprobe is closely related to the SEM, and in fact was in use five to ten years before the latter was in production. It also uses a fine electron beam to sweep over the specimen, but instead of image formation by secondary or scattered electrons from the specimen, the signal which the instrument picks up is the characteristic X rays which the electron beam excites in the specimen itself. Since the wavelengths of these X rays are specific for various elements, the electron microprobe can pick up the chemical elements. There are certain difficulties, however, mainly due to limited sensitivity, which reduce the resolution of the instrument to a cubic micron or so. The case is worse still for the light elements which are of interest in biology, since they are less efficient in X-ray production than the heavier elements. So far, the electron microprobe has been useful only in identifying discrete deposits of material such as calcium and phosphorus in bone, iron in hemoglobin of red blood cells or other iron pigments, and some metal-containing deposits in rare metabolic diseases. So far its sensitivity is insufficient to detect zinc, for example, in tissue which is known to contain zinc-binding metalloenzymes. The instrument is limited by the intrinsically low efficiency of X-ray production by electron bombardment, and also by the very low aperture of the crystal X-ray spectrometers used for the X-ray wavelength analysis. Solid state detectors may replace the crystal spectrometer in the future.

QUANTITATIVE MICROSCOPY

The future holds promise of further refinement of cell biology by direct and quantitative studies of cell function under the influence of various stresses such as external environmental conditions, chemical agents, toxins, metabolic and hormonal materials, microorganisms or viruses. At Battelle Institute eV., Frankfurt/M., Thaer (11) is assembling and utilizing equipment for investigating cells responding to changes in tonicity, pH, temperature and chemical environment by a variety of techniques (Fig. 6.5). For example, the water content of

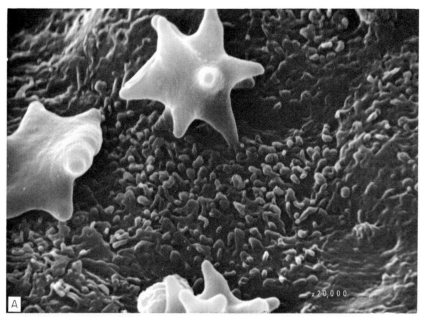

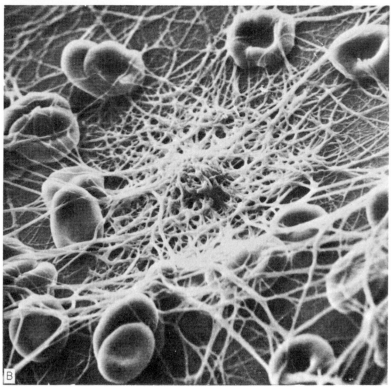

Fig. 6.4

single cells is studied by immersion refractometry supplemented by microinterferometry to obtain cell size. The mass and concentration of dry matter can be obtained by this combination. Enzymatic processes can be studied by determining the rates at which nonfluorescing substrates are split to produce measurable fluorescence. Capillary fluorescence standards have been developed using quartz capillaries 6 μm. inside diameter, checked by microinterferometry (13). As techniques of quantitative microscopy evolve and become standardized and automated, the information which can be gained from biopsied cell samples from various tissues should provide ever-increasing specificity, sensitivity, and quantification of diagnostic criteria.

NEUTRON ACTIVATION ANALYSIS

Gene L. Woodruff

Neutron activation analysis (NAA) is a technique whereby the quantity of an element present in a sample of interest may be determined. It is thus an analytical method that can be considered as a possible alternative when the use of atomic absorption spectroscopy, X-ray fluorescence spectroscopy, mass spectrometry or a similar technique is contemplated.

In the usual case the sample is placed in a neutron flux of high intensity. After an irradiation period the sample is removed and the induced radioactivity is measured. It is common practice to irradiate a standard having a known amount of the element of interest together with the sample and simply compare the ratio of the count rates produced.

The most significant feature of neutron activation analysis is the extreme sensitivity that can often be attained, especially when the intense neutron flux of a nuclear reactor of even moderate power is attainable. The interference-free sensitivities for 76 elements using a neutron flux of 10^{13} neut/cm^2 sec are indicated in Table 6.3. It should be noted that for many of these elements neutron activation analysis will ordinarily be the most sensitive method available.

Fig. 6.4. (A) The scanning electron microscope displays new dimensions of normal and pathological structure such as the inner lining, endothelium, of the normal pig aorta which appears rough because of numerous fingerlike projections, microvilli. The star-shaped cells are agglutinated red blood cells. (Photograph reproduced through the courtesy of A. R. Seaman, and S. A. Wesolowski and Rassegna; Medica e culturale XLVI-1969.) (B) The deposition of fibrin and red blood cells on a polyvinyl chloride surface illustrates how this research tool can be utilized to study processes such as blood clotting. (Photograph reproduced through the courtesy of N. E. Rodman and Rassegna: Medica e culturale XLVII-1970.)

Fig. 6.5. Equipment developed for quantitative microscopy for measuring excitation and fluorescence spectra as well as fluorescence polarization on single cells. This microscopic spectrofluorometer can detect primary fluorescing cellular substances (tryptophane, flavoproteins, etc.) and chemical binding between fluorescing molecules and macromolecules or configurational changes in macromolecules. (See Reference 11. Photograph through courtesy of A. A. Thaer.)

It is important to recognize the significance of the "interference-free" constraint on the values in Table 6.3. If more than one radioactive isotope is present in the irradiated sample, as will most invariably be the case, the activity from the various isotopes must be identified by using gamma spectroscopy. Highly sophisticated equipment and analytical schemes have been developed for this purpose and are widely used. Ordinarily a detector is used which is capable of producing electronic pulses whose height is proportional to the energy of the γ-ray. The resulting pulses are then sorted in a multichannel analyzer which produces a spectrum with peaks corresponding to gamma rays of a particular energy. In most applications the spectra are not overly complicated.

In summary NAA can be a powerful tool for elemental analyses. Some of the chief advantages of its use are

1. Usually high sensitivity is obtainable.
2. The method is often nondestructive.
3. Simultaneous multielement analyses are often possible.

TABLE 6.3

Activation Analysis Sensitivities

Element	Limit of detection[a, b]	Element	Limit of detection[a, b]	Element	Limit of detection[a, b]
Aluminum	0.001	Iodine	0.001	Ruthenium	0.002
Antimony	0.0009	Iridium	0.00002	Samarium	0.00009
Argon	0.0001	Iron	1 [c]	Scandium	0.002
Arsenic	0.0002	Krypton	0.001	Selenium	0.02
Barium	0.005	Lanthanum	0.0004	Silicon	0.009
Bismuth	0.05	Lead	2	Silver	0.0001
Bromine	0.0001	Lutetium	0.000009	Sodium	0.0004
Cadmium	0.009	Magnesium	0.05	Strontium	0.0009
Calcium	0.5	Manganese	0.000005	Sulfur	0.04 [c]
Cerium	0.02	Mercury	0.002	Tantalum	0.009
Cesium	0.0004	Molybdenum	0.01	Tellurium	0.01
Chlorine	0.002	Neodymium	0.02	Terbium	0.005
Chromium	0.2	Neon	0.4	Thallium	0.5
Cobalt	0.0009	Nickel	0.004	Thorium	0.007
Copper	0.0002	Niobium	0.1	Thulium	0.002
Dysprosium	0.00000004	Nitrogen	90 [b]	Tin	0.02
Erbium	0.0004	Osmium	0.004	Titanium	0.001
Europium	0.0000009	Oxygen	30 [c]	Tungsten	0.0002
Fluorine	0.2	Palladium	0.0002	Uranium	0.0004
Gadolinium	0.002	Phosphorus	0.02	Vanadium	0.0001
Gallium	0.0004	Platinum	0.004	Xenon	0.01
Germanium	0.00005	Potassium	0.002	Ytterbium	0.0004
Gold	0.0002	Praseodymium	0.0001	Yttrium	0.0002
Hafnium	0.007	Rhenium	0.00004	Yttrium	0.0002
Holmium	0.00004	Rhodium	0.00002	Zinc	0.02
Indium	0.000005	Rubidium	0.04	Zirconium	0.2

[a] Assume 1-hour irradiation in a thermal neutron flux of 10^{13} neut/cm² seconds.
[b] Interference-free limit.
[c] By fast-neutron reaction.

4. Good accuracy and high specificity are usually possible.

5. Analysis for more than one isotope of an element is sometimes possible.

Some disadvantages are

1. No information is provided about the nature of the chemical compounds involved.

2. Notwithstanding advantage (3) above, interferences are sometimes serious problems especially with biological samples.

3. Sometimes a considerable investment in equipment is required.

MEDICAL APPLICATIONS

Neutron activation analysis is inherently an elemental technique. As such it is somewhat limited in use as a routine clinical tool since it provides essentially no information regarding chemical structures. There are nonetheless many valuable medical applications already established and there will undoubtedly be developed many more. A few specific categories are worthy of note—toxicity determination, small sample analysis, stable tracer methods, and *in vivo* analysis.

Activation analysis *in vivo* has been performed both in a conventional manner and by using prompt gamma activation analysis. The conventional approach, whole body irradiation followed by whole body counting, offers a direct method for the determination of the elemental composition of a living organism. Such measurements with human patients will be severely limited by the acceptable dose accompanying the irradiation. Thus far experiments of this type have been performed to determine sodium, chlorine, and calcium.

Prompt gamma activation analysis is a variation of the basic method in which the prompt γ-rays, i.e., those emitted at the time of the neutron interaction, are detected in lieu of the delayed gammas produced by the decay of the product nuclei. In practice this means placing the sample in a well collimated beam of neutrons and then monitoring the gammas that are emitted. This method has been shown to be potentially very useful especially in the determination of elemental ratios in limbs when the radiation dose involved is more acceptable (14).

Finally it may be noted that some diseases can be well correlated by abnormal trace element concentrations. In these instances NAA may be useful either as a diagnostic tool or as an aid in performing basic research in the mechanics of the disease. Sodium analysis of the fingernails and toenails of children has been a useful supplementary diagnostic criterion for cystic fibrosis (15).

INTERNAL INFORMATION SOURCES

The functional state of internal organs is difficult to assess on the basis of energy or samples outside the body surface. The anatomical origins or geometrical relationships are frequently lost in the process. With increasing frequency, information is being extracted from the body by means of probes which will be considered in two main categories—(a) Probes or sensors inserted into the body, and (b) beams of energy directed into or through the body (Fig. 6.6).

PROBES INTRODUCED INTO THE BODY

Hollow catheters or needles have been used for many years for withdrawing samples of blood or body fluids. They are also employed for the measurement of pressures in blood vessels, heart, urinary tract, cerebrospinal canal, gastrointestinal tract, or elsewhere. Such applications are familiar and require no further comment. The technology is now sufficiently developed to expand greatly the scope of catheter applications. Catheter-tip pressure transducers of very small size, accuracy, and dynamic responses are now being used because they greatly reduce disturbing artifacts from movement or impact when pressures are recorded from the end of long fluid-filled catheters. Continuous flow velocity detection is now possible at the tip of catheters using ultrasonic Doppler techniques (see Fig. 6.9). Fiberoptic bundles are being used to transmit along catheters light which is reflected from the blood and transmitted back out to photoelectric devices for analysis. Such devices provide continuous determinations of oxygen saturation of blood at the tip of the catheter without the necessity for withdrawing or analyzing samples of blood.

FIBEROPTIC OXIMETRY

Curtis Johnson

The measurement of oxygen saturation, defined as the ratio of oxygenated hemoglobin to total hemoglobin is important in clinical and diagnostic medicine. Conventional measurement techniques involve the removal of blood from the cardiovascular system for subsequent laboratory analysis. Instruments have recently become available which can measure oxygen saturation at the tip of an intracardiac catheter.

A fiberoptic catheter-tip oximeter is commercially available and has been extensively evaluated on human subjects (16). In this instrument pulses of infrared and red light are alternately sent down the fiber by means of a chopping wheel which contains an infrared and red filter. Another instrument utilizes a lamp source with two separate photodetectors, one with an infrared filter and the other with a red filter. These instruments give continuous readings of oxygen saturation in the cardiovascular system.

Another oximeter system utilizes semiconductor light-emitting diodes at red and infrared wavelengths in place of the lamp optical source (17). A pulse generator triggers the semiconductor diode current pulser which alternately pulses the red and infrared diodes. The light pulses are sent down one branch of a fiberoptic catheter. Reflected light is brought out through another branch to a photodetector. The light pulses are preamplified, and processed by a sampling circuit. Synchronizing pulses from the pulse generator are used to time the pulse-sampling circuits, and an analog computer forms an analog output voltage from the reflected optical intensities proportional to oxygen saturation. Gallium-arsenide and gallium arsenide–phosphide light-emitting diodes are used with a spectral width of approximately 400 Å.

In vivo fiberoptic catheter oximeter systems are presently being used in anesthesiology for monitoring oxygen saturation values in patients during and after surgery. The instrument is also being used in pediatrics for detecting and estimating the magnitude of left-to-right shunts, and in cardiology for heart catheterizations.

Steady progress is expected in making catheters smaller for easier and safer penetration into the body. Also, progress is expected in developing a catheter which can make a series of measurements simultaneously and continuously, including oxygen saturation, pressure, sounds, flow velocity, pH, or blood gases (fig. 6.6).

REFLECTION SPECTROPHOTOMETRY

By installing fiberoptic bundles into a needle, the tissues through which it penetrates can be identified by reflection spectrophotometry by a method described by Edholm, Grace, and Jacobson (18). The process is very similar to that employed for oximetry except that a sequence of colored light pulses are transmitted into the tissue along one bundle and the other bundle returns the light from the tissues to a photomultiplier to obtain a reflection spectrum on an oscilloscope. The spectra recorded *in vivo* differed greatly for fat, cysts, and tissues such as muscle, liver, kidney, and white and gray matter in the brain. The ability to distinguish tissues under these circumstances promises information beyond mere localization of the needle tip. For example, the existence of atheromatous lesions on the wall of the aorta has been distinguished through an intravascular catheter by the same kind of technique (19). Further refinements in these techniques could add greatly to our ability to gather significant data from tissues *in situ*.

HYPODERMIC MICROSCOPE

Few locations on the body provide an opportunity to directly view living tissues except for the eye or mucous membranes. Fiberoptics are capable of conveying information in the form of images if they are properly organized (20).

For this purpose a hypodermic microscope has been developed to permit direct observation of living tissues through fiberoptic imaging bundles composed of some 10,000 fibers of about $5\,\mu$ in diameter leading to a microscope objective. Resolution is reported to be about $14\,\mu$ on the long axis of the bevel and $10\,\mu$ across the short axis. The capillaries and other vessels of the microcirculation can be visualized. Living tissues are quite lacking in distinguishing colors or details but distinctive characteristics might be brought out by vital staining as needed (Fig. 6.6).

CHEMICAL PROBES—ISOTOPE APPLICATIONS

The literature on applications of the many different radioactive isotopes to explore biological functions is much too voluminous to consider here. For completeness, one might mention some of the most obvious applications. The thyroid gland has great affinity for iodine which is found in a convenient isotope. The rate of uptake of ^{131}I provides not only a way to localize and scan the thyroid but also a fairly direct way of assessing thyroid function in terms of its rate of uptake, concentration, and elimination of this vital material. Since

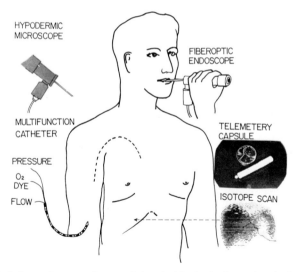

Fig. 6.6. Information regarding conditions inside the body can be obtained by means of various types of probes such as the hypodermic microscope, catheters with various functions, or endoscopes for direct vision. Energy can be transmitted from sources inside the body such as telemetry capsules in the gastrointestinal tract or isotopes in the liver. (Telemetry capsule photograph courtesy of L. Slater and *N. Y. State J. Med.*)

radioactivity can be readily detected and localized by appropriate columnation, it is possible to develop rather crude outlines of organs containing significant concentrations of radioactive materials. For this purpose, it is possible to provide visualization of lungs, liver, spleen, or pericardial effusions as part of diagnostic procedures. The blood flow through various tissues and organs can be estimated in accordance with either the Fick Principle or the Stewart Principle depending upon the conditions existing. A number of applications of isotopes for exploring the circulation of various organs was presented in a monograph edited by Sevelius (21).

INTERNAL CAPSULES—BIOTELEMETRY

The inaccessibility of many internal organs for study has led to great interest in the prospects of introducing microminiature transducers capable of transmitting information from within body cavities to the outside by means of telemetering techniques. Mackay (22) indicated the potentialities of biotelemetry from radiosonds by listing some of the applications which have already been utilized, including transmission of pressure, temperature, acidity, oxygen tension, radiation intensity, the presence of blood, and various bioelectric potentials. Despite the fact that these devices were extremely small nearly 10 years ago (Fig. 6), and further reductions in size have occurred since, little clinical application has developed. The gastrointestinal tract is a prime target for use of radiosonds because it is so very inaccessible for measurements except by exceedingly long and uncomfortable tubes or by X rays which are plagued by overlapping shadows from tortuous loops of bowel. Biotelemetry capsules are capable of registering changes in significant variables (i.e., pressure, acidity, hemorrhages) as they progress down the intestinal tract. Precise localization of the capsules along the tract remains a problem as does the recovery of the capsules from the stools. Neither of these obstacles appear to be insurmountable. Future prospects for further microminiaturization were suggested by Slater (23) by indicating the rate at which technology is pushing toward electronic elements approaching the size of living cells. The neuron is taken as a kind of standard and a timetable was constructed as follows

Neuron	10^0
Vacuum tube (1940)	10^9
Transistor (1950)	10^6
Thin-film (1960)	10^3
Microfilm (197?)	10^0

According to this table, Slater postulated that designers might place devices the size of neurons in the hands of investigators in a few short years. It is easy to envision the insertion of such small sensors into the blood stream and directing them by external magnetic steering devices into locations or organs of choice where they could transmit diagnostic information of value over extended periods of time by transponder techniques in which the energy requirements are met by electromagnetic sources from outside the body. It will be much more difficult to achieve these goals than to predict them.

EXTERNAL TELEMETRY

H. Fred Stegall

Information about an astronaut's status during space flight can be interpreted by a physician thousands of miles away listening for subtleties in his voice transmission or his facial appearance on a television monitor. These are often better clues to his overall status than his electrocardiogram, oxygen consumption, or other quantifiable physiological variables which may also be transmitted to the medical monitor.

There are two major applications for telemetry of biological signals— (1) Transmission of the information across some distance because direct connections would be unwieldy or otherwise impractical, or (2) isolation of the patient from electrical shock hazard or the inconvenience of cables connecting him to monitoring equipment.

Telemetry involves several separate steps.

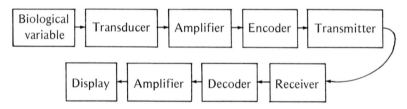

Energy representing a biological variable is converted into an electrical signal by a transducer. The electrical signal is amplified and encoded into a form suitable for transmission (e.g., by radiofrequency, light, or sonic waves) through the environmental medium to a receiver. The signal is generally decoded, amplified, and displayed.

A wide variety of biological data have been telemetered at one time or another: electroencephalograms and electrocardiograms, including fetal ECG's; phonocardiograms, electrical impedance of tissue, pressure, flow, and dimensions, temperature, pH, pO_2, etc., of body fluids; and many others. Recent developments in fabrication of miniature circuits have been largely responsible for the application of these techniques so that batteries are now the largest component in many of these systems.

Although telemetry usually refers to rf transmission, several other forms of telemetry are also employed. Each has particular advantages, and is used in fairly specific applications.

1. Telephone lines, for the transmission of electrocardiographs to distant consultants or in the transmission of data to and from a central computing facility

2. Sonic and ultrasonic transmission in underwater operations

3. Light transmission over very short range for electrical isolation

4. Electrical wires (e.g., power lines) in a building such as a hospital to carry the signals by carrier-current telemetry

All telemetry systems require electrical power from batteries, from biological sources (chemical, or mechanical via piezoelectric transducers) or by radiofrequency energy transmitted through the skin, rectified, and used for sensing biological variables which are then retransmitted back again.

Besides the familiar applications in monitoring astronauts in space, telemetry has been used to track animals in the wild (bears, birds), to obtain information about their behavior and their physiological function in their natural habitat. The devices to collect and transmit such data have become increasingly sophisticated and complex. Tortoises have swallowed temperature transmitters, baboons have carried backpacks in the wilds of Africa, seals have towed small floats with transmitters, dogs have towed sleds, and birds carried very light units for transmitting physiological information.

Thus far medical applications of telemetry have been exceedingly limited. Closed-circuit TV systems are sometimes used to transmit a view of the patient and oscillographic records of ECG, EEG, etc., to a central monitoring station as in a coronary care unit. In some hospitals carrier-current systems are used to transmit ECG's from the bedside to the nursing station, or patients have carried telemetry units which transmit to ceiling antennas when they became ambulatory after myocardial infarction.

Telemetry systems are likely to see increased use in future health care delivery. Small units could continuously transmit ECG's from a postinfarct patient anywhere in a city to a central monitoring facility and could sound an alarm. Tiny transponders (passive units which interact with an electrical field in their vicinity) which can be swallowed or implanted without the necessity for batteries might be used to monitor blood pressure, intraocular pressure, forces acting on the gut contents or the chemical composition of those contents, and the like, obtaining information about physiological function during disease. In particular, more extensive utilization of telephone lines might be anticipated because of their wide distribution, particularly in view of the crowding in the radiofrequency communications spectrum. Improved methods of coding and packing biological information are already being developed for aerospace and military applications, and will find their way into civilian fields in time.

ENERGY PROBES

The information which can be obtained by withdrawing samples or by inserting canulas, sensors, or capsules within the body is of extreme value in clinical diagnosis. Such techniques typically involve some degree of discomfort or hazard which interferes with routine application in the broad spectrum of medical care outside medical centers. Every effort must be extended to make available an expanding array of techniques for acquiring diagnostic data with minimal discomfort, distortion, or hazard. The advent and long history of X-ray techniques has indicated the great value which can accrue from directing a beam of energy into the body and analyzing the energy that comes out. With growing awareness of the cumulative effects of radiation, the exposure to radiation is being reduced to a minimum. For this reason, attention has been directed to other types of energy beams which can be generated, controlled, and recorded. Prominent among these are light and ultrasound. The various energy probes are supplementary rather than competitive.

TRANSILLUMINATION

Curtis Johnson

The most familiar and innocuous form of electromagnetic energy is light which conveys so much information to us about the external environment. Light transmission through a body, called transillumination, reveals much about the chemical and anatomical internal structure of that body. Transillumination in medicine has been used for examining sinus cavities, the scrotum, and fluid cavities in the head and abdomen. Instruments are available for transillumination of infants which utilize optical and electronic instrumentation to increase sensitivity and make quantitative measurements. For example, a cranial transilluminator has been developed principally for locating fluid cavities in the heads of infants. At present the optical density range of the transilluminator is 0 to 8 optical density units (optical density is defined as $OD = \log_{10} 1/T$, where T is transmittance). The optical density can be measured with an error of less than 0.1 OD units. The transilluminator consists of a light source which is externally triggered. The transmitted pulse heights are sampled and fed into a difference amplifier. The light intensity is read out on a scale which is converted to optical density. The output voltage is made independent of the ambient light by sampling the voltage from the photodetector just before and during the light pulse, or by using a high-ham filter.

Some tests have been performed with normal infants and infants with varying degrees of hydrocephalus. With light input at the frontal fontanel, OD-5 levels are observed normally at approximately 4 cm from the fontanel. A usual examination by means of a flashlight in a darkened room reveals light only within about 1 cm of the light source, thus the pulsed transilluminator is more sensitive. A series of tests has established that for normal term infants the OD per centimeter valued through the head averages 1.4. For premature infants the average may be as low as 1.0. Much lower optical densities are obtained from infants with hydrocephalus.

APPLICATIONS OF ULTRASONIC TECHNIQUES

Consider, for example, a local obstruction in an artery serving the leg which can be neither seen nor felt. The physician can detect changes in the pulse with his finger tips at various positions down the artery but he needs additional information. An X-ray photograph taken of the artery filled with a liquid which casts a heavy shadow will demonstrate the anatomical site and extent of the obstruction as illustrated in Fig. 6.7. An ultrasonic flow detector may be used to determine the functional effect on blood flow at various positions along the artery. Instruments of this type are described in greater detail in subsequent sections. Note that in this instance, the anatomical deformity can be assessed by one type of energy probe (X-ray) and the functional disturbance to blood flow by another energy probe (ultrasonic beams). These represent rather direct sources of information regarding the nature and severity of the disease process. The medical profession should aspire to this type of definitive information regarding the many different types of disease processes in various organ systems and the biomedical engineering community should actively respond to these needs.

A variety of crystalline structures, including quartz and certain ceramic materials have the property of emitting very high frequency sound waves when subjected to high frequency fluctuating potentials. In general, these crystals emit the most intense sound when the excitation frequency is at the resonant frequency of the crystal which is determined by its thickness. If the fluctuating voltage is more or less continuous, the crystal continues to generate ultrasound so long as the fluctuating voltage is applied. If a sudden burst of electrical potentials is applied, the crystal will emit a burst of sound much like the ringing of a bell. Such bursts of ultrasound travel at known velocities through tissues and body fluids and can be used to obtain information about structures and function inside.

A rather extensive review of the current status and future prospects of ultrasonic applications are included in this discussion for two reasons—(a) This is an area of investigation with which the authors have long and extensive experience, and (b) the very great number and diversity of applications of beamed energy are indicated to provide incentive to similarly explore other forms of energy for diagnostic applications.

PRINCIPLES OF ULTRASONIC ECHO RANGING

The distance between two points can be registered in terms of the transit time of bursts of ultrasound which originate at a transducer, travel to a reflecting surface and then return to a receiver at or near the point of origin (Fig. 6.8). This process is generally familiar in the case of sonar where a large piezo crystal generates pulses of ultrasound which traverse through the water and may be reflected from discontinuities in the water either in the form of free floating objects or the bottom of the body of water. The backscattered ultrasound returns to the receiver and the round trip time can be calibrated in terms of the distance between the transmitter and the point of reflection. Viewed on the face of an oscilloscope, the distance between the transmitter spike and the return signal from the reflected object can readily be calibrated in terms of the distance from the target. Note that the configuration of the back-scattered signal bears

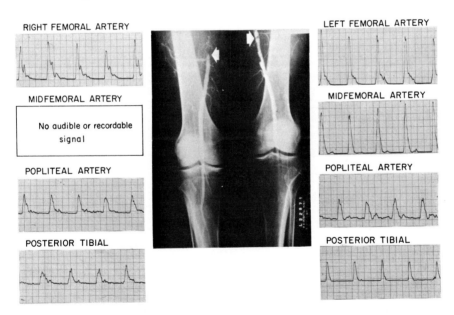

RIGHT FEMORAL ARTERY

MIDFEMORAL ARTERY

No audible or recordable signal

POPLITEAL ARTERY

POSTERIOR TIBIAL

LEFT FEMORAL ARTERY

MIDFEMORAL ARTERY

POPLITEAL ARTERY

POSTERIOR TIBIAL

Fig. 6.7. Blood flow detection by ultrasonic Doppler techniques indicate sites of obstruction or occlusion of arteries in the legs to supplement information obtained by X-ray pictures of radioopaque substances in the channels. The changes in flow signals above and below occlusive lesions facilitate diagnosis and evaluation of surgical therapy. (From photograph reproduced with permission of D. E. Strandness and *J. Amer. Med. Assoc.*)

no obvious relation to the shape of the target but may present somewhat characteristic identifiable patterns. Another application of this principle is clinically used with great frequency to determine the location of the midline echo in the skull. A transmitter pulse originating on one side of the scalp penetrates the head and produces signals on a cathode ray oscilloscope which indicates the distance between the transmitter pulse, the midline echo and the far side. By obtaining such signals from first one side of the skull and then from the other, it is possible to identify and measure displacement of the midline as may occur with hemorrhage or with tumor growth (Fig. 6.8).

The information displayed in Fig. 6.8A is called an A-scan and is represented by deflections on a time scale which is readily converted to distance knowing propagation velocity of ultrasound in tissues. If the ultrasonic beam is caused to traverse through angular rotation while echo ranging is going on, it is possible to develop signals which represent reflecting surfaces in two-dimensional space as indicated in Fig. 6.8B. This application is also employed for studies of the contents of the skull as illustrated in Fig. 6.8D. Circumferential movement of the ultrasonic source, around the head emersed in water produces pictures which represent a cross-sectional view of the skull and its contents (24).

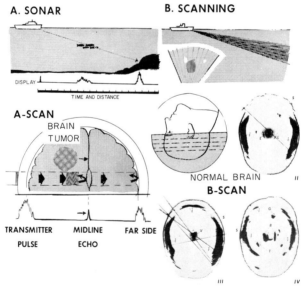

Fig. 6.8. A Transmission time of sound waves indicate distances to reflecting targets in sonar just as ultrasonic pulses can be utilized to indicate the distance to midline structures of the brain to detect displacement in the presence of a brain tumor. (B) Scanning techniques provide two dimensional displays of reflecting surfaces which indicate the location of structures within the brain which cannot be detected by X ray. (B-Scan pictures courtesy of D. Makow and Pergamon Press.)

Ultrasonic visualization has also proved possible in another extremely inaccessible site; namely the structure of the unborn child within the uterus. Robinson, Garrett, and Kossoff (25) reported echograms revealing the head, thorax, extremities. Sometimes the genitalia of fetuses can be seen and the placenta localized. The heart size of the fetus has also been measured (26).

A transmitter positioned over the precordium may emit pulses of ultra-sound through the chest wall and the heart walls with reflected or back-scattered ultrasound returning to the site of origin from the myocardium, from the heart valves and from other reflecting surfaces. In this way it is possible to identify the movement of individual structures such as the mitral valve leaflet or the ventricular walls in terms of the displacement of a deflection on a cathode ray oscilloscope.

ULTRASONIC ECHO RANGING

John M. Reid

It is in the area of soft tissue diagnosis that most of the applications of echo ranging ultrasound have been made. This is a welcome advance since X-ray techniques are of limited usefulness in soft-tissue diagnostic work.

Lung tissue scatters the ultrasound in an incoherent manner due to the high air content. In some cases a careful examination can detect propagation through portions of the lung where the air has been removed. The detection of embolism of the lung and the location of pleural fluid in the thoracic cavity can be done by this means now. The possibility of obtaining further information either about the lung or the heart by propagation through lung does merit some further consideration, although it is not clear at this time how to penetrate the barrier imposed by air-filled spaces.

Although the heart is accessible to ultrasonic waves only if they propagate through a fairly small area to the left of the sternum in the third through fifth intercostal spaces, a large amount of useful information has been gathered via this window. The motion of the mitral and tricuspid valves can be observed by recording the A-scope echo pattern in the form of lines representing the position of the valve. The technique is applicable to motion of artificial valves installed within the chest. The position, thickness and the motion of both anterior and posterior left ventricular walls can be observed. As more knowledge becomes available on the path taken by the ultrasound, the characteristic echoes obtained from various heart structures and the significance of the motions observed by this technique becomes clear, we can expect much greater utility from these methods. By monitoring movements of the near wall and far wall of the left ventricle, it should be possible to estimate changes in ejection rates and ejection

A. DOPPLER FREQUENCY SHIFT   **C. FM SONAR**

B. DOPPLER FLOW SENSOR (CW)   **D. PULSED DOPPLER**

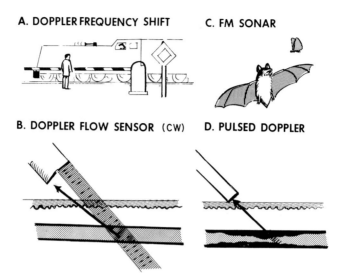

Fig. 6.9. (A) The Doppler shift of sound can be utilized to estimate the velocity of a sound source or moving reflector. (B) Ultrasonic waves undergo a Doppler shift when reflected from moving blood cells indicating changing flow velocity. (C) Bats employ a form of FM sonar while hunting prey. (D) The pulsed Doppler technique provides information regarding flow velocity in small volume of space a known distance from the transmitter.

fractions of the left ventricle in a convenient and completely nondestructive technique.

Pulse-echo techniques reveal information about the diameter of arteries or veins, either by scanning equipment or by fixed transducers positioned over the appropriate blood vessel. The fixed transducer offers the advantage of being able to follow the motions of blood vessel walls. One application being investigated is the detection of atherosclerotic deposits or plaques in the artery wall. The discovery that these deposits absorb the ultrasound has raised the possibility that they may be detectable by the presence of acoustic "shadows" cast by the plaque. The shadows behind the plaques coupled with greater reflections from plaque and the altered wall motion may make these deposits easily detectable.

The determination of internal structure of various organs may proceed beyond the present understanding of the relationship between the echo pattern and structure. Most of the present commercial equipment is effective for locating the margins of organs. This means that sizing and measurement of various tissues can be done. The sizes of the urinary bladder and kidney have been estimated. We can expect that other tissues may well be monitored by this means, particularly transplants. In liver the distribution of connective tissue can be shown to be responsible for the distinctive patterns obtained. A great deal of effort has been directed toward the differentiation of tumors from normal tissue.

PRINCIPLES OF DOPPLER SHIFT TECHNIQUES

When a locomotive passes an observer with its horn blowing, it is common experience that the apparent frequency of the horn has a higher frequency while approaching and suddenly shifted to a lower frequency while receding (Fig. 6.9). This familiar phenomenon is called the Doppler frequency shift and is a source of valuable information when used with ultrasonic equipment. For example it is possible to introduce a beam of ultrasound through the skin and subcutaneous tissues from a transducer positioned on the surface of the skin with a coupling jelly between the surfaces of the crystals and the skin. A small amount of the transmitted beam is backscattered toward the transmitter crystal by the acoustical interfaces which intervene along the beam. If these interfaces are stationary then the frequency of the backscattered sound is the same as the transmitted sound and if these are then mixed, no resultant beat frequency is observed. However, if there is a moving interface of substantial magnitude in the beam, the backscattered signal from this interface will have a different frequency based on the Doppler shift.

FM RANGING

A very sophisticated form of ultrasonic detection is utilized by the bat in pursuit of its prey (Fig. 6.9C). The bat is capable of emitting a series of chirps which consists of bursts of sound with a rapidly rising frequency, well above the audible level of man. If the target is relatively near, the initial portion of the chirp returns to the bats hearing mechanism while the remainder of the chirp is being produced. The difference between the transmitter frequency and the received frequency is an expression of the time lag between the transmission and the reflection back to its origin. By increasing the repetition frequency of the chirps, the bat can cone down on the target and rapidly identify its range and position. Techniques of this general sort are potentially applicable for ultrasonic application for data acquisition, but have not been widely used thus far.

PULSED DOPPLER TECHNIQUES

If bursts of high-frequency sound are emitted by an ultrasonic crystal and traverse through skin and subcutaneous tissues, backscattered sound from the bursts can be gated in a way such that the signals return only from interfaces a specified depth from below the skin surface. The application of these gated pulse Doppler techniques are clearly evident in Fig. 6.9. For example, by adjusting the range gate of the pulse Doppler system, it is possible to obtain information concerning the blood flow velocity in a stream emerging from a constricted portion of an artery. This velocity should be distinctly different from that upstream and downstream from the obstruction and helped to localize its extent and severity. By using the range gate and adjusting the sampling volume forward and backward, it is possible to get information regarding the dimensions from a single source, information regarding the velocity of blood cells, and at the same place, information regarding the dimensions of the blood vessels. These are important steps toward the identification of volume blood flow in arteries and veins under the skin by nondestructive ultrasonic techniques. The applications of these general principles are described in the remainder of this section.

FUTURE POTENTIAL FOR PULSED DOPPLER TECHNIQUES

Donald W. Baker

Pulsed ultrasonic Doppler techniques (Fig. 6.9) afford opportunities for making transcutaneous measurements on the cardiovascular system to provide information not available by any other known approach.

The first pulsed Doppler devices had poor resolution and low sensitivity but still provided signals identifiable as arterial wall movements, high velocities in the axial stream and velocity gradients across the blood vessel (27).

The principal application of pulsed Doppler to this date has been measurement of velocity profiles and vessel diameter. From these measurements it should be possible to compute the volume blood flow rate. To date this crucial measurement has been accomplished only in flow models under carefully controlled conditions. To make this measurement transcutaneously on man requires the determination of the angle between the sound beam axis and the flow velocity vector. When these problems are solved volume flow measurements will become feasible.

Future applications of pulsed Doppler go far beyond blood flow measurements. These possibilities can be grouped into two general categories. The first category involves measurement of the velocity of other interfaces apart from blood flow. They include

1. Valve velocity and displacement
2. Velocity of heart wall movements
3. Intraventricular blood flow velocity distribution

Motion of heart valves and heart walls can be detected and measured using a version of the pulse Doppler. In contrast to echo methods, which depend on specular reflections for maximum return, Doppler signals are detected from scattering regions. Thus pulsed Doppler motion detection should allow a wider selection of access points to the heart. The ultrasonic scattering process is generally not specular therefore the angle between the interface and the sound beam is not as critical. It is often possible to detect Doppler signal components even at grazing angles to the myocardium. The ability to cope with this type of situation may provide an opportunity to identify paradoxical movements of heart walls to diagnose the nature and extent of a myocardial infarct.

EXPERIMENTAL AND PROJECTED APPLICATIONS OF ULTRASOUND

H. Fred Stegall

The applications described above for pulsed or continuous-wave ultrasound involves generation of a one- or two-dimensional picture of reflecting surfaces (which correspond to tissue interfaces) and determination of the relative motion of reflectors and scatterers moving in an ultrasonic field. These are the most common applications for diagnostic ultrasound in experimental or clinical medicine today.

A number of other applications for ultrasound energy have been suggested or tried. Measurement of the transit time of a sound burst from one transducer to another has been used to estimate changes in dimensions of internal organs in chronically instrumented animals (see Fig. 6.10); including the ventricular diameters (29), liver and spleen size. These applications involve the physical displacement of the transducers fastened on the tissue surfaces to be measured. In such applications the transducers must be small and lightweight, and resolution requirements usually dictate use of relatively higher frequency sound (3 mHz or more). Fortunately transducers for generation of sound at this frequency can be quite small and thin, lightweight, and the fact that they are not physically linked to one another helps minimize constraint they offer to tissue movement.

Blood content of organs and hematocrit of blood have been estimated by measuring transit time of ultrasound between a transmitting and a receiving crystal a fixed distance apart as the transmission velocity is altered by changes in the tissue between the two crystals.

A very small pair of forceps has been equipped with an ultrasonic detector to allow an ophthalmologist to localize and extract foreign bodies from the eye which are nonmagnetic and even invisible.

The FM ultrasonic technique illustrated in Fig. 6.9 may be used to supplant pulsed ultrasonic techniques, particularly in those circumstances where power requirements, multiple reflecting surfaces, and the like make use of the pulsed techniques difficult (i.e., for measurement of flow velocity, dimensions, detection of reflecting interfaces).

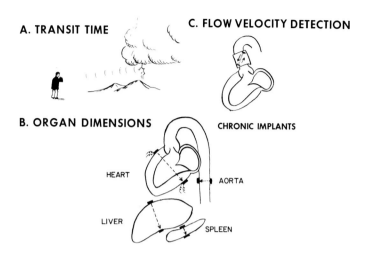

Fig. 6.10. Experimental applications of ultrasound include transit time measurements of ultrasonic bursts to indicate changing dimensions of internal organs (e.g., the heart, liver, or spleen). In addition, the difference in transit time of ultrasound diagonally upstream and downstream in an artery can be used to indicate changing blood flow from chronically implanted flow sensors in experimental animals.

VOLUME MEASUREMENT

Generally each echo from a series of reflecting surfaces is examined separately, but Lategola (30) has described a technique by which multidirectional bursts of ultrasound were generated at the tip of a catheter in the left ventricle and all the echoes were inversely proportional, in an unspecified way, to the volume of the ventricle assuming that the echoes would be smaller when the walls of the ventricle were more distant. It is technically feasible to monitor the changing internal diameters of the left ventricle from a ring of tiny transducers arrayed on the end of a catheter. Rather complicated electronics would be required to account for movements of the tip of the catheter in relation to the heart walls.

ULTRASONIC HOLOGRAPHY

The advent of lasers made three dimensional visualization of objects possible through the process of holography. One interesting application of this approach is the utilization of a liquid surface as a holographic "negative." A wide beam of ultrasound is transmitted through the part under examination and impinges on the surface of the water, creating a pattern of riffles. A reference beam of ultrasound plays directly on this same area of surface. A laser light source is directed through lenses to the water surface and reflected into a television camera to produce a continuous moving picture in real time displaying interfaces in both the hard or bony tissues and some of the soft tissues (blood vessels, fascial planes, tendons, etc.), which do not cast shadows on X rays (Fig. 6.11). To date, the technique is limited to relatively thin structures such as the hand through which the ultrasound can be readily transmitted. Rapid progress in both ultrasonic and electronic data processing techniques promise a progressively expanding array of new and useful recording techniques for research and diagnosis in animals and men.

Several potential areas for ultrasound application are still unexplored.

1. Measurement of transit-time along long bones as an index to osteoporosis, healing of fractures, etc.

2. Improved resolution of through-transmission ultrasonic imaging of the thorax or other body cavities

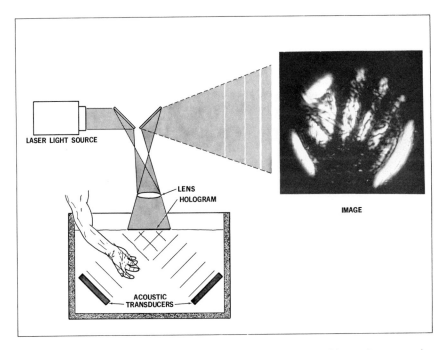

Fig. 6.11. Ultrasonic holography provides a moving picture of internal structures by television photography of patterns on a water surface produced by a transmitted beam and a reference beam of ultrasound. The internal structure of the hand is illustrated as the sample image. (Photograph reproduced through the courtesy of Holosonics, Richland, Wash.)

3. Measurement of movements of body surfaces by determination of transit time of lower-frequency, air-coupled ultrasound from fixed transducer to body surface

4. Catheter-tip sensors for determination of flow velocity, heart or vessel size, or visualization of deep structures in the thorax or abdomen where transcutaneous methods are inadequate

5. Gastrointestinal sonic visualization via sensors on tubes or even swallowed capsules

To this list could be added dozens of more speculations, but most have a common thread—introduction of ultrasonic energy into the body and using it in one of two ways, either to alter metabolism or to determine how the sound is changed (in frequency, amplitude, etc.), as it is reflected or back-scattered from deeper structures. Other applications will primarily depend on imaginative clinicians and bioengineers who can detect newer ways of employing these techniques in diagnosis or therapy.

References

1. Rushmer, R. F., Buettner, K. J. K., Short, J. M., and Odland, G. F. The Skin. *Science* **154**, 343-348 (1966).
2. Kenedi, R. M., Gibson, T., and Daly, C. H. Bioengineering studies of the human skin. *In* "Biomechanics and Related Bio-Engineering Topics" (R. M. Kenedi, ed.). Pergamon Press, Oxford, 1965.
3. Wennstrom, G. Medical engineering in Swedish medical service, *Proc. IEEE,* **57**, 1809-1819 (1969).
4. Rushmer, R. F., and Morgan, C. Meaning of murmurs. *Amer. J. Cardiol.* **21**, 722-730 (1968).
5. Rushmer, R. F., Morgan, C., and Harding, D. C. Electronic aids to auscultation. *Med. Res. Eng.* **7**, 28-36 (1968).
6. Reynolds, W. E., and Bazell, S. What medical instrumentation in 1974? *Med-Surg. Rev.* (First Quarter) **5**, 53-56, (1969).
7. Spencer, W. A., White, S. C., Geddes, L. A., and Vogt, F. B. The impact of electronics on medicine. *Part 1. Postgrad. Med.* **36**, 291-296 (1964).
8. Moss, D. W. Automation in clinical biochemistry. *In* "Advances in Biomedical Engineering and Medical Physics" (S. N. Levine, ed.). (Interscience), New York, 1968.
9. Dickson, J. F. Automation in clinical laboratories. *Proc. IEEE* **57**, 1974-1987 (1969).
10. Ledley, R. S. Automated pattern recognition for clinical medicine. *Proc. IEEE* **57**, 2017-2035 (1969).
11. Thaer, A. A. "Microfluorometric Analysis of the Reticulocyte Population in Peripheral Blood of Mammals" (D. M. D. Evans, ed.). Cytology Automation. Proceedings of the Second Tenovus Symposium, Cardiff Wales, October 1968. E. and S. Livingston Ltd., London, 1970.
12. Kamentsky, L. A., and Melamed, M. R. Instrumentation for automated examinations of cellular specimens. *Proc. IEEE* **57**, 2007-2016 (1969).

13. Sernetz, M., and Thaer, A. A capillary fluorescence standard for microfluorometry. *J. Microsc.* **91**, 43-52 (1970).
14. Comar, D., Crouzel, C., Chasteland, M., Riviere, R., and Kellershohn, C. The use of neutron-capture gamma radiation for the analysis of biological samples. *Nucl. Appl.* **6**, 344-351 (1969).
15. Woodruff, G. L., Babb, A. L., Wilson, W. E., Jr., Yamamoto, Y., and Stamm, J. L. Neutron activation analysis for the early diagnosis of cystic fibrosis. *Nucl. Appl.* **6**, 352-359 (1969).
16. Polanyi, M. L., and Hehir, R. M. *In vivo* oximeter with fast dynamic response. *Rev. Sci. Instr.* **33**, 1050-1054 (1962).
17. Johnson, C. J., Palm, R. D., Stewart, D. C., and Martin, W. E. A solid state fiberoptic oximeter. *Med Instr. J. Assoc. Advan. Med. Instr.* **5**, 77-83 (1971).
18. Edholm, P., Grace, C., and Jacobson, B. Tissue identification during needle puncture by reflection spectrophotometry. *Med. Biol. Eng.* **6**, 409-413 (1968).
19. Edholm, P., and Jacobson, B. Detection of aortic atheromatosis in vivo by reflection spectrophotometry. *J. Atherosclerosis Res.* **5**, 592-595 (1965).
20. Frommer, P. L. The principles of fiber optics and their clinical applications. *In* "Engineering in the Practice of Medicine" (B. L. Segal and D. G. Kilpatrick, eds.). The Williams and Wilkins Co., Baltimore, 1967.
21. Sevelius, G. (with 17 authors) "Radioisotopes and Circulation." Little, Brown, Boston, 1965.
22. MacKay, R. S. The potential for telemetry in biological research on the physiology of animals and man. *In* "Bio-Telemetry Proceedings of the Interdisciplinary Conference" (L. E. Slater, ed.). Permagon Press, London, 1963.
23. Slater, L. E. What's ahead in biomedical measurements? *J. Instr. Soc. Amer.* **11**, 55-60 (1964).
24. Makow, D. M., and McRae, D. L. Symmetrical scanning of the head with ultrasound using water coupling. *J. Acoust. Soc. Amer.* **44**, 1345-1352 (1968).
25. Robinson, D. E., Garrett, W. J., and Kossoff, G. Fetal anatomy displayed by ultrasound. *Invest. Radiol.* **3**, 442-449 (1968).
26. Garrett, W. J., and Robinson, D. E. Fetal heart size measured *in vivo* by ultrasound. *Pediatrics* **46**, 25-27 (1970).
27. Baker, D. W. Pulsed ultrasonic Doppler blood flow sensing. *IEEE Trans. Sonics Ultrasonics* **SU-17**, 170-185 (1970).
28. Stegall, H. F., Kardon, M. B., Stone, H. L., and Bishop, V. S. A portable simple micrometer. *J. Appl. Physiol.* **23**, 289-293 (1967).
29. Guntherorth, W. G., and Mullins, G. L. Liver and spleen as venous reservoirs. *Amer. J. Physiol.* **204**, 35-41 (1963).
30. Lategola, M. T. Ultrasonic echocardiography: a new vista. *J. Okla. State Med. Assoc.* **59**, 208-212 (1966).

Chapter 7

MEDICAL AND BIOLOGICAL APPLICATIONS
OF MODELING TECHNIQUES

Lee L. Huntsman and Gerald H. Pollack

INTRODUCTION

The formulation of models in order to better understand systems is a technique that is enjoying increasing use by investigators in the physical and biological sciences. In its broadest sense, modeling is an ubiquitous technique employed to help visualize concepts or interactions. However, the term "model" usually refers to a mathematical expression that is used to "simulate" a complex system. Such a model is a powerful tool because it can be used for controlled and quantitative analysis.

Well-formulated models have two key features which make them extraordinarily useful to investigators. First, models are simpler than the reality which they simulate and are therefore more tractable. Second, models are open to arbitrary manipulation and change, which is not necessarily possible for the real system. The first, for example, means that formulation of a model requires decisions about what characteristics of the system are really important in determination of the behavior being studied. The second means that the model can be tested, changed, retested, described.

Models prove to be useful both in design and analysis. Much of the modeling by engineers is intended to evaluate designs before actual construction in order to anticipate problems or discover unforeseen behavior. This application is now coming to be important in the health care system as engineering practices are increasingly applied to system design and to certain types of health care. Within the realm of biological investigation, however, models are used primarily as tools of analysis.

It becomes appropriate to move from reliance upon conceptual models to more formal ones whenever the mind boggles at complicated interactions. An obvious example is the analysis of a complex system. Formal models are also needed for simple systems whenever the need for quantitation becomes important. For example, the transition from static to dynamic analysis is often sufficient to necessitate some form of model. Understanding the relationship of subsystem behavior is another area in which models can become potent tools.

Though they may be formulated many different ways, most models are basically mathematical representations. As such, the considerable power of mathematics may be utilized. The extent to which one can make assumptions and approximations that will allow mathematical models to be solved analytically and yet remain useful is surprisingly great, even in the description of biological systems which are typically nonlinear. Yet modern simulation techniques employing digital and analog computers allow the model builder to circumvent the requirement that the model be one for which there is an analytical solution. This fact, together with the convenience of computer techniques, allows modeling to be done in response to needs for understanding and quantitation without undue consideration for the techniques of solution.

Warner has given a succinct statement of the steps in simulation of a biological system (1). They are

1. Review of present concepts and existing data
2. Construction of block diagram and system equations
3. Realization of model on appropriate media
4. Comparison of behavior of model and biological system
5. Modification of model (step 2), or extrapolation and design of new experiments (step 4)

The steps are essentially similar when modeling is used to develop an optimum design for a system which is to be constructed. A key feature of the sequence is its iterative cyclic nature (step 5). Note that the iteration can be either to improve the model or to improve biological understanding through further experimentation. For a design process this would correspond to modifying the design or altering the original design criteria. Fruitful results do not appear in this sequence solely at the end, but rather all through it. The simple mechanism for getting specific about one's thoughts (step 2) is often sufficient to expose inconsistencies and encourage innovative thinking.

A particularly important step in the sequence is step 4. Unfortunately, modeling a biological system has sometimes been done by people who felt no compunction to rigorously test their model against reality. If experimentalists were standing ready to test any model, this might be acceptable. However, unproven models are seldom picked up by others so that the model builder must assume responsibility for comparison through either direct experimentation, collaboration with experimentalists, or at least knowledgeable use of the literature.

A number of examples have been selected to illustrate the techniques and usefulness of modeling for the understanding of biological and medical systems. They have been specifically chosen to cover a wide range of complexity from single tissues to regional health systems.

A MEDICAL SERVICE REQUIREMENTS MODEL

Over the past decade, there has been rapid development of techniques for modeling large and complex systems which must by nature be described statistically. Developed principally by industry and the military to aid in quantitative understanding of such systems, these techniques are particularly useful in the prediction of probable responses of the system to a change or a perturbation. Only recently have such techniques begun to be applied to health care systems. Applicability to medicine is rather obvious because of the size of the systems involved and because statistical techniques are required in order to accommodate the unique constraints upon medicine by virtue of the fact that it is practiced by a diverse group of individual physicians.

A simple example has been chosen to illustrate the potential power and usefulness of such techniques when applied to the large scale aspects of health care. This model seeks to predict the demand for medical services within a specified population on the basis of health statistics for that population and medical judgments of services needed for specific disease states (2).

In order to achieve much-needed improvements in health system planning and design, such predictions are vital. This is especially true since the services studied, such as laboratory tests, special therapy, bed rest, and monitoring account for more than half the cost of hospital medical care. Presently the physician orders services according to the condition of the patient, but he is constrained by those services that are available and is often influenced by outside factors such as insurance coverage. The health system, on the other hand, seeks to provide needed services but often must simply expand services along traditional lines or in response to immediate needs. The result can be inefficient planning and duplication of services.

THE MODEL

In essence the investigators have modeled "the behavior over time of a pool of patients with medical conditions requiring health care services." Members of the population enter the pool when they seek medical care and remain in the pool so long as they require medical services. Each patient is classified according to disease and each disease is described by a branched array of disease states as illustrated in Fig. 7.1. These states are distinct in that they each require a unique set of services. For each state the average residence time is assigned as well as the probabilities for transition into the subsequent states. When coupled with a description of the disease distribution in the population, the model yields a statistical picture of the patient census of each disease and disease state. Since the service requirements for each state are known, the total demand for each service is known as well as the distribution of the demand among the different diseases and categories of patients. This information allows the development of possible strategies for optimizing the delivery of a particular service. By varying the description of disease in the population which is the input to the system, variations of service requirements, possible overloads or inefficiencies, etc., can be analyzed.

In formulating this model the investigators have had to develop adequate but manageable descriptions of three components of the model; (a) the population served by the health system, (b) the diseases and disease states which occur, and (c) the services which the health system is required to perform. All these descriptions were obtained largely by extensive personal interviews with physicians, other health professionals, and administrators. A description of the population was based upon available statistics. However, it is important to note that the model classifies incoming patients according to the disease they ultimately are shown to have, not according to the symptoms which they present. This is vital to keep the model manageable, since physicians will describe quite clearly the care required by a set of symptoms. With respect to disease, an important concept of the model is that a patient with a certain illness will be put through a few distinct care regimens corresponding to major stages of the disease during his illness. These regimens, requiring distinct services, are the disease states. The broad application of such regimens despite difference between individual patients allows the model to remain tractable. A specific regimen described for a patient will vary widely, but each regimen is composed of an identifiable set of tests and procedures, many of which are also common to other regimens. Thus the investigators have identified a "set of care elements

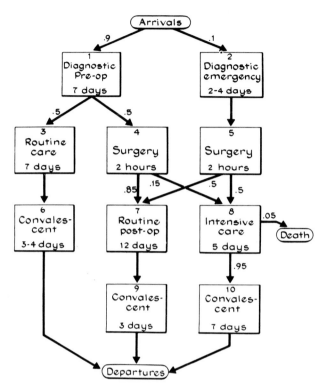

Fig. 7.1. Example of disease dynamics process for a hypothetical disease. [From Smallwood *et al.* (2).]

which are unit actions that the service system produces to form the care regimen for a patient." Importantly, these care elements are so fundamental as to be independent of their setting (hospital, office, clinic, etc.) provided sufficient flexibility has been designed into the model to allow additions and deletions of care elements as well as modifications of the input.

As an example of the application of this model, consider a hospital unit which supplies the services of blocks 3, 7, and 8 in Fig. 7.1. Suppose that the rate at which patients enter the health system with this disease is described by a Poisson process with an average of one per day. Then it can be shown that the number of patients in each disease state is statistically independent of the number of patients in any other state and that the average census in this unit will be about nine patients with a standard deviation of three. Similarly, if the survey of physicians had led to an estimate that the mean and mean square number of X rays for each of these states are 1, 2, and 4, and 1.5, 4.5, and 16.7, respectively, then the model would predict an average demand for X rays in this unit of sixteen per day with a standard deviation of six. While this example is hypothetical, it does illustrate the ability of models like this to generate quantitative information which can be extremely useful for hospital or regional health care planning.

COMMENT

Smallwood and his colleagues have put forward what seems to be a very sound approach to the prediction of health service requirements. This is true not only in the sense that the model is well conceived, but also more generally in that the area of medical service requirements is an appropriate one for modeling. The potential of this approach lies in the fact that it transcends present means of meeting health service requirements, and deals directly with requirements themselves. At the level of a single institution such as a large hospital, this means that administrators can begin to separate actual needs from the appearances given by existing structures, and seek to optimize the means of delivering the required services at the level of populations. As emphasized by the investigators, the model has great usefulness for long-range planning since adequate planning necessitates quantitation.

It is important to take note of those capabilities of the model which arise by virtue of its concern with service requirements rather than the nature of currently provided services. Statistics about current services are conditioned by availability, convenience, tradition, organizational structure, and many other influences in addition to more purely medical considerations. By appealing directly to medical opinion about the services actually needed, the investigators have circumvented these limitations to some extent. Another noteworthy feature of the model is that it allows the problem to be framed in probablistic terms, an approach which enhances its usefulness for long-range planning, since modifications and allowances for contingencies can be incorporated easily and quantitatively. In view of the apparent necessity for coming to grips soon with regional planning for health services, modeling efforts such as this are sorely needed.

AN *AD HOC* MODEL

An excellent example of a simple, successful model which has been designed specifically for generation of clinically useful information is that developed by Warner and his colleagues (3). Through several simple equations Warner has managed to leapfrog much of the terrifying complex analysis of arterial pressure-flow wave transmission which has preceded him, and has arrived at an entirely tractable model by which he claims to be able to calculate cardiac stroke volume from a single arterial pressure measurement.

The objective of Warner's model was to permit the acquisition of information which could not be obtained through current measurement techniques. No means are presently available by which beat-by-beat stroke volume may be measured in a clinical situation. The model then is simply a transfer function which grinds out one parameter—stroke volume—from input information concerning another parameter—central aortic pressure—which is easily measured with an arterial catheter.

THE MODEL

The model is implicitly a three component model, consisting of the heart, arterial tree and the periphery. The extent of the arterial compartment is defined as the aorta and large arteries from the aortic valve to a point downstream as far as the pressure wave will travel in 80-msec, a convenient definition. The peripheral compartment represents the remainder of the smaller arteries and the arterioles and capillaries (Fig. 7.2). The objective of calculating the volume of blood ejected from the heart to the arterial compartment is achieved indirectly by calculating the amount of blood flowing from the arterial compartment to the peripheral compartment. In a steady state situation the two values must naturally be equal over a full beat.

Flow into the periphery is broken into volume transmitted in two time intervals, diastole and systole. During diastole the calculation is easy since flow out of the arterial compartment can go in only one direction, and a change of volume in the arterial section is equal to the volume transmitted to the periphery. For an elastic body such as the arterial tree, the assumption is made that the volume in this compartment is proportional to the square root of the distending pressure, not an unreasonable assumption considering the shape of typical arterial wall stress—strain curves. Then it is necessary simply to know the mean arterial pressure at the onset of diastole and at the end of diastole in order to calculate the volume reduction in the arterial compartment and thus the flow

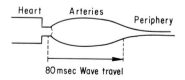

Fig. 7.2. Three compartment model of the systemic circulation used to evaluate stroke volume.

into the periphery during this interval. The measurement of pressure near the aortic valve coupled with the assumption of the 80-msec wave travel to the end of the compartment provides enough information to calculate pressure at the end of the compartment. Then average pressure throughout the arterial compartment can be calculated through diastole, and hence volume released to the periphery.

During systole, where there is inflow and outflow in the arterial section, the calculation requires a different approach. In this case the chief assumption is that flow out of the arterial compartment to the periphery is proportional to the exit pressure. Once again, by knowing the time course of pressure at the entrance to the arterial section, it is possible to calculate the pressure as function of time at the exit, so that the relative values of volume transferred during systole and diastole (proportional to relative exit pressures) can be calculated.

The final result is an equation which states that stroke volume is equal to a constant K times some integral function of central aortic pressure, a function which is sufficiently complex to require a calculation by digital computer, but which nonetheless is a function of pressure alone. Absolute values of stroke volume by the dye dilution technique permit a calibration of this function. But even if K is not calibrated—avoiding the necessity for dye dilution measurement—changes of beat-to-beat stroke volume can be ascertained simply by looking at this function of central aortic pressure and ignoring the absolute value of K.

Various studies and much experience have proved this technique to be a valid one. Despite threefold changes in peripheral resistance, and/or threefold changes in heart rate, good correlations have been found between stroke volume obtained using this technique and that obtained through dye dilution or electromagnetic flowmeter techniques.

This particular model is built upon a series of assumptions to which purists may legitimately object. The arterial tree, for instance, is broken into two segments, distensible large arteries and resistive smaller vessels. While this appears to be a rather arbitrary distinction *a priori,* such distinction has been employed traditionally in the form of a so-called "Windkessel" model and quite successfully at that. Its basis is simply that the larger arteries are characterized predominantly by their elastic properties, while the smaller vessels are less distensible and may be considered purely resistive in nature, i.e., flow is proportional to the pressure difference across the beds. But the limitations of accuracy imposed by such a simplification must be kept in mind.

There is also the assumption that the pressure wave travels at a fixed velocity without dispersion. Although other studies have shown that the arterial wall properties and the reflection of waves at bifurcations cause different components of the velocity wave to travel at different velocities, it is not clear that the errors introduced by assumption of constant velocity are of sufficiently large magnitude to void the model. Indeed the success of the model suggests just the opposite.

Finally the assumption that the volume is equal to the square root of pressure appears feasible at first but certainly may not apply in all cases, especially for those subjects whose arteries are under unusually high or low stress.

COMMENT

This model is of the *ad hoc* variety, designed specifically to obtain one piece of information. As such it represents a complete departure from the pathway paved before it. From the time of the famous mathematician Euler, the pulsatile properties of arterial systems have been investigated. Modern mathematicians such as Witzig and more recently Womersley in the early nineteen-fifties have expanded the analysis of arterial wave travel and provided some very sophisticated tools by which arterial properties could be investigated. But not surprisingly virtually none of this sophisticated information has found its way into the clinic. Instead clinical usage of circulatory models has either followed an entirely different pathway, such as Warner's, or has found itself at a dead end.

BLOOD GLUCOSE DYNAMICS

The system which controls the concentration of glucose in the blood is among the most important of the body's control systems because glucose is essential to the metabolism of many tissues. Although the control is "loose" in that glucose concentration can vary over a considerable range under normal conditions, it is of considerable medical importance because failure leads to pathologically high or low concentrations. Clinically it is necessary to assess the dynamic behavior of this control system, not just its static form. This is usually done by simulating normal perturbations with intestinal or intravenous loads of glucose or insulin. When the results of such tests are interpreted, the great complexity of the system makes diagnosis of a marginally abnormal system or localization of a malfunction within the system very difficult.

Application of modeling techniques to the control of blood glucose dynamics has proven very useful in coping with the complexity of the system. The particular model developed by the team of Gatewood, Ackerman, Rosevear, and Molnar has been extensively evaluated and so is a good example of the contributions of models to understanding of this system (4,5).

THE MODEL

Some indication of the known complexity of blood glucose regulation is shown in Fig. 7.3. Continued physiological study has revealed that even this may be a relatively simple approximation to the real system. Fortunately many of the relationships in Fig. 7.3 are usually of secondary importance so that a quantitatively useful description is possible with a simpler model. The form of the model used by these investigators is shown in Fig. 7.4.

This model is focused around the dynamics of two concentrations: glucose in the blood (G), and effective hormone in the blood (H). In the figure, solid lines indicate the flow of these substances between compartments, while dotted lines show mechanisms of control of these flows. The m's are activity coefficients, while J represents the addition of glucose to the blood via the intestine or intravenous injection, and K (not shown) represents the exogenous addition of insulin. By assuming the simplest linear relationships among the variables, and substituting $g = G - G_0$ and $h = H - H_0$, where G_0 and H_0 are the fasting levels of G and H, the model in Fig. 7.4 can be described by

$$\frac{dg}{dt} = -m_1 g - m_2 h + J$$

$$\frac{dh}{dt} = -m_3 h + m_4 g + K$$

Fortunately it turns out that one can go a long way toward the useful quantitative understanding of this system even with such a simple approximation to the complexity of Fig. 7.3.

When modeling the response to the glucose tolerance test, a rapid input of glucose, intravenous or intestinal, must be described. Depending upon the type of input, the accuracy required, and the technique for solution of the model, the authors have used pulses, exponentials, delta functions, and other forms to describe J. Except in the case of insulin infusion tests, K is 0.

Experience with comparison of this model to patient data demonstrates that G and g relate to the arterial blood glucose which is approximately the same as the venous glucose concentration usually sampled. The situation is not so simple for H, since H_0 seems to be a composite of several hormones. The dynamic component h, however, is related almost entirely to insulin changes, at least for the first several hours after a glucose load.

While the nature of the assumed model is specified by the equations above, some determination must be made of constants G_0, H_0, m_1, m_2, m_3, and m_4. This has been done by an iterative process of fitting the model to patient data. This process has demonstrated that the simple model will describe the dynamics of glucose and that some of the model parameters are characteristic of individual patients. The ultimate test of the model comes when constants chosen by

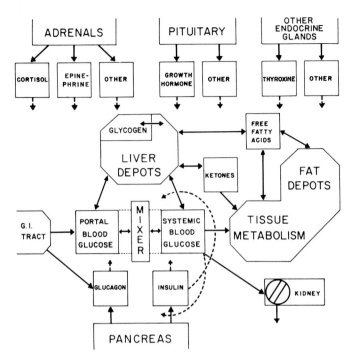

Fig. 7.3. Isomorphic model of the blood glucose regulatory system. Squares represent blood concentrations; solid lines indicate exchange; dashed lines indicate rate modification. [From Ackerman *et al.* (4).]

conformation of the model to one test are frozen and the model then compared without change to the results of another test on the same patient.

The fact that the model fits the usual response to the glucose tolerance test has allowed it to be used in the search for a single parameter which best indicates the normality or abnormality of the patient's response. While some parameters turn out to be very sensitive to sampling errors, it was found that the natural frequency, ω_0, defined by $\omega_0{}^2 = m_1 m_3 + m_2 m_4$, exhibited a low sensitivity to error but was a useful index of mild diabetes.

A number of additional comparisons of the model with patient data have been carried out in order to demonstrate its usefulness and define the limits of its applicability. Continuous measurements of blood glucose every 1.5 minutes for 5 hours indicate that the fit of the model is genuine, and not a result of the usually much slower sampling. Similarly, published data on the dynamics of immuno-reactive insulin following glucose load were shown to agree with model predictions. Attempts to fit the model to data from rapid intravenous glucose infusions failed, however, during the period of infusion. This revealed the

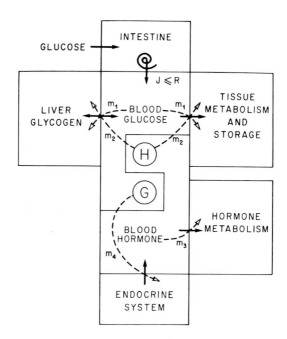

Fig. 7.4. Basic model of the blood glucose regulatory system. [From Ackerman *et al.* (4).]

limitation of an assumption inherent in the model, namely that the blood compartment is a single uniform volume. The fact that the model did conform to the data after the initial infusion period indicates that the assumption, and therefore the model, becomes tenable after such transient effects are over.

This model has also found application as a simulation tool used to perform experiments which are difficult or costly to perform with patients. For example, the manner in which the shape of the glucose input function J affects the response to a fixed total glucose load has been studied with the following interesting result.

1. If J was maximum in less than ½ hr and insignificant in less than 2½ hr the time course of G was unaffected.

2. If J declined slowly, G followed the decline of J and therefore did not indicate the true characteristics of the control system.

In the same way, it was found that only if the hormonal mass action constant m_3, and the decay time-constant of J are approximately the same will the glucose response show the often-observed damped sine wave type of response. However, for individuals in whom this equality does not apply, the oral glucose test does not offer a valid test of their regulatory system. All of these results indicate that the form of J, the manner in which glucose is absorbed from the intestine to the blood, can have a profound effect on the observed variations of blood glucose.

COMMENT

The model of blood glucose regulation which these investigators have formulated is an excellent example of the versatility and usefulness of a well-conceived and simple model. It is interesting that they apparently have managed to keep clearly in mind the original goal for which the model was developed. This was ". . .to design a simplified, lumped parameter model suitable for use in studies involving the oral glucose tolerance test." An especially noteworthy aspect of this modeling effort is the course of the model development. Starting with a goal calling for the simplest useful model, assumptions were made and identified, careful comparison of the model to experimental data was achieved, additional experimental designs were developed, model simulation of unavailable experimental conditions was used to broaden understanding, and the refinement of the model was carried out a step at a time, constrained by the original requirement for simplicity.

MODELS OF CARDIAC MUSCLE MECHANICS

An example of a group of controversial models which have stimulated wide interest are those purporting to represent cardiac muscle mechanics. These models are designed to serve primarily as a basis for gaining insight into contractile mechanisms. The theme is compartmentalization of the muscle behavior into several distinct mechanical components so that the effects of various mechanical perturbations or drugs may be attributed to one or another component. In this way the properties of each component and hence those of the muscle can be deduced. The concepts learned through this process can then be applied to the intact heart for use in the clinic.

The compartmental approach to modeling of cardiac muscle is, atypically, an historical one. When investigations of isolated cardiac muscle began intensively a decade ago it was thought that the similarities between cardiac and skeletal muscle were sufficient that the same approach to modeling skeletal muscle might also be applied to cardiac muscle. In this way the half-century required for skeletal muscle models to evolve to their present sophisticated form would not require repetition; the sophistication could be applied immediately to a parallel situation, thereby avoiding the stumbling blocks and inevitable time delays.

For example, in the case of skeletal muscle, a blatantly incorrect model flourished for years before the proof of its irrelevance was generally accepted. On the basis of data obtained from relatively primitive experiments early in the century it was originally thought that muscle could be represented by a spring-like element bathed in a viscous medium. This was called the "viscoelastic model" (Fig. 7.5). Stretching the muscle corresponded to storing potential energy in the elastic element. Active contraction was brought about by an abrupt increase in the elastic modulus of the elastic element, hence an increase of this stored potential energy. The release of this energy resulted in the tendency for the muscle to shorten.

Fenn proved in 1924 that this model was thermodynamically untenable, but Hill, who was its foremost proponent (7) and who accepted a Nobel prize in part for it, currently admits with some chagrin that the viscoelastic model had such a strong foothold that it took an unconscionable time beyond the appearance of Fenn's work for it to finally expire (8). Even today we are occasionally haunted by its ghost, for the appeal of considering muscle as a viscoelastic structure with phasic variations of elasticity and viscosity seems sometimes to outweigh the knowledge that this approach lacks physiologic significance.

It was not until 1939 that Hill (9) published his monumental work in which he characterized the muscle by an active contractile component (CE) in

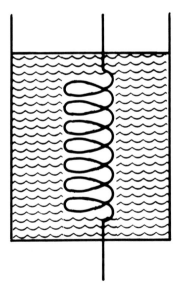

Fig. 7.5. The "viscoelastic model" conceived of muscle as a spring shortening in a viscous medium. It was determined to be invalid a half-century ago.

series with an elastic element (SE) replacing the viscoelastic concept and accurately predicting the mechanical behavior on the basis of thermodynamics (Fig. 7.6). With this, Hill ushered in the modern era of muscle mechanics.

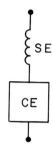

Fig. 7.6. The two-element model devised by A. V. Hill to represent muscle mechanisms. This model is widely used today.

Hill's approach is the jump-off point for models considered by cardiac muscle physiologists today. Although various elements have been added and changes suggested, the basic model and its attendant assumptions remain intact. In this model the contractile element is thought to be capable of generating power and exerting contractile force through (stretching) the series elastic element. This power is generated by the CE in the form of a trade-off between load force and shortening velocity: If the load on the muscle is small, the CE shortens with high velocity; if the load is high, the CE shortens with low velocity. Common experience confirms this; a small weight can be lifted rapidly while a larger weight cannot be lifted quite as rapidly.

After a decade of attempts to fit experimental data from cardiac muscle strips to this type of skeletal muscle model, the results remain inconclusive. Perhaps the only element whose properties are generally agreed upon are those of the SE. Data from many laboratories concur that the series elastic element has a stress-strain curve in the form of a "hard" spring, that is one whose elastic modulus increases as it is stretched. They also agree that the value of this elastic modulus appears relatively insensitive to various drugs and inotropic agents, suggesting that an anatomic basis for this element might indeed be a passive structure located within the muscle, perhaps totally independent of the contractile mechanism.

Other aspects of these models remain clouded. Although it is generally agreed that a parallel elastic element, or PE (Fig. 7.7) must be appended to the model in one form or another to represent the resistance to stretch of the unstimulated muscle, the stress–strain curves of this element differ depending on how they are measured. The force–velocity characteristics of the contractile element are found to be either hyperbolic—as in the original Hill model—or nonhyperbolic depending on the mode in which they are measured. The time course of the active state of the contractile element has either a slow or rapid onset, depending once again on the mode of measurement.

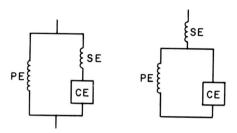

Fig. 7.7. The adaptation of the Hill model to cardiac muscle. A parallel elastic element is added to represent resting muscle elasticity.

In order to obtain better fit for their experiments some investigators have suggested tacking on extra elements to the basic framework. Some have proposed viscous elements in various arrangements to account for stress relaxation, diastolic creep, and changes of diastolic compliance. Others (10) have gone so far as to propose a complex multielement model (Fig. 7.8) in which the properties of each element could be determined in the future by extensive experimentation. Such complex versions of the Hill Model have not yet undergone serious study.

The current status of cardiac muscle models may be aptly characterized as a state of confusion. There are several possible explanations for this. First, there are some assumptions which carry over with the model from skeletal muscle such as the assumption that the resting elasticity of muscle can be considered to reside in a third, distinct, parallel elastic element rather than as part of the contractile element. Another possible basis for the difficulty may be that the CE-SE type of framework is not a reasonable one to begin with. It is possible that differences in properties of individual fibers within cardiac muscle and skeletal muscle and their geometric arrangements do not permit the extrapolation of a skeletal muscle model to cardiac muscle. For instance, it is well known that skeletal muscle fibers generally run parallel to one another, while cardiac fibers have been reported to be buckled or curled rather than straight. In addition, there is much inhomogeneity. These factors alone could preclude the use of such a model as a basis for interpretation of cardiac mechanics.

Apparently in the case of cardiac muscle the "historical" approach to modeling has not proved as fruitful as might be hoped. Few new insights have presented themselves through this vehicle. Rather, much effort has been expended trying to modify the basic model to fit the experimental data. By focusing so intensively on these models there has necessarily been a de-emphasis of the pursuit of new techniques for investigating muscle properties, thereby compounding the frustration. Moreover, these partially accurate models have served as a convenient but shaky crutch upon which further work has rested. For

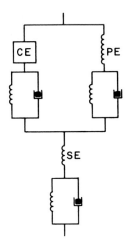

Fig. 7.8. One attempt at modifying the original two-element Hill model in order to fit cardiac muscle data.

example, considerable effort has been expended on deriving indices of contractile performance based on them. This work has proceeded despite the admittedly poor fit of these models.

One might rightly ask whether the ten years spent exploring these models has led to further understanding of cardiac muscle mechanics.

CONCLUSION

These examples of modeling applied to biological and health related questions, though only a few and limited in scope, serve to illustrate the ubiquity and usefulness of this approach. Some cautionary points are also evident, however. For example, the most complex or complete model is not necessarily the most useful one when one's objective is to increase understanding of a system, or to accurately mimic it in a specific situation. The modeling process should include at the beginning specifications of what an "acceptable" model would be. These specifications may well change with time, but they must be consciously and periodically determined. Another cautionary point is that the limitations under which the model was formulated must be kept in mind. Often misapplication of a model occurs because of a failure to appreciate the significance of the assumptions that were made, or because of a premature acceptance of the model. An ever-present danger is that a model—conceptual or formal—may prove to be a barrier to new insight and creative approaches because it has gained acceptance without adequate verification.

With proper allowance for its limitations, however, modeling provides a powerful tool for improved qualitative and quantitative understanding. As discussed in the introduction and illustrated in the examples, the development of models which improve quantitative understanding, even of simple systems, can be especially useful when dynamics are of interest, while for more complex systems, a model may be essential for even qualitative considerations. However, the applications of models extend beyond studies leading to "understanding" of the system being modeled, since they can be used to quantitatively summarize what is known about one system's behavior in order to understand or measure another system. This significantly extends the usefulness of models in diagnosis and therapy where a proven model may be applied in order to obtain clinically useful results. The blood pressure and glucose metabolism models described above illustrate this approach. An important aspect of improved understanding is the ability to identify the "hierarchy" of parameters in the system of interests. The usefulness of this for diagnosis is suggested, but the approach is also vital for the development of replacement therapy. This is especially so because in the design of prostheses it is often not possible (or necessarily desirable) to develop an exact replacement for the biological system. Therefore we must usually decide which of the natural systems' several functions are most important to simulate.

References

1. Warner, H. R. Simulation as a tool for biological research. *Simulation* **3**, 57 (1964).
2. Smallwood, R. D., Murray, G. R., Wilva, D. D., Sondik, E. J., and Klainer, L. M. A medical service requirements model for health system design. *Proc. IEEE* **57**, 1880 (1969).
3. Warner, H. R., Gardner, R. M., and Toronto, A. F. Computer based monitoring of cardiovascular functions in postoperative patients. *Circulation* (**Suppl.**) **37(II)**, 68-74 (1968).
4. Ackerman, E., Gatewood, L. C., Rosevear, J. W., and Molnar, G. D. Blood glucose regulation and diabetes. *In* "Concepts and Models of Biomathematics: Simulation Techniques and Methods" (F. Heimets, ed.), pp. 131-156. Dekker, New York, 1969.
5. Gatewood, L. C., Ackerman, E., Rosevear, J. W., and Molnar, G. D. Modeling blood glucose dynamics. *Behav. Sci.* **15**, 72 (1970).
6. Fenn, W. O. A quantitative comparison between energy liberated and the work performed by isolated sartorious muscle of the frog. *J. Physiol.* **58**, 175-203 (1924).
7. Gasser, H. S., and Hill, A. V. The dynamics of muscular contraction. *Proc. Roy. Soc.* B **96**, 398-437 (1924).
8. Hill, A. V. "First and Last Experiments in Muscle Contraction." Cambridge Univ. Press, New York, 1970.
9. Hill, A. V., The heat of shortening and dynamic constants of muscle. *Proc. Roy. Soc.* B 126-136 (1939).
10. Parmley, W. W., Brutsaert, D. L., and Sonnenblick, E. H. Effects of altered loading on contractile events in isolated cat papillary muscle. *Circ. Res.* **24**, 521-532 (1969).

Chapter 8

BIOMECHANICS AND BIOMATERIALS*

The structure and function of the body components at gross and microscopic levels have been of prime interest to anatomists for many centuries. In common with other disciplines of biology and medicine, the prodigious efforts of this army of anatomists have resulted in a vast storehouse of information which is almost completely descriptive and subjective in character. Many studies of the stress–strain relations of various tissues have been conducted by physiologists with a principal focus on very few types of tissues, notably contracting striated muscle, smooth muscle structures, and arterial samples. A very large proportion of this mass of data is of little value from the engineering point of view because so many of the measurements have been limited to static conditions. A rapid convergence of interest by materials scientists and engineers is responding to the crucial new requirements for biomaterials. For example, optimal mechanical properties of an implant material can be specified only with adequate knowledge of the material properties of the tissues with which it comes in contact. Biomaterials implies development, evaluation, and application of special substances to meet the specifications for research and practice in biology and medicine, with particular reference to materials which come into contact with tissues.

*Portions of this chapter are reprinted with permission from Rushmer, R. F., Biomaterials; an essential ingredient in bioengineering. *Mat. Res. Std.* **10**, 9-13 (1970).

Materials scientists can provide a wealth of quantitative information about materials tested under various conditions, but application of this information is difficult without corresponding information about the living tissues. The enormous number of voids in our knowledge of dynamic properties of tissues came to light when the biologists found they were totally unable to provide specifications for the various moduli of tissues when stressed or flexed in various directions and at different rates. The relevance of physical measurements which have been collected previously on tissues is highly questionable since a very large proportion has been conducted on specimens excised from their normal environment, destroying the normal relations and imposing serious functional disturbances. The mechanical and physicochemical properties of tissues are most in need of elucidation where there are new applications of nonliving materials as tissue substitutes in the human body.

The medical community has been singularly blind and unresponsive to the need for specialized materials to meet the demanding specifications of medical diagnosis and therapy. Few materials have ever been produced specifically for medical applications. In contrast the dental profession has maintained a thirty-year relationship with the National Bureau of Standards and derived significant benefits in terms of new materials and machines. A corresponding focus of attention on materials science has not been developed by medical investigators or practitioners. Now that many essential projects are being delayed and arrested by inadequate specifications and unfulfilled requirements for appropriate materials, we can confidently anticipate explosively expanding fields of biomechanics and biomaterials. In this sense biomechanics may be defined as the quantitative and dynamic description of the physical properties of tissues.

The diversity of requirements for materials for replacement or supplementation of tissues and functions in the body is indicated in Fig. 8.1. As an extreme example, artificial skin would be most useful for reconstructive surgery in a large number of individuals from among the 4,000,000 people with injuries from highway accidents and the 2,233,000 people suffering burns each year. Replacements for clouded corneas require optical transparency, durability, and tissue compatibility as prime requisites. The physical properties required of materials are quite different for reconstructing noses, ears, cheeks, chins, and breasts. The tough, durable, and flexible materials are required to take the pounding imposed on heart valves opening and closing more than 100,000 times each day. Inert tubes to serve as replacements for a variety of ducts and channels are in demand as substitutes for segments of the digestive tract, circulatory system, glandular ducts, and urinary tract. Replacement materials or substitutes for tendons must have great tensile strength with firm attachments at both ends and free sliding motion between. Structural materials are needed for implants as substitutes for cartilage in the ear or nose. Appropriate materials for reliable intrauterine contraceptive devices are required to help control the population explosion. The development of artificial bones and joints is being retarded by a lack of appropriate materials. Also, artificial internal organs (kidney, heart, lung), and pacemakers) and artificial extremities all require specialized materials which are currently unavailable. Not the least of these requirements is one for compact energy sources to provide power for various functions, such as artificial hearts, pacemakers, and prosthetic devices.

BODY-BUILDING BLOCKS

Each tissue or functional unit in the body consists of cells which are held together by meshworks of long-chain fibers in various orientations and combinations. Groups of cells are generally contained within envelopes of fibrous connective tissue through which the arteries, veins, and lymphatics course to come into close proximity to the individual cell clusters. Tissues are enclosed

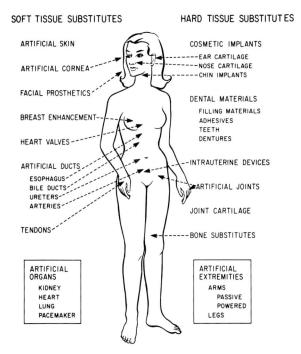

Fig. 8.1. The diversity of materials requirements for substitutes or supplements for either soft tissues (left) or hard tissues (right) is illustrated schematically. Artificial internal organs and artificial extremities represent many additional unmet needs for specialized substances.

and confined in a geometrical relation to other tissues and organs by strands and sheets of connective tissue. The supporting tissues, which hold the tissues together, maintain spatial relations between organs, support the weight of the body (bone and cartilage), and serve as the packaging material are composed of various combinations of a very few building blocks (Table 8.1). The most common supporting material is collagen which is exceedingly tough and is found in fairly pure form in scar tissues and tendon. Elastic fibers provide both flexibility and extensibility. The skin is covered by an inert layer of keratin that serves to protect the delicate cells of subcutaneous tissues from the rigors of the hostile external environment. Most of the spaces between cells are occupied by a loose gel structure with a framework of mucopolysaccharide. An astonishing array of supporting tissues are made up from these few common long-chain building blocks.

COLLAGEN: NATURE'S UNIVERSAL STRUCTURAL BIOMATERIAL

When nature conducts an eminently successful experiment, a human effort to compete is just a bit presumptuous. Collagen is a case in point since it proved such a successful supportive material early in the evolutionary development of lower forms (e.g., the sponge and jelly fish). In the intervening millions of years, it has persisted as one of the most common and consistent materials to be found in the bodies of animals with little change in basic structure or composition. If there is one material which confines and supports the tissues of animals of all sizes and complexity, that material is collagen. It is a protein with a characteristic composition of amino acids formed into a triple helix of three long peptide chains entwined around one another like a twisted rope. Although the basic structure is similar, variations occur in detail between samples from different animals and within individual tissues. The structural unit of fibrous collagen is not the molecule but a regularly recurring sequence of molecules.

TABLE 8.1

Structural proteins	
Collagen	Highly ordered, tough fibrous proteins
Elastin	Crosslinked, Elastic fibrous protein
Keratin	Crosslinked, stiff fibrous protein
Ground substance	
Mucopolysaccharides and glycoproteins	Hydrophilic polymers filling the space between the fibrous structural proteins and cells
Mineral	
Calcium hydroxyapatite	Stiff mineral phase

Fibrils composed of long-chain, intertwined molecules constitute the major structural support of skin, tendon, ligaments, cartilage, and bone. Distribution of collagen has been estimated for mammals at 20% of the total protein in the body and about 5% of the total body weight concentrated in tissues with major mechanical functions. About 70% of the dermis of the skin or of tendon consists of collagen excluding water with which it is always associated. About half of the total collagen is in the skin and is the source of leather, one of man's earliest "industrial" materials. It is responsible for the tenacity of glue, the viscous quality of gelatin, and the toughness and suppleness of leather.

A great deal of knowledge has been assembled regarding the constituents of collagen (1,2). Glycine, one of the smallest amino acids, is also one of the most prominent. Proline is also common, forming about one-third of the residues. Most other amino acids have been found in varying concentrations in different situations. The molecular arrangement consists of three polypeptide chains entwined around each other like a three-strand rope, each with a twist in the opposite direction. The strands are held together primarily by hydrogen bonds between adjacent CO and NH groups and also by covalent bonds, with each molecule about the same length, about a thousand residues. Raising the temperature disrupts the helical arrangement and the individual chains separate and become soluble. (Collagen means "glue–forming.") Single chains, pairs, and strands of three can be separated. Two types of chain can usually be distinguished indicating that each molecule is composed of two chains of one type and one of another.

The molecular chains are found in most tissues in the form of fibrils ranging in diameter from 100–200 Å of fairly uniform diameter along the length. The fibrils characteristically demonstrate cross-banding regularly repeated at 600–700 Å along the length of the fiber as viewed by electron microscopy (Fig. 8.2A). For years this banding was regarded as an indication of the length of the molecule. However, collagen can be dissolved in acid and reconstituted spontaneously by neutralizing the solution (2) to produce bands at 2800 Å or four times as widely spaced. This collagen with "fibrous long spacing" (FLS) can be produced with almost 100% yield from purified collagen. Still another arrangement can be produced (without the characteristic beltlike appearance, emerging) as short segments (about 2800 Å long) consisting of threadlike units lined up in parallel. It is now believed that the collagen molecules are about 2800 Å long and are normally arranged in parallel array and overlapping about one-fourth their length to produce the banding at 700 Å. The regularly ordered structure is interspersed at intervals by "disordered regions" along the length of the fibers. Detailed knowledge of the molecular composition and fine structure of collagen is of the greatest importance in approaching the problem of developing materials of optimal characteristics for medical application on two counts—(a) Specially prepared forms of collagen may have great potential usefulness; (b) tissue reactions, compatibility, and binding of synthetic materials are closely related to the formation of collagen in production of scar tissue. The durability of collagen is dramatically displayed in the survival of shoes and other leather goods for centuries in graves, particularly those with very low humidity. Even more impressive is the persistance of the ultramicroscopic structure with cross banding displayed in an electron micrograph of collagen from the skin of a Ptolemaic mummy some 2000 years old (3). (See Fig. 8.2B.)

The natural process by which collagen is formed at the surface of fibroblasts remains highly conjectural. The fibroblasts evidently synthesize complete collagen molecules (triple strands) and extrude them into the space outside cells, where they polymerize into fibrils (Fig. 8.2C). For example, collagen from young animals can be dissolved in cold neutral salt solution and reconstituted merely by warming to body temperature. It is believed that this "youthful" collagen is not tightly aggregated. As the collagen ages, it dissolves in dilute acid and on further aging it becomes insoluble in acids. A similar "aging" process can be demonstrated *in vitro* with collagen extracted in cold neutral salt solution.

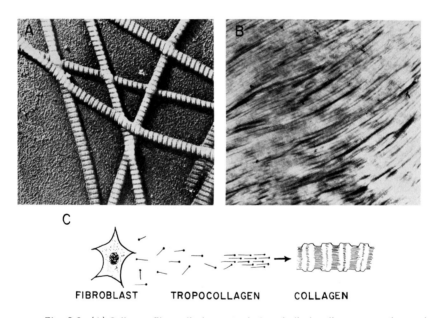

Fig. 8.2. (A) Collagen fibers display a typical periodic banding representing periodicity in the molecular arrangement of the component molecules as viewed by electron scan microscopy. (B) Period banding is visible in collagen fibers obtained from the skin of a Ptolemaic mummy some 2000 years old as examined by electron microscopy (prepared by Dr. D. D. Reichenbach). (C) The mechanisms have not been established by which fibroblasts fabricate tropocollagen which is, in turn, highly organized in to the many molecular geometries found in supportive structures (e.g., tendon, collagen, bone, corneas, etc.)

Reconstituted by rewarming it produces a gel which will immediately redissolve on cooling. After 24 hours it will not redissolve on cooling and after two weeks it will resist solution in weak acids. These experiments suggest that the collagen molecules extruded from fibroblasts must aggregate into the highly ordered patterns outside the cell and form into the sheets or parallel arrangements by forces or mechanisms which are currently obscure (see Fig. 8.2C). For example, it is not clear how collagen may form as alternate layers (like plywood) to form such an ordered structure that the cornea of the eye is as transparent as pure water. The sclera of the eye is also made of collagen arranged as a feltwork so that the structure is white and quite opaque. The dermis of the skin is a feltwork of fibers. In the walls of blood vessels the supportive layers of collagen intertwine about the smooth muscle cells which contract and adjust the caliber of the blood channel. Tendons are composed of collagen in parallel array. We have little concept of how these diverse forms of collagen arrangements are formed when extruded by the fibroblasts. The way in which the fibrous tissue is formed is basic information needed for a logical process of developing specifications for synthetic structural materials with appropriate properties and compatibility. In many different circumstances optimal supplements for supportive structures may depend upon developing artificial meshwork or foam of polymers into which connective tissues can be induced to grow in the most appropriate arrays to provide the necessary strength and durability. Ultimate success in producing optimal materials may depend upon developing proper combinations of synthetic plastics, specially reconstituted collagen, natural collagen, and living cells designed to produce supportive materials which bond tightly and permanently to tissues through normal tissue reactions.

ELASTIN

Elastin forms in long fibers from cells like fibroblasts but differs in many respects from collagen. The fibers are branched, yellow in color, and reversibly elastic so long as they remain moist (dry elastin is brittle). The distribution of elastin differs from that of collagen, having low concentration in skin and tendon, about equal to collagen in lungs and arteries, and representing about 80% of the material in ligaments (6). It confers the rubberlike elasticity to these structures and is capable of rapid extension under small stresses with prompt and complete retraction when the stress is removed. Despite the differences in physical properties, elastin has an amino acid composition very like that of collagen with proline, glycine, alanine, and valine comprising about 80% of the amino acids, in addition to small amounts of hydroxyproline. The yellow color is commonly attributed to two unusual amino acids, desmosine and isodesmosine. These and similar intermediate substances apparently form the bridges between the polypeptide chains of elastin in which almost 95% of the side chains are nonpolar. As a general rule, those tissues with substantial amounts of both elastin and collagen tend to display very large deformations with small stress, which are completely reversible owing to the elastin content. During this phase little of the stress is borne by the tortuous collagen fibers which become straightened out along the lines of force. When collagen fibers uncoil and align along lines of force, further distortion is strongly resisted by these strong fibers acting in parallel (5). (See also the section on skin, page 305.)

KERATIN

A fibrous protein is synthesized within the cells forming the skin, hair, and nails. Delicate fibrils appear in the dividing cells and thicken and coalesce to form fibrils which are visible in the light microscope. Granules of amorphous material appear in the cell layers and become associated with the fibrils to form the final product—keratin—which endows the skin with a highly protective covering.

MUCOPOLYSACCHARIDES

Although many diverse substances have been grouped with the mucopoly-saccharides, the term is usually applied to substances which contain uronic acid residues and hexosamine, common among them being the hyaluronic acid and the chondroitin sulfate. The structural formulae for these two acidic substances are quite similar and they tend to be bound firmly to protein in nature (6). These materials appear in small quantities in tissues, the highest content being in skin and bone (2.02 and 0.88 mg/gm, respectively), but they tend to trap water in their structure and occupy very much larger volumes than would be suggested by the concentration in the tissues. The mucopolysaccharides constitute hydrous gels usually called the ground substance. The gel structure contains protein fibers producing a material with appreciable tensile strength but readily deformable in response to pressures or stresses.

COMBINATIONS

In general, a more rigid structure is produced by chondroitin sulfate (e.g., in cartilage), and more flexibility is produced with hyaluronic acid intermixed with fibers of collagen and elastin. The solid rigidity of bone is attained by deposition of an inorganic crystaline material, hydroxyapatite, on a matrix composed of collagen and mucopolysaccharide. Various combinations of these five materials produce an amazing array of natural supporting materials. Cartilage displays rigidity, elasticity, and low friction and contains a large proportion of chondroitin sulfate. Synovial fluid serves as a fine lubricant with its high content of hyaluronic acid. Tendons have a tensile strength in the range of 17,000 lb per square inch and are composed primarily of collagen in parallel array.

The principle load-bearing members of the body are composed of bone and cartilage. During normal walking, forces three or four times the body weight develop in the hip shortly after the heel strikes the ground. Although bony structures of the legs and spine support the body weight, the ligaments holding the joints and the tendons transmitting forces from muscles across joints are load-bearing structures.

BONE AS A SOLID STRUCTURAL MATERIAL

Bones consist of a framework of collagen which serves as the organizational substructure for the highly complex crystalline mineral called calcium apatite. The rigidity and apparent permanence of bone provides a misleading concept of a stable material that is not subject to change. On the contrary, bone is an exceedingly changeable material, for it represents an important store of calcium as a pool from which the blood can be replenished or into which calcium can be deposited if excess appears in the blood. In addition, bone is a remarkably well-engineered structural member, composed as it is of relatively smooth external surfaces which enclose highly organized spicules of bone oriented along the lines of stress. The vacant spaces between the spicules are not wasted space but are occupied by the blood-forming tissues of the body which produce the red blood cells and most white blood cells.

The upper end of the femur, which constitutes part of the hip joint, is a remarkable structure in that it clearly shows the radial array of spicules of bone arranged along the lines of force generated by supporting the weight of the body, both during passive and dynamic weight bearing (Fig. 8.3A). The spicules of bone orient along the line of force because bone tends to grow in the direction of stress. If this bone is fractured and is reset at a new angle, the spicules of bone tend to become quickly reoriented and follow the new lines of stress.

Clearly, there is no immediately available substitute for materials such as bone, particularly when viewed in relation to the long-range permanence and the self-restorative properties of bone (see Fig. 8.3B). At the present time, active work is going on to produce substitutes for bone, which not only support the weight and resist the large forces that are developed, but which also tend to become firmly adherent to the bone at its areas of contact. These requirements

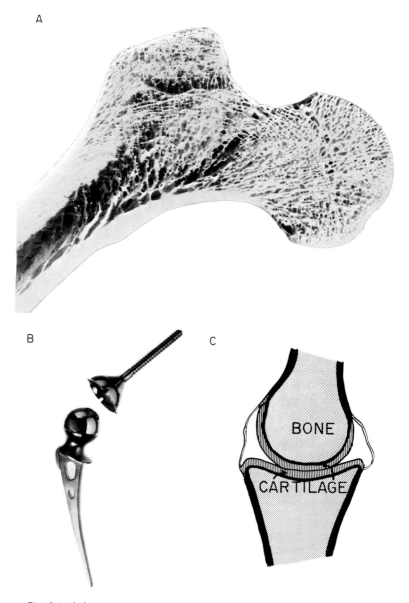

Fig. 8.3. (A) The complex architecture of bone is exemplified by the head of the femur in which spicules are oriented in intersecting radial lines along the lines of force to provide great strength per unit weight. (B) Artificial hip joints made of metal are rather crude substitutes for the original partly for lack of the lubricating and shock absorbing properties of the joint cartilage illustrated in (C).

have not yet been met and will challenge the ingenuity of materials scientists for some time. For example, requirements for a replacement part for the upper end of the femur pose some very serious problems. The lower end of the prosthetic device must be firmly adherent to the shaft of the bone, allowing neither longitudinal nor rotational displacement, while the upper end must move with minimal friction against a joint surface. The joint surfaces are lined with a remarkable material called cartilage, which is composed of another form of collagen organized in a matrix capable of imbibing the joint fluid when the stresses on the joint are minimal (Fig. 8.3C). This fluid is expressed into the joint space when the stresses are increased. The joint fluid has the remarkable thixotropic property of exhibiting less viscosity with more rapid shear by virtue of high concentration of the long-chain mucopolysaccharide molecules. The coefficient of friction in a normal joint is competitive with the best that can be achieved by any known engineering technique.

LOAD-BEARING BIOMATERIALS

The biomaterials requirements for hard and soft tissue replacements have been the central theme of a rapidly growing number of scientific meetings in the past few years. One extensive review addressed the two questions—"What's needed?" and "What's available?"—at a Battelle sponsored seminar-workshop (4). The use of metal alloys for nails, plates, and formed substitutes for bones and joints (see Fig. 8.3) are decidedly widespread. Virtually all types of the common metals and alloys have been tried in prosthetic devices; gold, silver, copper, lead zinc, cadmium, tin, iron, nickel, aluminum, magnesium, vanadium, bronze, brass, steel alloyed with other metals, ticonium. Vitallium, titanium, zirconium, and various types of stainless steels have been employed at various times and in various shapes in hopes of improving bone healing. The most common requirements are corrosion resistance and general inertness.

The three most common metal systems currently employed are the cobalt–chromium base alloys, some of the stainless steels, and unalloyed titanium. None of these materials duplicate the low-friction-bearing surfaces at the joints, and they often fail to remain firmly attached to the shaft of the bone. In addition, the electrolytes in body fluids can attack the surfaces of the prostheses made of stainless steel or vitallium. The differences in physical properties between the metals and bones tend to produce loosening of screws and separation at interfaces. Hurlburt (4) has studied implants of ceramic materials with varying pore size to determine optimal conditions for tissue ingrowth to provide positive bonding of supporting implants to bones. For example, calcium aluminate ceramics with pore sizes of 75–100 μ pore size with

70% porosity revealed tissue ingrowth in addition to the laying down of new bone at the external surface. Vascular connective tissue was replaced by osteoid tissues within the pores of the samples to a depth of about 2 mm. The ceramics with pore sizes in the range of 150–200 μ were penetrated by vascular connective and osteoid materials to even greater depths. Judged by histological examination, calcium aluminate produces little or no tissue reaction. The mechanical strength of porous ceramics may not be adequate to provide the necessary durability and impact resistance. For this reason consideration has been given to the coating of a porous ceramic material on a strong weight-bearing member or core of metal to provide the necessary durability and impact resistance.

New techniques for developing porous metal may have applications for production of bone prosthesis. For example, powdered metals can be tamped into molds containing spherical balls of a material with a lower vaporization temperature. The powdered metal can be converted into porous metal by high-temperature treatment so the material in the spheres is vaporized leaving connected pores of predetermined size and distribution. By such techniques a transition from porous metal to solid metal at the core could be achieved, avoiding the problem of tight bonding of two different materials (e.g., ceramic and metal as above). Still another approach to the problem of stabilizing prosthetic devices within bone cavities is the use of adhesive materials as mentioned in a subsequent section.

Within the last few years, a number of new physical forms of carbon of high chemical purity have been developed to meet requirements of aerospace and nuclear industries. Vitreous graphite has a combination of very great strength and very favorable tissue compatibility. Filaments have been prepared with tensile strengths of around 300,000 psi. and many different forms with tailored properties can be produced. These graphite composites are extremely resistant to corrosive attack by body fluids. They have been studied experimentally in the form of bone pins, and nails and have been proposed as possible low-friction bearings for the joint surfaces of prosthetic devices. There seems little prospect of developing a material that can compete with natural cartilage for both self-lubrication and impact energy absorption.

SHOCK-ABSORBING PROPERTIES OF CARTILAGE

Cartilage forms a thin covering over the opposing joint surfaces and serves not only as a self-lubricant of the moving surfaces but also acts as a shock absorber. In the substance of the cartilage, collagen fibrils traverse a gel substance with large amounts of stiff chondroitin sulfate to form a three-dimensional network of fibers and interconnecting pores. The pores are too small to appear on histological preparations but allow penetration of liquid through the matrix at low flow rates. When cartilage is under compressive load, it exudes synovial fluid into the joint cavity and on release it expands and takes up the fluid. This process of decreasing the liquid content of a porous material is called consolidation. At the onset of a compression load, the forces appear as excess hydrostatic pressure producing a sequence of liquid flow through the pores, absorbing the impact by both consolidation and by deformation of the cartilage at their areas of contact (7). The bone is spared direct transmission of dynamic loads from one unyielding structure to another.

In addition to the weight bearing and mobility functions of bone and joints, delicate structures are protected from external forces by enclosure within bony cavities. For example, the hollow channels and voids within cancellous bone are occupied by blood-forming tissues. The vital functions of the heart and lungs are enclosed with the rib cage. The brain and spinal cord are enclosed within the cerebrospinal cavities, cushioned by a surrounding layer of cerebrospinal fluid (8).

SOFT TISSUE SUBSTITUTES

The specifications for materials to be used as substitutes or supplements for soft tissues vary with the function of each of the many different tissues involved. For example, the essential properties of an artificial skin are obviously entirely different from internal reinforcing sheets, or channels or conduits in the body or the space occupying materials for cosmetic implants in the breast, ear, nose, or chin. The diversity of materials which have been used in these many applications defies any concise review (4). The complicated types of compromise required to optimize various types of nonliving implants is indicated by the variety of patterns of specifications suggested by the tabulation in Table 8.2. Several of the many factors are presented, including the expected duration of the implant, some positive physical requirements and functional properties needed for the particular application. From such a listing one can confidently predict that the basic studies of the biomechanics of tissues, in terms of both static and dynamic stresses, will be greatly stimulated as the empirical trials of biomaterials are superseded by a more scientific approach. But it seems desirable to indicate the scope of the problem by presenting representative examples of the various applications and some of the materials that have been employed. In those instances in which a large number of materials are listed for a single type of tissue there is justification for suspicion that none of them approach the ideal. The large number of tissues that require nonliving implants gives an indication of the quantity of detailed information needed to provide the necessary specifications.

TABLE 8.2
Requirements for Biomaterials

POSITIVE REQUIREMENTS

Column groups — Duration: Degradable (Rapid, Controlled, Slow), Permanent · Physical Characteristics: Strength (Tensile, Compress), Flex, Fatigue Resist., Compliance, Softness, Elasticity · Functional Properties: Tissue Ingrowth or Seal, Non-thrombogenic, Permeability (H_2O, O_2, CO_2, ions)

	Rapid	Contr.	Slow	Perm.	Tensile	Compress	Flex	Fatigue Resist.	Compliance	Softness	Elasticity	Tissue Ingr. or Seal	Non-thrombo-genic	H_2O	O_2	CO_2	ions
ARTIFICIAL SKIN																	
1. First Aid	X				X		X		X		X			X			
2. Temp.			X		X		X		X		X			X			X
3. Semi-p.			X	X	X		X		X		X			X			X
4. Perm.				X			X				X			X			X
COSMETIC IMPL.				X				X	X	X	X						
SUPPORTING SUBSTITUTES																	
Fascia							X										
Dura			X	X	X		X	X	X								
Tendons				X	X		X										
DUCTS, CHANNELS																	
Blood Vessels				X			X	X	X			X	X				
Ureters				X		X	X	X	X			X					
Bile Ducts				X		X	X	X	X			X					
Esophagus				X					X			X					
BONES																	
Weight Bearing			X?	X		X	X	X				X					
Supportive				X		X	X	X				X					
Mobile				X		X	X	X				X					
JOINTS																	
Weight Bearing				X		X	X	X	X		X						
Mobile				X			X	X									
ADHESIVES																	
Surface Organs	X	X					X					X					
Blood Vessels				X			X					X	X				
Bones				X	X		X										
Teeth				X	X												
MEMBRANES																	
Art. Lung				X	X								X	X	X		
Art. Kidney				X	X								X	X			X
O_2 Enrichment				X	X										X		

BIOLOGICAL PACKAGING MATERIAL: THE SKIN

The skin is the largest organ in the body and is a study in contrasts in various regions over the surface. The scalp is thick, and hairy, and skin of the palms is very thick, tough, and hairless. The skin of the back, abdomen, scrotum, soles, and eyelids are as different in a single man as though they came from different animals (9). Biomechanics is such a new field that the physical properties of the skin at various sites is in a primitive stage of knowledge. The deficiencies of our knowledge regarding the physical properties of this most accessible tissue are most painfully apparent when faced with the need to provide an artificial substitute for skin lost from exposed regions of the body.

ARTIFICIAL SKIN

Cosmetic Requirements

The social significance of skin is apparent from the overwhelming emphasis on skin presented by media for mass communications. The appearance of facial skin has undeniable importance to man as a social animal. Consider the plight of the patient who has been disfigured by loss or excision of portions of the face following trauma or surgery. Patients whose lives have been prolonged by excision of portions of the nose, lips, or cheeks as a result of cancer cannot bear the mental strain of social contact. They may suffer such severe psychological strain that they become hopeless recluses unless they can be rendered presentable by prompt replacement of the lost tissues by artificial inserts or prosthetics. The essential requirement for more precise knowledge regarding the properties of these tissues is apparent when one considers the problem of matching shape, texture, color, deformation, and resilience of facial skin under the widely variant conditions of normal everyday life. Presentable substitutes for skin are also required as coverings for artificial hands and arms, but this is a simple matter compared with the requirements for materials that can serve the role of a protective barrier for regions of the body from which the skin has been lost.

Functional Requirements

The outermost layer of skin is a horny layer composed of dead epidermal cells that are continuously being replaced from below as they are lost by contact or abrasion (Fig. 8.4). It serves as an essential mechanical buffer, cushioning the impact of the body against all manner of obstacles, but this is far from its most important role. The loss of the outer layer of skin by burning and blistering is an experience shared by everyone. The red, painful, weeping surface of delicate tissue over a very small area is a minor inconvenience; corresponding loss of skin by burning over large surface areas threatens life by escape of body fluids from the surface. The normal skin is an essential barrier to the evaporation of body water into the surrounding environment and to the invasion of the tissues by infectious organisms. The healing process of large areas is very slow but can be greatly accelerated by covering these denuded areas with patches of skin layers removed from other portions of the body. More effective therapy for burn victims awaits the development of artificial skin particularly applicable to the problems that arise in sequence during therapy. The immediate requirement after a burn is a sheet of material that will serve the barrier functions of the skin against loss of water, proteins, and other body constituents and protect against infection. When the acute emergency has passed, artificial skin is needed that will adhere to the surface and stimulate regrowth of epidermis in the same way that patches of excised skin layers do at present. Finally, permanent artificial skin is needed which can be used to cover surface areas that failed to generate replacement skin by regrowth. The materials that could serve these purposes would be an enormous boon to patients who are currently facing the prospect of repeated and exquisitely painful restorative operations and plastic surgery.

Ultimate Requirements

The specifications cited above will be difficult to realize but even they will fall far short of conforming to the real functional role of natural skin. The protection of the delicate underlying tissues against environmental threats such as impact and abrasion, heat, and irritating materials may be reasonably achieved by the kinds of plastic materials which are likely to subserve the needs of burn patients. It should be recalled that normal skin effectively absorbs ultraviolet light and is an effective insulator against electrical currents, and a barrier against a wide variety of irritant and toxic substances. Deadly poisons and venoms can be safely applied to intact skin. The heat-regulating function of the skin may be difficult to reproduce by artificial skin. Similarly the sensory interface between the skin and the environment is an important protective mechanism against excesses of heat, cold, pressure, impact, and distortion.

Attachment

There is little prospect for total replacement of the skin's function in the foreseeable future since we have barely begun to develop the most rudimentary substitutes for its simplest and obvious functions. We can anticipate serious problems in developing a skin substitute which will be both as flexible and as adherent to the underlying tissues as the normal skin. The skin must exhibit a high degree of stretch and mobility in all directions to accommodate the remarkable flexibility of the human body. At the same time the skin must be very strong and firmly attached to withstand the very large shear stresses that

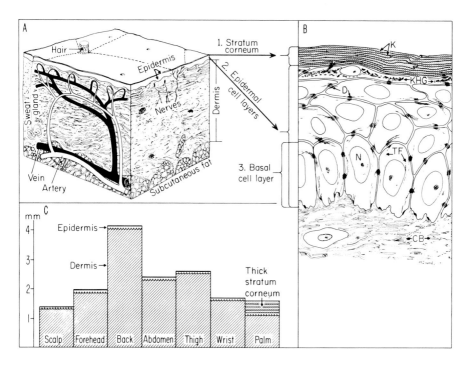

Fig. 8.4. (A) The skin layers contain samples of many types of tissues including blood vessels, connective tissue, nerves, fat, and muscles. (B) The epidermis: The barrier functions of skin are highly dependent upon superficial layers of keratin, specialized remnant of dead epidermal cells. K, Keratin; D, desmosomes; TF, tonofibrils; N, cell nucleus; KHG, keratohyaline granules; and CB, collagen bundle. (C) Skin layers vary greatly in thickness and composition in many different surface areas of the body, reflecting regional differences in function.

may develop. In normal skin these properties are embodied in a dense meshwork of flexible fibers called collagen, which approach the tensile strength of fine steel wire. Stress–strain curves for skin (excised strips or *in situ*) characteristically reveal two distinct regions—(a) large extensions at low loads, and (b) greatly decreasing extension with high loads as reported by Kenedi *et al.* (5). The structural counterpart of these properties may be described as follows: The loose meshwork of collagen fibers offers little resistance to distortion as they become progressively oriented in the direction of stress. When these fibers are oriented in parallel and packed, they strongly resist further distortion (5). Consider the enormous shear stresses that must develop between the skin of the palm and the underlying tissues of the hand when one is exerting maximal torque on a screwdriver or a jar lid. We must develop a material that will assure firm attachment by the ingrowth of collagen, that ubiquitous material which holds virtually all tissues together.

INTERNAL SUPPORTING LAYERS

The fibers of collagen which account for the great strength of skin and leather are to be found as essential supportive layers interspersed between tissues, cells, and bundles or enclosing organs in fibrous capsules. In many individuals, weakness of such supportive tissue at critical sites in the walls of body cavities may result in extrusion of internal organs as hernias in the groin, umbilicus, at the site of previous surgical scars, or even through the diaphragm into the chest. The surgical repair of such defects by sewing up the hole frequently provides only temporary restoration of the integrity of the wall because the fibrous layers were weakened at the outset. To support previously weakened fibrous tissues may require the use of patches composed of materials that not only exhibit intrinsic strength but that will also encourage the growth of tough collagenous fivers into and around the patch. For such purposes, special plastic materials in the form of woven plastic fibers or sponges can interact with the normal tissue repair mechanisms to provide the necessary strength and flexibility for more permanent repairs in difficult situations. The crucial requirement is not confined to its particular properties in isolation because of potential deterioration of materials exposed to corrosive body fluids. It may be more realistic to strive for the optimization of its interaction with the living tissues of the body.

POWER SOURCES

The most familiar biological sources of power are the skeletal muscles which provide mobility and manipulative motility by contractile forces acting on connections to bones across the various joints in the body. Skeletal muscles occur in a wide variety of sizes, shapes, and locations, but they consist of individual fibers with diameters between 10 and $100\,\mu$ generally running longitudinally along the muscle. When contracting the individual muscle fibers can shorten at speeds several times their length per second and can generate tensions of some 40 lb per square inch at its cross section. It can contract or relax in a small fraction of a second. Striated muscle contraction is dependent upon interactions between thick (160 Å) and thin (60 Å) filaments in diameter which are packed in sequential arrays within the muscle fibers. The contractile mechanism has been described in terms of a sliding filament concept by Huxley (10). In addition to the obvious effects of muscle contraction, the excitation process which initiates or stimulates contraction also requires energy expenditure, a very small but important power requirement for both skeletal muscle and heart muscle. The contractile properties of skeletal muscle and heart muscle are quite similar but their functions are quite different because of major differences in their excitation and control mechanisms. Muscle fibers without visible striations (smooth muscle) are embodied in the walls of many hollow organs, channels, or contractile structures over which we cannot exert voluntary control. Some smooth muscles serve as peristaltic pumps and others as constrictive valves. Engineering efforts to duplicate or supplement the mobility of the body or pumping action of the heart have resulted in gross caricatures of the original. Duplication of smooth muscle activity in general has not even been attempted in any concerted effort.

MOTIVE POWER

A patient who loses the function of his weight-bearing muscles must either adapt to a passive artificial leg or be sentenced to a wheelchair. No engineering solution has been developed which can compete or substitute for the power development of the legs for purposes of walking. The principle engineering contributions to artificial legs have been improvement in the interface between the stump and the socket for distribution of forces and assurance of proper fit. The most likely future improvement is better control by appropriate locking and release devices to extend the flexibility and diversity of its uses. There seems no immediate prospect of developing a power source by which the artificial leg would be powered and controlled to actively contribute significantly to walking or running. This is partially because a man can live a reasonably satisfactory life with a passive artificial leg, particularly if he retains the use of his hands and arms.

MANIPULATIVE CONTROL

Man is extremely dependent upon his ability to execute fine manipulative movements of his upper extremities in the course of everyday living and employment. The loss of an arm is a tragedy and great effort has been exerted toward supplying powered prosthetic devices which replace to some small degree the wide range of movement, in terms of degrees of freedom, forces, velocities, and coordination that whole individuals take for granted. The most recent trends have been directed to providing as close a duplicate of the human arm and hand as "possible." An artificial substitute for the full range of hand and arm movements is clearly beyond the present state of the art, so rather extreme compromises are being struck based on limited data regarding the functional requirements. For example, the specifications for artificial hands needed for the occupations most common or most appropriate for amputees have not been clearly decided. It is quite possible that for certain occupations, amputees could be provided with special attachments which would actually give them advantages over the intact normal. A selection of prosthetics (i.e., for home and work) might be more appropriate than current efforts toward a more "universal" solution.

Many power sources for arm prostheses have been tried out including electric motors, hydraulic actuators, pneumatic devices, and contractile materials. Nitinol is an alloy which can be fabricated as wire and will contract reversibly when heated by flow of an electric current and exert considerable tension. Such approaches are intriguing but are not likely to replace standard mechanisms like battery-powered motors or pneumatically activated pistons. The most crucial requirement is improvement in control mechanisms.

The many degrees of freedom of movement in arms, hands, and fingers challenge the most authoritative control systems engineer. The coupling between the intentions of the patient and the movement of the prosthesis has been the subject of many very sophisticated feasibility studies. The most promising depends upon using to the maximum the external manifestations of the normal controls. For example, the electrical potentials recorded from the skin surface overlying contracting muscles can be used to operate switches and actuating devices. The feasibility of chronically implanting electrodes into muscles with wires running out through the skin to provide control has been demonstrated by Reswick and others. A future prospect is to implant microminiature telemetry equipment completely beneath the skin with rechargeable batteries providing the power. The power requirements for such devices are undoubtedly attainable with batteries or external sources, but the future will undoubtedly witness continued efforts toward development of effective biological fuel cells. The electric eel has a highly specialized organ capable of repeatedly discharging 200–400 V. Mechanisms will be explored by which naturally occurring potentials in the body can be utilized for power to control devices implanted in the body.

CARDIAC PACEMAKERS

The benefits of technology are clearly displayed by thousands of patients whose heart rates are sustained by self-contained pacemakers chronically implanted under the skin. In the simplest form, a constant rate is impressed on the heart, regardless of the level of activity. Others provide for a variable heart rate dependent on the continued function of the normal atrial pacemaker. The reliability of these devices and their connections to the heart muscle has progressively improved to the point that the most significant remaining problem is replacement of the batteries at intervals of 2–4 years or so. Power supplies based on radioisotopes are designed for extended life up to 10 years but are extremely expensive at present (about $2000 apiece). In contrast with energy densities in regular batteries of less than 800 Watt·sec/gm, isotopic thermionic power supplies of small size can deliver energy at 1600–160,000 Watt·sec/gm. To date, efforts at utilizing biological energy sources such as biological fuel cells have not yet proved successful. The contraction of muscles or distension of arteries can generate electrical power in small quantities by flexion of piezo-crystals, but this approach has not proved practical. Considering the enormous difficulties encountered in providing the very small quantities of power required for pacemakers, energy sources to supply pumping action of a mechanical heart appear beyond reach.

ARTIFICIAL HEARTS

A massive research and development program has been mounted by the National Heart and Lung Institute utilizing the contract mechanisms to tackle the problem of providing artificial hearts for temporary assists and chronic implantation on a large scale. The mission of the program was interpreted to encompass ultimate provision of total heart replacements in numbers up to 200,000. Many potential sources of energy and pumping mechanisms and other components have been studied (Fig. 8.5). The contractors of the Artificial Heart Program have met in annual conferences. In 1969, the reports totaled 92 covering an enormous range of subjects beginning with 23 papers on materials and their interaction with blood (11). Without question, the most important single obstacle to progress is the unfulfilled requirement for surfaces which do not damage the blood in any way nor encourage clotting. Techniques for coupling antithrombogenic agents like heparin to various suitable materials such as silicone rubber have been shown to provide substantial improvement in nonthrobogenicity through controlled test procedures. These efforts are bound to have salutory effects on many other developments, but the goal of a chronically implanted total heart replacement appears to be completely unattainable in the foreseeable future. Implanted power sources, for one, are a great problem. Fortunately, the emphasis of this program is shifting toward temporary assist devices such as the intraaortic balloon pump mechanism or heart-lung bypass procedures. Nontraumatic pumps, control systems, and quick, easy attachment methods are still problems.

A number of assist devices have been proposed ranging from pumps inserted between the left ventricle or ascending aorta and the descending aorta, or from the left ventricle or atrium reached through an appropriate vein or artery to the femoral artery, or mechanical compression devices surrounding the heart itself or pressure suits resembling those worn by high altitude pilots. Even

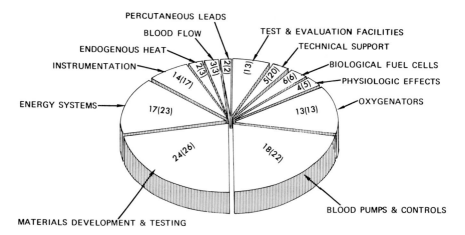

Fig. 8.5. The Artificial Heart Program represents a coordinated attack on a wide variety of problems which must be solved to produce an effective mechanical substitute for the human heart. Numbers in parentheses represent the total number of contracts. (From Artificial Heart Program Conference, National Heart and Lung Institute, NIH, Frank W. Hastings, M.D., and Lowell T. Harmison, Ph.D., Washington, D. C. June 9-13, 1969.)

inertial methods have been used to assist blood flow. At this time, balloon assist devices have shown the most promise clinically. It is of interest to note that this device was first proposed in 1961. Obviously the successful development of any assist device or artificial heart will depend on the close cooperation of specialists in the fields of biology, medicine, and engineering, and materials scientists.

EXCHANGERS

Heart-lung machines have fully established an essential place in making it possible to perform reconstructive surgery on the nonbeating heart.

HEART-LUNG DEVICES

The use of the mechanical devices to bypass the heart and lungs for open-heart surgery began in 1953 and attained maturity in the 1960's with the availability of many different combinations of blood pumps and oxygenators (12). For example, gas exchange with the blood requires a large surface area of exposure. For this purpose the blood could be spread out on bubbles of oxygen when the foaming problem was solved by use of silicone compounds.

Alternatively, the blood could be spread out while flowing down as films on screens or sheets of plastic-covered material. Rows of rotating discs which dip into a shallow reservoir of blood spread the liquid out over a broad surface for oxygenation. Contact of the blood with gases has been found to lead to denaturation of proteins with a consequent limitation to the time a patient can be kept on such a heart-lung machine. To eliminate this source of blood damage, effort is being directed toward development of membrane oxygenators which avoid immediate contact between blood and gases. Here the blood flows in a thin layer between plastic sheets and the gases flow on the opposite sides of the membranes. The most important requirement is the selection of a satisfactory material for compatibility with blood and permeability for the respiratory gases. The design of the oxygenator must be such as provide minimal impedence to blood flow, the least possible resistance to both oxygen and carbon dioxide diffusion, and optimal distribution of the area of exposure, hence the emphasis on designs that reduce the blood film thickness and increase mixing within the blood layer. Both of these lead to increased rates of gas diffusion. Another approach has been the use of capillary exchangers wherein the blood is passed through many small capillaries with a large total exchange area in a physically small space. Excessive pressure drop (resistance) is avoided by running the many capillaries in parallel. Materials with relatively high permeability to oxygen and carbon dioxide and relatively innocuous surfaces for blood contact include silicone rubber, and Teflon; each has its advantages and disadvantages. With continued research on materials and their fabrication new prospects for continued development emerge.

ARTIFICIAL KIDNEYS

The development of permanently implanted arteriovenous shunts made it possible for chronic patients with little or no kidney function to be maintained for long periods (years) on artificial kidney machines. This stimulated research and development on improved, more efficient artificial kidneys. Many of the requirements of an artificial kidney are common with those of a membrane oxygenator (e.g., design for optimum flow, nonthrombogenicity, and membranes of desired permeability). In fact the two devices are so similar that designs have been proposed that are interchangeable using either oxygen or dialysate depending on the need.

Much effort has also been devoted to increases in both reliability and simplicity, as well as reduction in both initial and operating costs, so that artificial kidneys can be operated in the home by the patient and/or members of the family. Happily, definite progress has been made and a number of home units are in use.

Although progress has been made in the engineering design of artificial kidneys, corresponding progress in membrane materials has been slight. Cellulose, the original membrane material, is in common use despite its thrombogenicity. Improvements are needed as borne out by the fact that uremic patients often are subjectively worse for a period after dialysis even though their blood chemistry is apparently near normal. Possible explanations are the nonremoval of an unknown "uremic factor," or more likely the unwanted removal of a needed factor from the blood, or perhaps some subtle injury to the blood by the kidney machine. When this question is resolved, materials specialists will be called upon to design membranes or materials with the required properties.

The ideal artificial kidney system would include a long-lasting A—V shunt that is relatively trouble-free and a compact, reliable, and inexpensive kidney machine that can be patient operated. Construction materials (including the shunt and associated blood tubing) must be nonthrombogenic and noninjurious to the blood components. Even more desirable would be an implanted kidney so that only dialysis lines would have to be connected for each use. This means that the blood would be flowing through the artificial kidney passages at all times, a condition which poses even more stringent material requirements.

VALVES

The normal human heart valves are light, strong, durable, with minimal obstruction to forward flow and essentially leakless backward flow occlusion. A mechanism capable of withstanding unremitting stressing for 60 or more years represents a challenge to engineering talent which has stimulated production of a myriad of mechanical designs. The Second National Conference on Prosthetic Heart Valves (13) provides a comprehensive and critical review of the various approaches which have been attempted in many different institutions, along with the biological, hydrodynamic, and functional problems that must be confronted. Prosthetic valves had been implanted to replace about 1800 aortic valves and 1500 mitral valves. The operative mortality was reported as 18.5%. In addition, the most common complications in 3620 patients were the formation and release of blood clots (8% of patients, 4% fatality), infection (10% of patients, 2% fatality), and mechanical problems (13% of patients and 2% fatality). The combined mortality was 29.2% one year after surgery and may be as high as 50% after four years. It must be recalled that many of these patients were operated because their short term survival appeared in doubt without replacement therapy. The wide variety of different designs for prosthetic valves totals 50 or more. The caged ball and caged disc generally reveal substantial pressure drop and backflow but have greater durability. However, one of the most discouraging experiences was the discovery that caged silastic balls in aortic valve prosthetics degenerated over time with variations in the color, shape, and consistency accompanied by surface damage, abrasion, or fracture as the cause of death of many patients after a period of about 2 years. This serious example of material failure has required replacement of the artificial valves in many patients. The combination of blood clotting and material failure represents extremely challenging problems for the materials scientists.   .

Woven materials are being used with increasing frequency, taking advantage of new natural and synthetic materials and new types of knitting, weaving, or felting to produce a wide variety of fabrics with a wide choice of surfaces. For example, thin porous fabrics of Dacron, polypropylene, or Teflon have been used to cover the nonmoving parts of rigid valves to allow spontaneous ingrowth of tissues to diminish the tendency for clot formation. According to Braunwald and Bull (14), the apparent sequence is the initial deposit of blood clotting deep in the interstices of the fabric lattice, incorporating red cells and fibrin. The superficial portion is replaced by a deposit of platelets and fibrin; fibroblasts appear within tunnels in the structure and produce connective tissue while endothelium grows over the vascular surface. The quality of ingrown tissue appears to be improved by loose weaving so that the cells can be well nourished. Unfortunately, flexible Teflon fabric proved unsatisfactory as a replacement for aortic valve leaflets because of fatigue fracture despite their development of connective tissue ingrowth and a smooth glistening surface. This may have been due to the use of a tightly woven material. It seems entirely possible that cells required for tissue growth within the fabric (fibroblasts) or on its surface (endothelial cells) might be seeded prior to implant.

TUBES AND CONDUITS

A moment's reflection makes one wonder if man-made materials can be expected to display the kind of durability under unremitting impact and stress that many body tissues sustain. Effective combinations of living tissues cultivated on nonliving frameworks are undoubtedly appealing as a future prospect. It first developed in the use of loosely woven Dacron tubes as replacements for seriously diseased arteries. In this application, the woven Dacron tubes were sutured in to replace segments of diseased arteries. Blood could be observed to migrate into and through the meshes of the fabric, but quickly clotted and sealed the interstices so that blood loss was minimal. Fibrous connective tissue replaced the clotted blood and the combined strength of the collagen and Dacron proved capable of supporting the pulsatile arterial pressure for long periods of time. The development of combined living and nonliving materials for such prosthetics appears to have great potential as an example of biomedical engineering teamwork. It seems quite possible that the best antithrombogenic surface in contact with blood might well be achieved by seeding endothelial cells that might be induced to spread and cover the internal surface completely. Tissue-culture techniques have clearly developed to the stage that supplies of such cells could be developed.

In addition to the arterial and venous channels of the cardiovascular system, the human body is equipped with a large number and variety of channels and hollow organs for which nonliving replacement may be required. In young children developing hydrocephalus, excess cerebrospinal fluid can now be siphoned through a fine tube down the jugular vein to the level of the heart. The ureters conduct urine from the kidney to the bladder and the urethra drains the urine to the outside. The gall bladder concentrates bile and the bile ducts convey bile from the liver to the small intestine. Secretions are conveyed by ducts from various glands such as the salivary and lacrimal (tear) glands or pancreas. The full length of the gastrointestinal tract from the esophagus to the rectum is a series of specialized conduits with multiple functions. The walls of all these channels are invested with smooth muscles which either produce phasic contractions to produce propulsion of the contents or smooth muscle rings which serve as occlusive valves. Replacement materials for all these applications consist of inert tubes which serve as passive conduits. Little or no effort has been directed toward incorporating either the propulsive or the valve functions in these substitutes. Some examples of the materials applications collected by Ruth Johnsson-Hegyeli (4) are summarized in Table 8.3.

OPTICAL TRANSPARENCY: ARTIFICIAL CORNEAS

The eyeball is a spherical shell of collagen fibers with two different arrangements. The sclera appears white and opaque because randomly oriented fibers scatter the light. The cornea of the eye is composed of layers of collagen of fibers with each successive layer at right angles to those above and below (2). The uniform diameter and spacing of the fibers suppresses any scattered light except that which passes straight through (see the section on body building blocks, page 293).

Normal vision is highly dependent upon the crystal-clear cornea of the eye through which light must pass undistorted to be focused on the sensitive retina at the back of the eye. When the cornea becomes cloudy, clear vision is lost until a substitute can be sutured into place. The cornea from the eye of a living or dead person can be transplanted as a substitute for a clouded cornea with a high success rate, and eye banks have been organized to supply the needed tissues.

TABLE 8.3

Conduits and Channels

Application	Materials used	Advantages
Ureterostomy Cystostomy Nephrostomy	Polyethylene	1. Chemical inertness 2. Natural flexibility 3. Not softened by blood, urine, gastric fluid 4. Does not show presence of crystalline urinary salts 5. Can be kept in place for months
Lacrymal intubation following stricturotomy	Polyethylene	1. Does not become blocked by mucus, aggregates of sequestered cells, or coagulates of organic fluids
	Polyvinyl	1. Water repellant 2. Prevents adhesions to tissue 3. Therefore permits easy and painless manipulation
Drainage of cephalorachidial fluid	Polyvinyl	1. Can be left in place for hours, even days, without risk of trauma or infection 2. Patient need not be immobilized
	Silicone rubber	1. Low surface tension 2. Inertness 3. Excellent behavior at temperatures

Artificial corneas fabricated from plastics have an unfortunate tendency to be rejected or to become cloudy. There is vital need for a source of artificial corneas which are cheap, durable and fully compatible with surrounding tissues while conforming to the rigid requirements for normal vision.

ADHESIVES

Closure of wounds and cessation of bleeding may be achieved by rapidly polymerizing monomeric cyanoacrylate adhesives which are capable of setting up while in contact with moist tissues. By spraying these compounds on bleeding surfaces, blood loss is rapidly diminished. They are also generating interest as a means by which incisions or tears of highly vascular solid organs like the liver and spleen might be repaired since sutures are very likely to tear out. Tests have been applied to the homologous series of alkyl α-cyanoacrylates, and the higher homologs were found to spread and rapidly polymerize to produce stronger bonds on tissues while the lower homologs spread and polymerized more slowly and weakly bonded. Tissue tolerance is apparently improved by use of the higher homologs. Some of these materials are biodegradeable so that they will completely disappear as healing becomes well established. The day has not yet come when these substances are widely used in surgery or in the treatment of injuries or burns. However, the widespread interest that the basic experience gained thus far appears to warrant optimism for successful outcome in the future.

Orthopedic prostheses of the sort illustrated in Fig. 8.3 may be seated in the marrow cavity of long bones by means of rapidly curing acrylic polymer compounds. Materials like methy methacrylate monomer and elthylene dimetha-crylate are mixed with accelerators and stabilizers to form a paste that solidifies in a very few minutes with the liberation of considerable amounts of heat, perhaps attaining temperatures near the boiling point at the interface with the bone. Thermal necrosis of bone and escape of the material into the bloodstream represent potential sources of complications in the use of currently available materials. It seems likely that the future will see greater emphasis on the use of porous metals or even ceramic materials into which connective tissue can actively grow to provide a firm bond between the prosthesis and the bone made up of living tissue rather than a potentially destructive artificial material.

Dental adhesive materials have been under development and study for many years and many different types are available. An important future objective is a material which can be installed within the prepared cavities of teeth and solidify to become sufficiently hard and durable to act as fillings of great permanence. Interest has been generated in the fact that barnacles are capable of synthesizing cements which fasten them firmly to rocks or ships' hulls in the presence of dirty seawater, a condition not too different from the oral environment. The applications of comparative physiology for the purpose of simulating nature's techniques for the development of an effective dental cement is an extremely attractive prospect.

FUTURE REQUIREMENTS FOR BIOMATERIALS RESEARCH

It is currently impossible to specify with confidence the materials requirements for either soft or hard tissue implants, primarily because studies of physical properties of tissues are grossly incomplete. Tissues are virtually all viscoelastic materials that are neither homogenous nor isotropic. Although a large mass of data has been collected regarding the strength of biological materials (14), this is an insufficient base from which to specify the essential properties which will permit mechanical compatibility between the tissues and an implant. According to Fung (15), three-dimensional stress–strain history laws have not been determined for soft tissues; this necessitates a large experimental effort to study the relationships in three dimensions at various rates. He proposed three steps for the future.

1. Study the natural, relaxed state and its dependence on physical and chemical parameters (e.g., temperature, pH, osmotic pressure, etc.).

2. Study the stress–strain history with respect to the natural state.

3. Describe the physiological conditions in terms of strains measured in the natural state.

The last requirement is of extreme importance and frequently neglected. In the first place, the natural state of tissues in the body is not in the unstrained state. Furthermore removal of samples from the body is virtually always accompanied by changes in state and function that would not be predicted, particularly by the investigator. As a specific example, an enormous body of information has been collected on the mechanical properties of arterial walls studied on excised samples and in exposed vessels. It is generally conceded from such studies that the arterial pulse causes a distension of 2 or 3% under these conditions. Studies of the distension of arteries *in situ* by ultrasonic techniques have disclosed pulsatile distension as great as 15%. In a similar way, an excised or exposed heart functions at or near its minimal dimensions while the normal heart in its normal environment functions at or near its maximal dimensions. Experimental studies of muscular contraction often neglect the fact that there is no physiological counterpart to the electrical stimulus delivered to a motor nerve supplying a muscle. Possible changes in physical properties of other tissues, including bone and connective tissue by virtue of removal from the body cannot be ignored.

With a growing knowledge of *what is needed*, it should be possible to establish more specific criteria for evaluating *what is available*. For example, Lyman (4) proposed the following properties for the *initial* screening of polymers as candidates for tissue substitutes: tensile strengh, flexural strength, moduli, solubility, melting temperature, deformation temperature, hydrolytic stability, hydrophobicity (water absorption), and permeation.

Needless to say, this list is far from being complete enough to provide full scale specifications for a material needed to perform a specific function (see also Table 8.2). Other factors such as fabrication, sterilization, durability, biological compatibility, and many other considerations will be required as more is learned of the problem.

References

1. Gross, J. Collagen. *Sci. Amer.* **204**, 120-134 (1961).
2. Harkness, R. D. Collagen. *Sci. Progr.* **54**, 257-274 (1966).
3. Rushmer, R. F. Biomaterials; an essential ingredient in bioengineering materials. *Mat. Res. Std.* **10**, 9-13; 33-34 (1970).
4. Bement, A. L. (ed.). Biomaterials; bioengineering applied to materials for hard and soft tissue replacement. "Proceedings of a Seminar-Workshop, Battelle Seattle Research Center, Nov. 12-14, 1969." Univ. of Washington Press, Seattle and London, 1971.
5. Kenedi, R. M., Gibson, R., and Daly, C. H. Bioengineering studies of human skin II. *In* "Biomechanics and Related Bioengineering Topics" (R. M. Kenedi, ed.). Pergamon, New York, 1965.
6. Bartley, W., Birt, L. M., and Banks, P. The biochemistry of structural proteins, mucopolysaccharides and bone. "Biochemistry of Tissues." Wiley, New York, 1971.
7. Zarek, J. M., and Edwards, J. Dynamic considerations of the human skeletal system. *In* "Biomechanics and Related Bioengineering Topics" (R. M. Kenedi, ed.). Pergamon, New York, 1965.
8. Caveness, W. F., and Walker, A. E. Head injury: conference proceedings. Lippincott, Philadelphia, 1966.
9. Rushmer, R. F., Buettner, K. J. K., Short, J. M., and Odland, G. F. The skin. *Science* **154**, 343-348 (1966).
10. Huxley: The mechanism of muscular contraction. *Sci. Amer.* **213**, 18-27 (1965).
11. Hastings, F. W., and Harmison, L. T. Artificial Heart Program Conference National Heart Institute, Washington, D.C., June 9-13, 1969.
12. Galleti, P. M. Advances in heart-lung machines. *In* "Advances in Biomedical Engineering and Medical Physics" (S. N. Levine, ed.). Wiley (Interscience), New York, 1968.
13. Brewer, L. A. (ed.). "Prosthetic Heart Valves, Second National Conference." Thomas, Springfield, Illinois, 1969.
14. Yamada, H. *In* "Strength of Biological Materials" (F. Gaynor Evans, ed.). Williams & Wilkins, Baltimore, Maryland, 1970.
15. Fung, Y. C. Biomechanics. *Appl. Mech. Rev.* **21**, 1 (1968).

Chapter 9

TECHNOLOGICAL TRAINING OF MEDICAL MANPOWER

The problems of nature do not yield to one approach or to one type of investigator. It seems wise to consider the contributions to scientific knowledge by the widest conceivable spectrum of talent. The individual who is fully trained in both engineering and medicine can identify problems and pursue investigations of the greatest importance. Training of such "hybrids" must receive support and encouragement. However, there are many examples of effective collaboration and productive research through cooperative efforts of engineers and life scientist with neither having any special training in the other's discipline.

Existing mechanisms for interdisciplinary training in the field of biomedical engineering are limited in number. Although the students all come from the same population pool in high school, they diverge widely when they choose between engineering or biomedical areas on entrance into colleges or universities. The division was formerly so complete that engineering students received no course work in biology and students of biology and medicine were exposed to no contact with engineering. The intellectual barriers between these two vitally important areas are now being breached at a few sites in the academic structure. First, a growing number of students with engineering training are applying as candidates for advanced degrees in basic medical sciences, medicine, and dentistry. The numbers are still small but very significant. Most of the formal training of hybrid bioengineers is taking place within engineering colleges in doctoral programs with information input from the basic medical science departments (or clinical departments). Hybrid bioengineering training is designed to prepare its graduates for academic positions to serve independently in teaching and research. In contrast, collaborative teams represent a very wide spectrum of backgrounds, extending all the way from the thoroughly trained bioengineer to the physician with little or no training in engineering, physical sciences or mathematics. Engineers with no formal training in biology or in medicine can also contribute effectively in collaborative relationships. Some engineers increase their competence by postdoctoral training in biological or medical environments. The future of biomedical engineering depends in large measure on our ability to build and strengthen the interconnections between the diverse disciplines of engineering, biology, and medicine.

GRADUATE TRAINING IN BIOENGINEERING

Lee Huntsman

The rapid growth in bioengineering which took place during the late 1950's and early 1960's, was concentrated in the universities rather than in industry or in the health care system. Such a pattern was dictated by the rapid rise in interest and funding within the university, while obstacles detailed in other chapters served to limit the growth outside the university. An emphasis on advanced study and doctoral training was compatible with this pattern, and the result was a quite rapid proliferation of bioengineering research and training programs in many universities across the country. From its inception, bioengineering was heavily oriented toward fundamental research, and the training of bioengineers was accomplished predominantly at the doctoral level. The most common pattern of development was a graduate training program established in a Department of Electrical Engineering with some degree of affiliation with the Department of Physiology. With growing interest in the field, the diversity of programs and the scope of involvement by different disciplines has expanded greatly.

Many existing bioengineering training programs share a common statement of philosophy; namely, to offer graduate-level training in engineering together with a broad exposure to the biological sciences. In practice, however, these programs differ significantly in emphasis and time devoted to the study of biological sciences. These discrepancies are much less at the level of Master of Science than in programs leading to the Ph.D degree because the time available for nonengineering course work is very limited in the former case. Within doctoral programs, however, the emphasis ranges from one or two physiology courses, to the entire first year or two of medical school and additional courses. This range reflects different levels of total effort which these programs require of their students. More fundamentally, it reflects attitudes about whether the training is designed for engineers who can collaborate with health scientists, or independently competent biomedical investigators. Most of the older and larger training programs have chosen to train research scientists who are bilaterally competent in an engineering discipline and in at least one aspect of biology. This objective has been compatible with their emphasis on fundamental research and the academic market for their graduates.

We can now see evidence of saturation of this initial need for academic personnel and the transition to a new pattern. A reduction in available educational funds from all sources has quickly curtailed the growth of bioengineering activities within universities, and reduced academic positions available for those who are being trained. Simultaneously, the application of engineering technologies is becoming emphasized throughout the health care system with a growing need for bioengineers, particularly at the M.S. level, to implement these applications. This new trend is sustained by the growing acceptance of bioengineers into the health care system. This acceptance, in turn, is largely motivated by the crisis of cost-effectiveness in health care, and the publicity which that crisis has received. The situation in the health care sector almost directly determines the utilization of bioengineers by supporting industries.

The nature of this evolution and the number of bioengineers which are trained will depend heavily on the manner in which the health care industry utilizes engineers. As pointed out elsewhere in this book, the opportunities for engineers to make important contributions to the quality and effectiveness of health care are great. The extent to which these possibilities will become realities, remains to be seen. However, it is clear that a very wide variety of engineering techniques and talents will be called for. This probably means that bioengineering will continue to be poorly defined as a discipline, but it also means that the spectrum of training opportunities must become quite broad. "Market" requirements for such diversity make present choices difficult regarding the structure and support of training programs individually and collectively. A further complication is the fact that the compartmented structure of the modern university is not well suited to such interdisciplinary activities in either research or training.

For an individual bioengineering training program, there are two possible solutions. A specialized program may be established to train bioengineers in a particular area. Presumably not very many of these programs would exist at one university. Alternatively a broad, flexible program could be established which offered opportunities for students with various backgrounds and interests. Hard realities indicate that specialized programs are probably most easily attainable and perhaps the only possible program in most universities. These realities include the departmental and school structure of universities that makes a broadly based program difficult if not impossible; the lack of interested faculty with sufficient knowledge to implement a broad program; the strong faculty inclination towards specialization in a well-defined area of excellence; and a desire to produce a known product through a preset and efficient program. One unfortunate consequence of specialized programs is the "definition" of bioengineering on a particular campus as being equal to the specialization practiced there, ruling out other legitimate areas of activity. This phenomenon is real and may have been partially responsible for the disproportionate activity of electrical engineers in the bioengineering field. A major disadvantage of the more specialized program is the need for students to select a program with limited knowledge about the program area and little confidence that the professional opportunities actually lie in his major area of interest. There are a number of advantages to the concept of having a broadly based training program within a single university. Principal among these is the possibility of attracting and training students from a wide variety of backgrounds in an atmosphere of stimulation and cross-fertilization. Also, in view of the diverse and rapidly changing nature of the health care industry, it may be argued that breadth of exposure is what a student most needs, as he will likely get into many diverse areas of activity during his career.

MULTIDISCIPLINARY TRAINING IN UNDERGRADUATE ENGINEERING

Curtis Johnson

The training of undergraduates in engineering has traditionally been accomplished in the established departments. The four mainstay departments are electrical, chemical, mechanical, and civil engineering, but recently many institutions have broadened their departmental scope to include nuclear engineering, ceramic and metallurgy, bioengineering, and a host of other areas of specialized interest. Another approach to undergraduate training has been to provide a broad undergraduate training in engineering techniques with no departmentalization. This approach has not gained wide acceptance, but is a trend in some institutions. Presently there are experiments in mixing these two approaches; where the departmental structure is maintained as areas of proven interest of the students, faculty and potential employers, while broad interdisciplines interlace and cut across many departmental barriers. These interdisciplines may have formal administrative structure, based on their research or training functions, and often represent areas of strong research support from government and industry. Examples of interdisciplines of this type are bioengineering, ocean engineering, computer sciences, environmental engineering, etc.

Historically, bioengineering programs have commonly been set up as unique departments in the traditional academic structure, and have developed their own curricula to produce a "hybrid" engineer, with major time commitments devoted to both engineering and medical sciences. Such a departmental faculty may provide departmentally organized courses, with students obtaining a bachelor of science degree in bioengineering. The advantage of this system is administrative.

Another approach is to train the bioengineer in an existing department, recognizing that each department has many areas of important interest in bioengineering. For example, electrical engineering departments have great interest in bioinstrumentation, the uses of ultrasound, electromagnetic, and optical processes for therapy and diagnosis, and in the application of computers. Similarly, mechanical engineering departments have great interest in biomechanics, biomaterials, fluid dynamics, control systems and instrumentation. Chemical engineering departments are interested in biochemistry, diffusion in fluids and through membranes, fluid mechanics, and biomaterials. Training bioengineers within existing departments can be stimulated by providing research funds and by identifying new applications of the various engineering disciplines. A bioengineering subspecialty in an existing department is a convenient way to train bioengineers, as the departmental administrative structure is already in existence. This approach may suffer from a lack of fertile interdepartmental ties. Most departments have from 10 to 20 credit hours of technical electives in addition to several credit hours of free electives. If the student desires to specialize in bioengineering within his department, judicious choices of free and technical electives can prepare him satisfactorily to pursue a bioengineering career as a participant in a multidisciplinary team.

Introductory bioengineering and zoology courses can be included as free electives in the lower division. At the junior–senior level, technical electives may be utilized to take a human physiology course, plus a selection of other courses with some specialization. As bioengineering interest and involvement on the part of the departmental faculty increases, departmental bioengineering courses can be offered. This opens up new opportunities for the departmental student, and these courses fill technical elective requirements within the department.

In some institutions, a degree of a Bachelor of Science in engineering is available, unspecified as to department. Interdisciplinary training in bioengineering can be achieved in such programs.

Such an interdisciplinary approach, is distinctly different from the two pathways previously mentioned. The curriculum is designed so that premedical requirements are satisfied and the student is prepared for medical school if he so desires. Such an undergraduate training program, strong in the engineering basics, will be very important in the background of a modern medical practitioner. It will prepare students to enter graduate programs in bioengineering, physiology and biophysics, and related health sciences. With an emphasis on application, such a program can prepare an engineer to satisfy the increasing demand for technology in medicine by industry, hospitals, and the government.

TRAINING PHYSICIANS FOR APPLICATIONS OF NEW TECHNOLOGIES

Fred Stegall

Two central themes have repeatedly recurred throughout this monograph—(a) the quantity and quality of health care must be rapidly expanded to meet the growing demand, and (b) the cost-effectiveness of medical care must be improved to arrest the spiraling costs. In either case, it is clearly necessary to conserve the precious time of the physician for his essential decision-making role by increased reliance on specially trained allied health personnel supplemented by new technology. The role of the physician in the outpatient clinic and hospital settings will shift toward leadership of a team of qualified personnel using the most modern equipment to gather precise data, to display it to the physician in readily interpreted form, and to administer appropriate therapy, often employing modern techniques manned by other trained personnel. To carry out his leading role, team physicians must become much more familiar with the general principles of diagnostic and therapeutic equipment than at present. Some will also play a key role in training a new breed of paramedical personnel—the physician's assistant (see next section). New techniques and equipment also offer some hope of relieving the physician of routine portions of history-taking and physical examinations.

The changing nature of medical practice, and the expanded role of technology in helping free more of a physician's time and energies for patient contact, support the conviction that physicians must have more extensive exposure to the principles of technology as an appropriate, perhaps essential, part of the medical education. If new devices are to extend his senses to obtain information about a patient or to allow an assistant to obtain information, he must understand the principles and limitations of these new tools to use them intelligently.

The form and function of medical schools are in the process of a major overhaul characterized by increased time for elective courses and reduced emphasis on the required "lock-step" sequences. In the past, a single medical curriculum was utilized as a training ground for a very wide spectrum of careers, ranging from basic medical scientists to general practitioners but including investigators, administrators, public health officials, and many others. Now it is becoming possible for a student to direct his educational experience toward long-range professional objectives beginning early in his undergraduate college years. The need for training in physical sciences, engineering, and quantitative techniques will vary among individuals progressing through medical education. With the current rapid incursion of technology into all branches of medical practice, the proportion of physicians who could benefit from broader backgrounds for science and technology can only become rapidly greater.

DURING PREMEDICAL TRAINING

Over the last decade or two, college preparation for medical school has shifted away from a relatively structured course concentrating on "essentials" to a far less rigid set of course requirements. The specific entrance requirements for acceptance in medical school are generally minimal. Freshman medical students with undergraduate majors in fine arts, philosophy, urban planning, music, or engineering are becoming much more common.

Many educators recognize that the growing impact of technology in this age of "Technological Revolution" is producing a more sweeping change in our social fabric than the Industrial Revolution of the last century. Appropriate adaptation is not obvious in the education of young men and women on American campuses. While engineers and physical scientists are exposed to the principles of technology as a matter of course, today's students of liberal arts, education, fine arts, or other academic discipline lack even a minimal exposure to technical knowledge. Programs of study for the premedical student might well include at least an introduction to the general principles of technology that future physicians will need.

A number of disciplines could contribute the essentials of modern technology needed by the college undergraduate, i.e., chemistry, including physical chemistry, and physics, including basic information on isotopes, mathematics (at least through calculus) and engineering. Among important engineering topics might be mentioned fluid dynamics, control systems, mass and heat transfer, basic electronics, materials sciences, structural analysis, and many others depending upon the particular inclination of the student.

Each contribution might be in the form of lectures and demonstrations, with minimal laboratory exposure, and approached as a survey of each discipline and its investigative tools rather than with the goal of making the undergraduate technically competent. A logical case can be made for exposure to technology a requisite for all college education, as mentioned above, and perhaps only slightly more extensive for those (including physicians) who will need more familiarity with technology in their career fields.

DURING MEDICAL SCHOOL

Assuming that future physicians will use tools and employ paramedical personnel more widely than today, what sort of training should be added to his years in medical school? An undergraduate student with at least minimal exposure to the fundamentals of technology and physical sciences can identify a somewhat more extensive special training than would be possible today. The course content appropriate for some medical students in preclinical years might include bioengineering or biotechnology, and would encompass many areas of specialization now identified under this "umbrella" such as bioinstrumentation, biomechanics, biomaterials, information theory, and applied controls systems concepts, data processing and display of biological data, and the like. Trained in biotechnology, tomorrow's medical student could attain better understanding of the machines and materials he will encounter in clinical practice.

The format for this training would undoubtedly be varied in different institutions. In some, an elective in "bioengineering" would serve as a suitable vehicle. In others, material added to the traditional lectures and laboratories for basic medical sciences would be employed. Course material might include some of the following

   a.   Basic principles of instrumentation
   b.   Fundamental electronics (fields, charges, current, electromagnetism)
   c.   Transducers for biological measurements: biological reactance
   d.   Data processing: amplifiers, filters
   e.   Display systems: oscilloscopes, meters
   f.   Pulse techniques: digital logic; digital displays
   g.   Energy sources for diagnosis and treatment: light, diathermy, ultra-
        sound
   h.   Bioelectrical potentials: ECG, EEG, EMG; biological impedance: electrodes
   i.   Interaction of materials and tissues
   j.   Principles of analog and digital computers

Somewhat more specialized course material, which could be included in the preclinical course or reserved for the clinical years of medical school, might further include

   a.   Engineering applications to health care delivery
   b.   Criteria for selection of instrumentation systems
   c.   Coding, transmission, and recording of information
   d.   Patient safety
   e.   Prostheses and implanted pacemakers
   f.   Hum, noise, and other forms of interference

These could be presented in separate courses or included in the usual curriculum (prostheses and control in lectures on orthopedics, for example, and patient safety in anesthesiology).

The wide range of material suggests that it might be best incorporated into current course material, particularly in the upper division. The usual medical student focuses on those courses which seem most important to his future functioning as a physician. He follows advice from practicing physicians, house staff, and his upperclassmates to make such decisions. He will elect to use some of his newly freed time to learn more details about the interactions of technology and medicine only after seeing technology's impact on his professional future. Thus feedback from the wards and clinics in medical schools will influence the response of the medical students to these elective offerings. The stimulus to learn about these developments is already growing and medical schools must plan now to meet this need in their preclinical curricula, organizing course material and faculty and pressing for development of suitable textbooks.

POSTDOCTORAL TRAINING

Most medical students, receiving their doctorate in medicine, complete an internship for a year and proceed to residency training in a medical or surgical specialty or family practice. For many, this is an ideal time for additional grounding in those areas of technology important to their specialty. For example, intensive care demands better understanding of data collection, manipulation, and display. Respiratory and circulatory assistance is particularly important to surgical specialists, anesthesiologists, and cardiologists. Biomaterials is especially important to the orthopedist. Patient safety and selection of instrumentation systems is important to those headed toward careers in medical administration and public health. Postdoctoral programs in bioengineering or biotechnology may encourage the development of a professional for the setting in which bioengineers might serve as a consultant on the applications of technology to clinical care, and conduct training programs for clinical staff at professional and subprofessional levels.

The relatively loose preceptorship approach to most postdoctoral training in clinical specialties offers another possibility for presenting technological training in parallel with traditional medical topics. During this period it seems particularly desirable to organize relatively short, intensive courses in principles of technology for house officers and staff who have not had such exposure earlier in their careers and now recognize the need for it. The obvious impact of technology on the day-to-day clinical care of patients will be the motivating factor in acceptance of these offerings, and successful application of engineering principles to health care delivery will insure their success.

Practicing engineers or graduate bioengineers will have an important role in the training of physicians at this level, too. Engineers actively involved in clinical problems will become members of health care teams which advise the physicians and help them make decisions. The interaction between engineer and physician in postdoctoral training will generate the sort of informal teaching environment in which a great deal of information can be abosrbed quickly, and greater utilization of technology (and technologists) realized. For those in training as bioengineers with medical background this might amount to intense, daily cooperative research or planning effort with an engineering group, and such trainees should serve as a particularly useful route for feeding technological developments into clinical care.

CONTINUING EDUCATION

Among all professional groups, physicians are especially aware of the need for continuing education in their special area of competence. Advances in medical practice have tended to outstrip a practicing physician's ability to keep up with medical literature. Professional meetings, continuing education seminars, and short courses have been widely utilized to fill this need. Courses in instrumentation, materials, hospital systems, and others are already offered. Such courses will reflect practitioner's desire to keep abreast of developments essential to his professional practice growth.

TRAINING OF PHYSICIAN'S ASSISTANTS

The demand for health care is increasing faster than the rate at which physicians can be trained. In some areas the problem is becoming particularly acute, and patient–physician ratios in rural and central city areas are unreasonably high. In these areas physicians increasingly recognize the need to delegate increasing responsibility to paramedical personnel (like physician's assistants).

The concept of delegating responsibility for patient care is hardly new. Decades ago it was not uncommon for a physician to draw a blood sample and analyze urine. In contrast, a physician in practice will usually delegate this responsibility to a technician trained especially for this limited role, and he has learned to trust within limits the data obtained this way. Similarly, technicians obtain X rays, electrocardiographs, and electroencephalographs; carry out radio-isotope scans; operate heart-lung machines; conduct ultrasonic or diathermic treatments; and handle a variety of other duties once reserved to physicians themselves.

Chapter 10

CLINICAL ENGINEERING:
Future Engineering Requirements by
Medical and Surgical Specialties

H. Fred Stegall (Editor)

The new and novel diagnostic data sources summarized in the preceding chapters represent opportunities for expanding sources of information about patients without specific reference to the unmet needs of the various medical and surgical specialties. Obviously the wide variety of problems presented by the various organ systems demand a correspondingly wide spectrum of instrumentation to meet the stringent requirements imposed by the many abnormalities and disease processes affecting tissues and organs in different locations in the body. This problem was indicated superficially and schematically in Chapter 6. A more realistic assessment of the current status and future requirements for technology in medical diagnosis, therapy, and evaluation of therapy can be obtained by directing specific inquiries to investigators who are intimately involved in medical and surgical specialties. A comprehensive inquiry into the future needs for new technology by all the components of clinical medicine would be most valuable for future planning, particularly if it were undertaken on a large scale with an effort to assemble the considered projections of selected representatives of the medical community known for their vision and foresight. However, this would be a much more ambitious and extensive effort than is appropriate for this text. Instead, we decided that representative examples of needs could be obtained from collaborators and colleagues on the local scene who are affiliated to a varying extent with the Center for Bioengineering of the University of Washington. To attain a more or less consistent and comparable reaction, a standard format was proposed to these selected investigators consisting of four parts.

a. A general statement defining the sphere of interest of the specialty for the benefit of readers without a medical background

b. A listing and brief description of the techniques which are currently available for routine practice (i.e., by physicians in solo practice or in small local hospitals)

c. A somewhat more extensive description of techniques currently available in medical centers (i.e., in 40–60% of major medical installations)

d. Last, but most important, the specification of needs and opportunities for new technology in the future, including items currently under development or evaluation and forecasts of future needs as long as 10 years from now.

The contributors to the chapter responded without being informed of the material in the other sections, so there is obvious duplication in the description of specific instruments or requirements. Not all clinical specialties were included because we did not intend to make this an exhaustive survey. For the benefit of readers who are not familiar with the details of medical specialties, somewhat more familiar terms have been substituted for technical ones used in the original manuscripts prepared by the contributors. The very large number of new and innovative ideas presented in the individual sections clearly indicates the extent to which engineering can be applied to the medicine of the future—if the academic, medical, and industrial segments of our society can be induced to respond to the many opportunities available.

CARDIOLOGY

James S. Cole

DEFINITION

Cardiology is the area of medical practice concerned with the diagnosis and treatment of diseases affecting the heart and blood vessels. To practice this specialty effectively, a physician must appreciate the normal structure and function of the heart and blood vessels and how a variety of diseases can alter this normal anatomy and physiology. He must also be competent to gather and interpret the data from techniques for evaluating normal and altered anatomy and physiology and understand the limitations of the systems used to make these measurements.

ROUTINE PRACTICE

The general practitioner or internist arrives at many clinical decisions on the basis of information obtained from the patient's verbal history and physical examination. The latter includes palpation of the precordium and peripheral pulses with the hand, percussion of the heart and lungs (Fig. 6.1), auscultation of the heart, lungs, and peripheral vessels, and determination of blood pressure. The standard 12-lead electrocardiogram (ECG) provides information about abnormal cardiac rhythms and injury to heart muscle, while the chest X-ray and fluoroscopy provide information about cardiac chamber and great vessel size, calcification of heart valves, and areas where heart muscle contraction is depressed or abnormal. The major limitation in all of these sources of information is their subjective nature. Only the amplitude and time relations of the ECG and certain specific X-ray measurements can be regarded as quantitative information.

Techniques in Medical Centers

Cardiology in medical centers has emerged as one of the most quantitative of the clinical specialties as a result of two mutually dependent developments. Surgeons demonstrated their ability to correct certain types of deformities or defects in the great arteries, valves, and heart chambers, beginning in the early 1940's with surgical correction of abnormal shunts between arteries, removal of local obstruction in arteries and opening up malfunctioning heart valves. As the treatment became more definitive, the need for more accurate diagnosis became sufficiently important to justify accepting the risk to the patient involved in getting the necessary information. Inserting tubes along blood vessels to the heart (cardiac catheterization) provided information about blood pressure, oxygen saturation, cardiac output, and shunt flows for functional and clinical evaluation of candidates for surgery (see Fig. 6.6). Improved X-ray techniques with radiopaque dyes outlined the heart chambers and valves, indicating more precisely the abnormalities of structure in the heart. This increasing accuracy in diagnosis uncovered a wide variety of congenital and acquired deformities of the heart which became suitable targets for surgical correction. Hypothermia, and later, artificial heart-lung machines, permitted stopping the heart and exposing defects for repair. Thus improved treatment and more quantitative detection developed more or less in parallel with important engineering contributions to both diagnostic techniques and therapeutic intervention (Table 10.1). As a result of such progress, physicians practicing in medical centers have at their disposal more quantitative information than is available in most other specialties. For example, the physical examination may be supplemented by graphic recordings of heart sounds and murmurs, local vibrations on the chest (apex cardiogram), and pulse tracings from arteries (1). Additional electrocardiographic information is obtained by vector analysis and exercise stress testing (2). Direct intravascular measurements of pressure recorded through fluid-filled catheters with external strain gauges (3), cardiac output determinations using the Fick or dye dilution principles, and angiographic demonstration of cardiac chamber size and contraction are established diagnostic procedures performed by clinical cardiologists. More recently, the anatomical distribution and caliber of coronary arteries and great vessels can be visualized by special injection techniques (4).

The problems posed by rheumatic fever and subsequent deformation of heart valves have largely succumbed to effective antibiotic administration, rendering chronic rheumatic valvular disease a diminishing threat to the public. The direct surgical attacks on congenital malformations of the heart have greatly alleviated that important category of heart disease. Despite these remarkably effective advances in both diagnosis and therapy, heart disease remains the major cause of death in the United States, largely because the great mortality from heart disease stems from diseases affecting the heart muscle (mainly the result of atherosclerosis in coronary arteries) for which direct therapy is inadequate. Our most significant problems are an inability to identify chronic degenerative disease processes before they progress to significant dysfunction or catastrophic events, and our inability to fully correct the dysfunction once it has become apparent. Two fundamental problems have significantly impeded progress in this area. The noninvasive routine diagnostic techniques (i.e., the physical examination, ECG, chest X-ray, and external graphic recordings) yield semiquantitative information which may show statistically significant differences between normal and abnormal groups of patients (see also Chapter 5), but the information is often too imprecise to allow a valid conclusion in a specific patient. Invasive techniques do provide much more quantitative data, but the application of these techniques is limited by the risk of complications from these procedures and by the limited number of situations in which they can be employed. Engineering solutions should be found which (1) improve the resolution and accuracy of noninvasive techniques to a point at which the data are precise enough to permit valid conclusions in specific situations, and (2) improve invasive techniques so that the morbidity and mortality associated with them will be reduced, the range of their applications extended, and the number and quality of data sources increased.

Rapid progress is apparent in improving the usefulness of catheterization and other invasive techniques. Currently, these invasive techniques are usually restricted to the diagnostic cardiac catheterization laboratory and special situations in the intensive care facilities. One significant recent development is the use of fiberoptics (5) and projected development of multifunction catheters (see Fig. 6.6). The ultimate objectives of these invasive techniques should include continuous or repeatable measurements of a wide variety of variables over a prolonged time with ample accuracy and responsiveness, consistency and

TABLE 10.1 *Technology in Cardiology*

A. In routine practice	B. Additional techniques in most medical centers	C. Projected developments
Cardiology technique		
History	Phonocardiography	Automated history
Physical examination	Apex cardiography	Spectral analysis (heart sounds)
ECG	Vector cardiography	Portable arrhythmia detector
X-ray	ECG under stress	Subtraction techniques (films, TV)
Fluoroscopy	Angiocardiography	Invasive techniques (catheters)
	Cardiac catheterization	oxygen, dye concentration
	Mixed venous O₂	pressure sensors
	Cardiac output	dimensions (vascular, cardiac)
	Fick	flow (velocity, volume flow)
	Dye dilution	surface irregularities
	Shunt detection	Noninvasive techniques
	Pressure	Instantaneous ventricular wall
	Intracardiac	movement
	Intravascular	Valve movements
	Across valves	Geometry (e.g., B scan)
		Instantaneous velocity and
		acceleration
		Volume flow and stroke volume
Therapy		
Prophylaxis against rheumatic fever	Intensive care	Coronary surgery
Control of hypertension	Coronary care	Infarct excision
Supportive treatment of	ECG monitoring	Artificial heart assists
coronary occlusion	Defibrillators	Hypertension control
	Extra cardiac repair	Prevention or reversal of arteriosclerosis
	Open heart surgery	
	Septal defects	
	Valvular reconstruction	

ease of maintenance and operation, and above all, minimal risk to the patient. Fiberoptics provide a means of transmitting light through a catheter, and extends spectrophotometric techniques directly into blood vessels. Pressure and sound energy can be recorded through either miniature piezoresistance or fiberoptic transducers and miniature ultrasonic flowmeters can provide measurements of flow velocity, and acceleration of blood (6,7). Isotope techniques will be expanded to provide measurements of flow rate and distribution of blood flow and blood content in various regions of the body.

One of the major limitations of invasive techniques is the threat of complications. Current materials used in catheter construction stimulate production of blood clots and irritate the lining of blood vessels, resulting in floating clots threatening obstruction of cerebral, coronary, and peripheral arteries. Acute inflammation of veins (thrombophlebitis) may also occur with prolonged use of catheters. Developments in biomaterials science must include synthetic materials which are inert to extend the use and reduce the threat of these procedures. Improved catheter material and design will also provide a means of positioning the catheter at the bedside when x-ray visualization is unavailable (8).

Improvements in the therapy of heart disease should take several forms. The improved diagnostic and monitoring techniques discussed above can provide a basis for improved medical therapy, by providing more precise information regarding responses to medication in acutely and chronically ill patients. Facilities should be extended to start treating acute myocardial infarction within minutes after the onset of the attack, either through rapid ambulance service (9) or helicopters (see also Chapter 4). Cardiopulmonary assist devices (artificial hearts and lungs) will be improved by the development of improved biomaterials, control systems and oxygenation systems (10) plus extensive physiologic studies identifying both beneficial and deleterious effects of assisted circulation.

A few specific applications have indicated the feasibility of obtaining more precise information by noninvasive techniques (Table 10.1). Initial studies have already indicated that pulsed ultrasound may be used to estimate the size of the mitral valve area. Other studies have indicated that pulsed ultrasound may also estimate left ventricular stroke volume and cardiac output (11). This would permit continuous monitoring of a function related to cardiac output during periods of stress testing, in response to specific therapeutic manipulation and during periods of acute circulatory dysfunction. The major problem with this technique is limited resolution and imprecise localization of the origin of the returning echoes. Studies needed to identify the origin of echoes arising within the cardiovascular system will necessarily be performed initially in the animal laboratory, later in the diagnostic cardiac catheterization laboratory, and finally during open-heart surgery.

Flowmeters based on the Doppler shift or electromagnetic principle may be implanted on arteries or veins and accurately determine volume flow through a vessel (see Fig. 6.10). Similar instruments can detect blood flow velocity in blood vessels under the intact skin, but these measurements do not indicate volume flow at present. Quantitation of the transcutaneous signal depends on knowledge of the caliber of the channel and the geometrical relationship between the vessel and the source of ultrasound. Solutions to these problems appear straightforward but are not now available.

One of the major advantages of noninvasive techniques is the opportunity to follow a specific variable continuously for several hours or days. However, transducers applied to the skin surface must usually be directly connected by wires to their appropriate amplifiers and recording systems. This greatly restricts the mobility of both hospitalized and ambulatory patients. Short-range telemetry systems could be used to transmit the desired signal from the patient to an appropriate amplifier (see Chapter 6). This would eliminate wires connecting the patient with the display unit, thus providing freedom of movement during prolonged continuous observations, isolation from extraneous noise, and reduction of electrical hazards. This should be particularly applicable to electrocardiograms, apex cardiograms, phonocardiograms, external pulses, ultrasonic and pressure signals.

In summary, advances in the diagnosis and therapy of heart disease will depend on physicians and physiologists identifying significant parameters on which to make clinical judgments while bioengineers develop improved biomaterials and techniques to obtain this information.

PERIPHERAL VASCULAR DISEASE

D. E. Strandness

Peripheral vascular disease can interfere with the function of the arterial, venous, or lymphatic systems. By tradition this field is largely concerned with those entities which influence the large and medium-sized arteries, the microcirculation, and the major draining veins. In practice the most common disease on the arterial side of the circulation is atherosclerosis which leads to arterial occlusion and aneurysm formation. On the venous side of the circulation, inflammation with clot formation (thrombophlebitis) often leads to additional complications such as pulmonary embolism and the postphlebitic syndrome (obstruction of veins with malfunctioning valves). Accumulation of fluid in the tissues (lymphedema either congenital or acquired) accounts for the majority of problems in the lymphatic system.

PRESENT STATUS

The diagnostic procedures available for blood vessels are summarized in Table 10.1. While there have been remarkable technological advances in instrumentation, the routine testing procedures applied to patients are little changed over the past 20 years. The current practice in the community hospital in assessing arterial, venous, and lymphatic disorders is to make a tentative diagnosis based upon the history and characteristic physical findings. The verification of the clinical diagnosis is made in most cases by the appropriate X-ray procedure. This includes, in nearly every instance, the introduction of contrast material into the appropriate vascular channel to define in anatomic terms the location and extent of the abnormality (see Fig. 6.7). From these results all therapy is planned, usually without employing any intermediate physiologic testing procedures that might be of value.

In the larger medical centers, particularly those associated with medical schools, a fairly wide spectrum of diagnostic tests is often available (Table 10.1). These more sophisticated procedures usually occupy a place between the initial clinical appraisal and angiography and are designed to provide tentative anatomic and physiologic data concerning the extent of the disorder (12). These applications can briefly be summarized as follows.

1. Plethysmography to measure flow, limb blood pressure, and sympathetic activity (13)

2. Flowmeters to assess qualitatively and quantitatively arterial or venous flow transcutaneously (ultrasonic) or with the vessel surgically exposed (electromagnetic) (14, 15)

3. Isotopic studies, using dye-dilution methods to measure total limb blood flow or clearance methods to measure nutritional blood flow in tissues (16)

4. Ultrasonic echo ranging to measure the dimensions of abdominal aortic aneurysms

5. Thermography, to assess cutaneous thermal patterns

6. Isotope scanning, for the detection of deep venous thrombosis

The diagnostic techniques available in large centers have not been generally accepted or applied in the routine practice of medicine. The lack of wide acceptance of these techniques is due to several factors, some of which include (1) cost, (2) complexity, (3) need for ancillary personnel, and (4) reluctance on the part of the medical community to acknowledge the need or desirability for such studies. The training of physicians at both the medical school and postgraduate level has only recently included much emphasis on either the value or use of these well-established techniques.

On the other hand, therapeutic approaches to peripheral vascular diseases have been widely accepted and applied by both physicians in private practice and those in university centers. Reparative surgery is now widely used. Engineering technology has not contributed materially to nonoperative methods of management. Surgical procedures (plaque removal or shunt installation) for the complications of atherosclerosis are often successful in selected cases, but they can be applied to only a small percentage of patients with the problems. These therapeutic limitations revolve largely about the location and extent of the disease. It is not likely that surgery, by itself, will be able to expand its vistas much beyond the current therapeutic approaches. At the present time therapy in nearly all types of peripheral vascular disease can only be considered as palliative and not curative.

The problem of thrombophlebitis from a therapeutic standpoint is largely stopgap. Since the cause of this disease remains obscure, treatment is directed only at the complications of the problem. While some of the newer diagnostic tests (e.g., the ultrasonic velocity detector and isotope scanning) have permitted an earlier, more accurate diagnosis of venous thrombosis, additional experience with them is needed to tell whether they can contribute materially to a reduction in development of pulmonary emboli or the postphlebitic syndrome.

The treatment of lymphatic disorders is generally discouraging. Lymphedema is usually secondary to a congenital absence of lymphatics or their interruption during extensive surgery. No effective measures other than controlled external support have been successful in reducing the tissue swelling due to this problem.

FUTURE PROSPECTS

In the next two decades the greatest potential for bioengineering is in the area of improving the diagnostic technology available both to the community physician and the academically oriented scientist. There should be some attempts to make the currently available diagnostic methods simple, inexpensive and convenient for the physician interested in this field. This has already been accomplished to some degree in the case of the ultrasonic velocity detector which is being used more widely in the practicing community.

The great emphasis for the future must be directed toward perfection of noninvasive methods of measuring those critical variables essential to the description and understanding of the disease processes that affect the vascular system. These variables include (1) volume flow of blood, (2) blood-velocity profiles, (3) vessel dimensions, (4) stress–strain relationships of vascular walls, and (5) wall composition.

The most promising type of energy for instruments which may provide this data is ultrasound. Preliminary work using more sophisticated continuous wave and pulsed systems indicates that information concerning many of these variables can be obtained. Though each generation of instruments increases complexity by a variable factor, their utility justifies these steps. The new developments will, in all likelihood, follow a pattern similar to breakthroughs in the past—the initial evaluation and use will remain in the large centers for a period of 3–10 years. As information and confirmation of the value of the methods becomes widespread, the diagnostic system will be modified and simplified and its cost reduced so it may be applied in the smaller hospitals.

The therapy of the future for diseases of the peripheral vascular system will largely fall in the area of prevention and nonsurgical management. The greatest potential contribution that engineering can possibly make, in my view, is to permit disease detection in its earliest stages before irreversible changes occur. At the present time the diagnosis is only made after severe and chronic anatomic changes have occurred within the vascular system. Improved techniques would provide serial assessment of the value of diet, drugs, and other nonsurgical therapy.

One major area which warrants an intensive, combined biomedical engineering effort is prosthetic arterial grafts. The major problems, which still need solution, are the development of a nonthrombogenic intimal lining and walls whose properties match those of the host vessel. For example, all of the currently available Dacron or Teflon prostheses are extremely stiff in comparison to the host arteries, and the differing compliances of the implant and the artery to which it is joined cause late disruption of the union. When the anastomotic union ruptures, a false aneurysm results, necessitating another operation. Without surgery these aneurysms can rupture or thrombose, leading either the the patient's demise or loss of the affected limb.

RESPIRATORY MEDICINE

J. Hildebrandt and C. J. Martin

Respiratory medicine is concerned with estimation of lung function and determination of mechanisms underlying respiratory insufficiency and disability. Inadequate respiratory gas exchanges may occur when the work of breathing is increased (obstructive pulmonary syndromes), often expressed as shortness of breath (exertional dyspnea) which can accompany heart failure as well. Respiratory insufficiency is evident in inadequate arterial oxygenation (arterial hypoxemia) or increased blood carbon dioxide (hypercapnia). The growing emphasis on function and pathophysiologic mechanisms stems from the fact that respiratory disease may have multiple etiologies and fail to respond to traditional approaches of treatment.

Inflammatory diseases of the lungs (pneumonia, tuberculosis, bronchiectasis, etc.) were once important to the practitioner, but are less commonly seen in the current age of antibiotics. The chest physician today is faced with a different spectrum of lung disorders. Two "diseases of civilization" have become more prevalent. Lung cancer and irreversible obstructive diseases of the airways (emphysema) have risen alarmingly in their incidence, doubling every five years or so. In all probability these diseases do not have a single cause; their frequency is related to repeated insults applied to a susceptible lung. On the one hand, personal pollution (tobacco), environmental and occupational forms of air pollution and infectious agents, in concert with genetic conditioning, serve to distort normal lung properties and function (clinical emphysema), or to provoke new growths (cancer). A third condition of severely deranged function, the respiratory distress syndrome, occurs in the newborn (hyaline membrane disease) and in adults after burns, injury, or following surgery. Finally, respiratory care has achieved prominence in the management of patients having diseases that affect other organ systems. Intensive care units in which injuries, heart disease, and myriad other illnesses are treated use the aids of the modern chest physician. Anesthesiology and postoperative care are devoted to intelligent use of respiratory therapy. Thus the scope of chest medicine has both changed and expanded.

AVAILABLE TECHNIQUES

One must distinguish between the diagnostic and treatment mechanisms available to the solo practitioner from those found in urban medical centers and from those available in specialized research facilities. Many general practitioners may have no technical apparatus for pulmonary diagnosis, function-testing and therapy beyond the chest X ray (Table 10.2). Most do have access to equipment providing oxygen therapy and assisted ventilation. Some use a gas volume recorder (spirometer), a timed vitalometer or a peak flow meter, and have access to blood gas apparatus for the measurement of partial pressures of oxygen and carbon dioxide. Few private practitioners use pulmonary function testing routinely to screen for presymptomatic abnormalities. On the other hand, larger medical centers are able to measure additional variables such as diffusing capacity and alveolar capillary membrane permeability to carbon monoxide; uniformity of gas distribution in lung (single breath or multiple breath nitrogen washout); residual volume and functional residual capacity (nitrogen dilution techniques); blood gases at rest and after standard exercise; lung compliance (using an esophageal balloon); shunt and physiological dead space; and sometimes airway resistance with an applied sinusoidal air flow (Table 10.2).

At the highest level of specialization (universities, federal hospitals, and research centers) one may find equipment which is still under evaluation and improvement or which is too expensive or too complex for wide duplication. This would presently include diagnostic apparatus employing radioactive iso-

TABLE 10.2   *Techniques for Respiratory Function*

	Routinely available	Available in medical centers	Projected technical developments
Diagnosis	History	Pulmonary volume determinations	Isotope evaluation of
	Stethoscopy	Tidal volume, vital capacity	Blood perfusion
	X ray	Residual volume	Gas distribution
	P_{O_2}, P_{CO_2}, sometimes available	Functional residual volume	Pulmonary embolism
	Spirometer	Dead space	Whole-body plethysmography
	Timed vitalometer	Maximum breathing capacity	Airway resistance
	Peak flow rate meter	Diffusion capacity	Lung volumes
		Alveolar membrane-capillary	Capillary blood flow (from N_2O uptake)
		permeability	Instantaneous multiple gas analysis
		Uniformity of gas distribution	O_2, CO_2, N_2, etc.
		Blood Gases	Blood gases
		At rest	Respiratory monitoring devices
		After exercise	CO_2, O_2
		Lung compliance	Hyperventilation
			Hypoventilation
Therapy	O_2 therapy	Assisted respiration	Portable oxygen supplement
	Antibiotics	Intensive care facilities	Respiratory pacemaker
	Symptomatic treatment		
	Lung resection for cancer		

topes. Radioactively tagged macroaggregates of albumin may be injected intra-venously for transportation to the lung and allow detection of regions with poor perfusion. A radioactive gas, ^{133}Xe, may be inhaled to visualize the regional distribution of inspired air in the lung, using either fixed scintillation counters over the chest wall or the Anger γ-camera. The same gas (^{133}Xe) may also be dissolved in saline and injected as a bolus to map the distribution of blood flow in the lungs. Another class of measurements involves enclosing the body in a box or chamber (plethysmograph) which is capable of precise determinations of airway resistance and lung volume. A whole-body plethysmograph is also used to measure instantaneous pulmonary capillary blood flow by measuring the rate at which nitrous oxide enters the blood. Structural studies may employ inhaled lead dust or finely powdered tantulum to study the outlines of the airways.

The modern chest physician must aspire to detect lung disease before gross symptoms appear. Techniques to detect signs of early obstruction to airways of the lung, early carcinoma, and potential pulmonary distress syndrome in infants may make these diseases amenable to successful therapy that would be impossible at a later stage. Current knowledge of air pollutants, although incomplete, is sufficiently alarming to the population that they may demand its level be reduced in the near future. The importance of modern respiratory problems has also prompted investigative effort into the genetic factors involved. Until these problems are solved, however, the primary concern of the chest physician is that of presymptomatic screening. The patient having fairly early asymptomatic obstructive syndrome can be identified with simple measures of expiratory air flow. This may enable the physician to use early preventive measures and to greatly delay progress toward serious disability. No such early detection techniques are available for cancer and distress syndromes. Identifi-cation of those tests (physiological, biochemical, radiological, etc.) that are useful in presymptomatic identification of disease is important to modern medicine. However, we are currently at a stage where mass screening and prospective epidemiological studies may be premature since highly sensitive and reliable techniques for diagnosis are not always available and therapy is not fully effective.

PROJECTED TECHNICAL DEVELOPMENTS

During the next ten years we may anticipate a number of practical developments in studies of lung function. The most promising instrument for continuous analysis of multiple gases is the mass spectrometer adapted for respiratory use. Current models are able to analyze O_2, CO_2, and N_2 continu-ously and therefore replace three separate meters. Mass spectrometers are not at present economically competitive since the cost is still several times that of individual meters. However, they have the capability of analyzing any other gas, from H_2 to SF_6. Mass spectrometry may also be utilized for blood gas analysis using a diffusing Teflon membrane that separates the vacuum line (catheter) from the liquid phase. These catheters are theoretically implantable in arteries and veins, allowing continuous arterial monitoring of respiratory gases in seriously ill patients, in exercising subjects and in other research applications.

The development of respiratory monitoring for intensive care units is under way as well. Alarm systems can be coupled to carbon dioxide meters so that an alarm is triggered when the patient fails to breathe, or when he hypoventilates or hyperventilates. This system can also monitor the integrity of pressure breathing systems. Arterial O_2 levels may be continuously monitored using transmission oximetry (earlobe) or mass spectrometry.

Treatment of respiratory disorders should also change in the years to come. Methods of increasing the inspired oxygen concentration for the chronically ill are expensive and/or unacceptable to the patient. Many patients could become ambulatory with improved oxygenation and need only a convenient, portable inexpensive and rechargeable oxygen supply for a few hours of daily light exercise. An artificial portable blood oxygenator and CO_2 exchanger should be possible using the techniques developed for the artificial kidney including dialysis and arteriovenous shunts. This would be an ancillary (temporary) device only, since lungs appear to have vital nonrespiratory functions. Certain individuals with hypoventilation secondary to abnormalities of central neural control need a respiratory pacemaker. These individuals fail to breathe with sufficient frequency and depth, especially when asleep, to maintain adequate gas exchange. Respiratory pacemakers could be as beneficial to these unfortunate individuals as the cardiac pacemaker is to heart patients.

As a final item in this brief summary we must list standardization of clinical and testing procedures and respiratory equipment. Cooperative clinical trials of drugs and therapeutic regimes should be promoted on a regional or national level to establish the usefulness of existing and proposed practices. Second, pulmonary testing equipment should be standardized so that data and normal values may be compared between laboratories. Calibration facilities for the various pieces of test apparatus are essential to relieve the nonresearch oriented practitioner from uncertainties about measurements and limitations of equipment. The need for standards of calibration and evaluation is immediate. It is difficult to imagine an adequate procedure established by other than some federal agency, for manufacturers have not been able to do so.

What of the future in respirology? The answer must be highly subjective and limited. In the area of irreversible obstructive pulmonary syndromes one is concerned with techniques that measure the scleroprotein abnormalities responsible for changes in tissue elasticity. These changes also occur with age but are more marked in lungs with obstructive syndromes. Following identification of the molecular change and its correlation with physical properties, treatment approaches can be designed. With regard to lung cancer, our hopes are periodically raised when claims are made for new screening factors, such as an X-ray change, a viral agent or serum factor, or a change in sputum cytology. To date, early detection techniques seem as remote as determination of etiology.

GASTROENTEROLOGY

Charles E. Pope, II

Gastroenterology is that subspecialty of internal medicine which focuses attention on the digestive tract from the pharynx to the rectum. The mouth and pharynx are the preserve of dentists and otolaryngologists, and the superspecialist proctologist stakes his claim to the rectum. The digestive tract absorbs, synthesizes, detoxifies, transports, and excretes various substances. Topologically, the gut interior is continuous with the external surface of the body, so that investigations of its interior is simply a matter of extending one's senses into relatively inaccessible locations. The gut can serve as a site for bleeding, for inflammation, and for cancer. In terms of technological innovation, the gut and its appendages lag far behind the cardiovascular and pulmonary systems.

In the community practice of medicine, one finds fiberoptic esophagoscopy (Fig. 6.6) and fiberoptic gastroscopy used to visualize the interior of the upper portion of the gut in a search for bleeding sites, ulcers, tumors and the like (17). Roentgenograms, fluoroscopy, and cineroentgenograms are available in most hospitals for visualizing more remote portions of the gut (Table 10.3). Far less information is obtained from these black-and-white shadows than would be observed if the structures could be seen directly. Injection of dye into the spleen (18) or the umbilical vein (19) allows radiological visualization of the venous drainage of the gut. Selective arterial contrast injection permits visualization of the gut's arterial tree (20).

Frequently available, too, are intraluminal pH electrodes, and in more and more hospitals, the capability to obtain a tissue sample of the small bowel and gastric lining by means of a remotely controlled, spring-loaded biopsy capsule (21). Esophageal dilators for treatment of obstructions are crude, but effective applications for technology. Liver scanning after isotope injection has become more widespread (Fig. 6.6) with greater availability of the necessarily complicated and expensive equipment (22).

On the next step up the scale of complexity, techniques generally restricted to medical centers (with associated medical schools and research facilities) include fiberoptic viewing of the colon or appendages of the gut. These techniques require much greater skill on the part of the operator. Perhaps the

TABLE 10.3 *Techniques for Gastroenterology*

Available technology in community	Medical school technology	Possible future technology
1. X ray, fluoroscopy	1. Fiberoptic colonoscopy	1. Use of computer-assisted pattern recognition for detecting abnormal gastrointestinal cytology
2. Esophagoscopy	2. Fiberoptic choledoscopy	2. Swallow device to detect fresh gastrointestinal bleeding and its location
3. Gastroscopy	3. Multiple retrieving hydraulic biopsy capsule	3. Ultrasonic study of liver consistency including fat and fibrous tissue
4. Intraluminal pH electrode	4. Peritoneoscopy	4. Methods of pancreatic visualization
5. Esophageal dilators	5. Pancreatic scans with liver subtraction	5. Transport studies using microspheres detected by ultrasound or by radioactivity
6. Mosher bag	6. Esophageal motility	6. Control of ionizing or thermal damage to parietal cells in an attempt to produce therapeutic achlorhydria
	7. Colonic motility	7. Quantitative gastric emptying studies
	8. Detection of electrical activity of gut	8. A directional intraluminal flow-probe for studying movement of fluid in the gastrointestinal tract
	9. Duodenoscope	9. Catheter tip flowmeter for study of vascular flow to arteries supplying small and large bowel
		10. A monitoring system for shock due to gastrointestinal bleeding
		11. Improve methods for dialysis of patients in hepatic coma using a normal donor as part of the dialysis system, but separating their bloodstreams
		12. A peritoneoscope small enough to be introduced through a No. 14 needle
		13. Palliative cryotherapy for esophageal or rectal cancers involving careful monitoring of dose levels
		14. A stimulating device to correct metabolic ileus in the small and large bowel

ultimate is the direct intubation of the pancreatic duct by direct visualization with a fiberoptic instrument having a controllable tip (23). Multiple retrieval of biopsy specimens by means of newer capsule techniques has been described (24). Pancreatic scans (which require more complicated equipment to subtract the contribution of the liver to isotope uptake) are also available in the centers. Studies of motility of esophagus and colon by means of mechanical transducers to determine forces or pressure by means of radiotelemetering capsule (25) would help determine whether the gut is capable of carrying out its assigned function of transportation of its contents at a reasonable speed. Newer techniques of determination of electrical activity of the gut (26), which shed light on either its secretory or its smooth muscle function, also show promise of assisting in the diagnosis of gastrointestinal dysfunction and assessing the results of treatment.

There appear to be a number of significant areas of development in diagnosis or treatment to which modern technology may address itself in the next few years, some of which are widely applicable in medicine and some of which are specific to gastroenterology.

DIAGNOSTIC NEEDS

a. Screening of cytologic slides obtained from the gastrointestinal tract as well as other sites is time-consuming, laborious work with high rates of observer error. Computer-assisted pattern recognition would be of great assistance

b. A simple, small device which could be swallowed easily and pinpoint the site(s) of fresh gastrointestinal bleeding

c. Better visualization, perhaps by ultrasound, of liver consistency with estimates of the ratio of hepatic to fatty or fibrous tissue (to detect, quantitate, and follow the course of treatment of cirrhosis, for example)

d. Improved methods for visualization of the pancreas, again perhaps by ultrasound, for detection of neoplasia

e. Techniques for studying transport of gastrointestinal contents, particularly using a tracer (perhaps microspheres) whose movement down the gut could be followed with ultrasound or radiofrequency energy

f. Quantitative measurements of the timing and rate of gastric emptying

g. A directional intraluminal flow-probe for studying the movement of fluid in the gastrointestinal tract

h. Better methods, perhaps by catheter-tip flowmeters, of determining blood flow through arteries supplying the large and small bowel

i. A technique to detect and monitor the physiological evidence of shock associated with gastrointestinal bleeding might include some method of estimating changes in blood volume repeatedly. Corresponding methods are needed to follow the course of treatment of patients in shock and would be applicable in other areas of intensive care

j. A peritonescope small enough to be introduced via a No. 14 or smaller needle to reduce the hazard of peritonescopy and allow its use in more patients.

NEPHROLOGY

T. Graham Christopher

INTRODUCTION

Nephrology is a clinical specialty which deals with the function of the kidney in health and disease based on an understanding of the renal physiology, particularly with reference to the regulation of sodium, potassium, water, and pH of the body. Nephrology deals with the mechanisms for identifying abnormalities of kidney function, instituting appropriate therapy and ultimately evaluating the success of the therapy. Since high blood pressure is attributed by many to malfunction related to the kidney, nephrology also includes, by implication, the renal participation in the control of blood pressure and the hypertensive state. Since failure of kidney function results in metabolic toxicity, the recognition and therapy of the accumulation of toxic materials resulting from kidney failure is appropriately included in the field of nephrology.

DIAGNOSTIC TECHNIQUES

The common and age-old approaches to the diagnosis of abnormal renal function have included urinanalysis of casual specimens as well as collected 24-hour urine specimens, with the determination of the specific gravity, protein, glucose and cellular content of urine including white blood cells, red blood cells, and casts. The urinanalysis is commonly supplemented by determinations of the blood chemistry to determine if any of the critical elements in renal excretion are accumulating in the blood. As an indication of overall urinary capability, the creatinine clearance is the most reliable and popular since the urea clearance has fallen into disrepute. The creatinine clearance is a relatively simple determination which contains valuable information with little cost in time or energy. In the presence of suspected renal pathology X-ray films taken during the concentration of radiopaque materials within the kidney and in the excreted urine will commonly identify the presence of renal calculi (stones), tumors, cysts, and obstruction of the urinary tract. An estimate of the blood supply to the kidney can often be obtained using this intravenous pyelographic technique.

Additional Techniques in Medical Centers

The more comprehensive study of urinary function often accomplished in medical centers includes an extension of the urinary analysis, to measuring osmolarity of the urine, the amino acid content, bicarbonate concentration, the titratable acids, and total acidity. In some cases the character of protein in the urine is studied by separation of the protein components using electrophoresis. The determination of the blood clearance of paraamino hippuric acid provides an indication of the blood flow through the kidney in a quantitative manner. The clearance of inulin provides information regarding the quantity of fluid which filtered from blood in the glomeruli per unit time. The clearance of vitamin B_{12} contributes additional information regarding active transport in the kidney.

Supplementary tests to study specific functions of the kidney can be obtained by intentionally loading the patient with acid, by determining the tubular reabsorption of phosphate, by establishing the concentration and dilution capacity of the patient's kidneys and the excretion of intravenously administered mannitol. The effects of salt restriction can also be studied in terms of urinary function. The kidneys, ureters, and bladder can be outlined by intravenous injection of radiopaque dyes. Such studies can be extended by catheterization of the renal veins to obtain measures of the renal secretion of selected materials and also a more accurate estimate of renal blood flow. Renal artery catheterization allows injection of contrast media to outline the dimensions of the renal artery and its branches; this will often reveal the existence of obstruction or stenosis of the renal artery and structural abnormalities in the region such as tumors and cysts. Isotope studies include clearance and renograms displaying isotopes concentrations in the kidney.

Samples of kidney tissue can be obtained by inserting appropriately designed needles through the skin and muscles of the back and into the kidney substance for later studies by light microscopy and electron microscopy supplemented by special stains such as those used for fluorescent antibodies. Extended tests of the blood composition can help to determine the degree and nature of any disorders of electrolyte balance and specific tests of diseases that may involve the kidney can be applied to identify the presence of systemic lupus erythematosus, hemolytic uremia, Henoch Schonlein purpura, and amyloidysis, etc.

TABLE 10.4

Techniques in routine use	Techniques available in medical centers	Opportunities for the future
Diagnostic techniques		
Urinalysis	Urinalysis extended	Analysis of urinary sediment
(Samples 24 hr)	Blood chemistry	Urine protein analysis
Creatinine	Clearance	Urine enzyme studies
Clearance	pah, inulin, vitamin B12	Improved radiographic techniques
Blood chemistry	Catheterization-renal veins	Gamma camera studies isotopes
X ray	X rays	Ultrasonic studies of renal blood flow
Intravenous	Renal arteriograms	Renal biopsy plus improved microscopy
Pyelograms	Isotope renograms	electron, light
	Renal biopsy	
Therapy		
Renal disease	Dialysis—artificial kidneys	Improved tissue typing for
Diuretics	peritoneal	transplantation (i.e., cell cultures)
Antibiotics	Renal transplantation	Improved artificial kidneys
Steroids		Reduced expense
Dietary limitations		Shorter dialysis times
Peritoneal dialysis		Smaller dialysis units
		Ultimately a portable artificial kidney
		Identification of toxic substances in uremi

Future Requirements for Diagnostic Techniques

Analysis of chemical composition and sediment of the urine have been extremely valuable in the past. Techniques are now being developed whereby specimens of urinary sediment can be appropriately stained and preserved to greatly improve the value of urinary analysis as it is now performed in the routine laboratory.

The fractionation of the proteins present in urine has demonstrated that not only the quantity, but also the type of protein provides indication of both the nature of the pathological lesion and also the expected response to therapy. Improved fractionation techniques together with correlation of the findings with diseased states will undoubtedly occur and may provide early and more specific evidence of renal disease.

Abnormal concentrations of enzymes in the urine have been demonstrated in various pathological conditions of the kidney. The diagnostic value of these tests has been limited by the lack of specificity of the enzymes measured.

More specific enzyme tests will be developed so that recognition of damage to specific cells in the kidney may then become possible (27).

Improved sensitivity of X-ray imaging apparatus will lead to further reductions in the amount of radiation used. Significant improvement both in film quality and in image processing techniques will result in better resolution of the details of the vascular supply and functioning areas of the kidney (28). The use of image-magnifying techniques will increase their value as diagnostic tools.

For radioisotope studies the stationary imaging systems such as the Anger camera and Ter-pogossian camera lend themselves to both computer analysis of the images and to sequential study of the time dependence of the distribution of radioisotopes in the organ. Thus with the Ter-pogossian camera, images representing time intervals as short as 1/60 second can be obtained. Isotopes with such short half-lives that they decay to low levels during a single pass through an organ, are available, and can provide steady images of blood vessels, or other structures. Washout curves (^{133} Xe) are being used to measure the renal cortical and medullary blood flow. It is possible to project that additional isotopes and labeled compounds will be combined with the imaging techniques to provide both specific and quantitative information about tissues which cannot at present be visualized. This will also give information about blood flow distribution to individual compounds of renal tissue. It may also provide information about the metabolic performance of individual functional units of the kidney.

Doppler ultrasound monitoring has been reported to be effective in detecting the slowing renal blood flow associated with rejection of a renal transplant (29). The further development of pulsed Doppler techniques will permit improved discrimination of the target vessel and more precise flow measurement. This may then prove to be the method of choice for studying renal blood flow and arterial narrowing (see Fig. 6.9D).

Ultrasound holography is capable of producing well-defined images within continuous media (see Fig. 6.11). Considerably more research must be undertaken so that this technique may develop to the point where soft tissue imaging techniques can effectively supplement existing radiological techniques.

Treatment of Renal Disease

Progress in the treatment of renal disease is dependent on the achievements of physiological and pathological research, combined with the development of specific therapeutic agents. At the present time the immunological and vascular attributes of renal disease are being investigated without assurance that the fundamental mechanisms will be elucidated by this approach.

The exact mechanisms by which sodium, potassium, water, calcium, and phosphorous are regulated by the body are not known. There is continuing speculation about the existence of additional hormonal controls. The functional relationship between parathyroid hormone, calcium regulation and vitamin D are under intensive investigation. Intrarenal control systems for the regulation of blood flow distribution and nephron function are also being studied. The elucidation of these mechanisms will significantly enhance the understanding of disease processes, and will eventually lead to improved therapeutic techniques.

Renal pathology is to be studied by more quantitative electron and light microscopy techniques for measuring the changes in renal histology associated with disease (see Chapter 6). Stains with increased specificity, coupled with improved instrumentation, will be employed to measure the nature of both the renal deposits and cytochemical changes within the kidney. These results could be correlated with the measured clinical parameters of renal damage and provide improved understanding and recognition of the early phases of the disease process (see the section on quantitative cytology, Chapter 6, page 247).

Tissue culture techniques for separating individual components of renal tissue will be developed. These are necessary for the development of more specific and more sensitive tests of tissue compatibility and antigen content. Thus, tissue typing will become more specific, and knowledge of the process of graft rejection will be improved. The association of antibody deposition in the glomular basement membrane with the lesions of glomerulonephritis indicates that increased understanding of the antigenic structure of the individual components of renal tissue will be of considerable importance in the investigation of the cause of glomerulonephritis.

Hemodialysis with artificial kidneys was a therapeutic technique applicable to a select few patients but is now considered suitable for many patients with chronic renal failure, particularly those with a prospect of renal transplant at some future time (see below). Biomedical engineering research has been directly responsible for considerable cost reduction in kidney centers (30) and for equipment suitable for use in the home (31). Therapy with artificial kidneys remains time-consuming and uncertain since the 8-year survival is reported to be only 55% of patients (32).

Future Therapy

The major research requirements for improvement in hemodialysis include (a) identification of currently recognized toxic factors occurring in patients with chronic kidney failure (33), (b) development of truly nonthrombogenic surfaces to avoid blood clotting in canulas, and dialysis units, (c) the development of absorbent columns for regeneration of the dialysis fluids to provide satisfactory removal of toxic materials from the blood with small volumes of fluid (i.e., for portable artificial kidneys), and (d) new dialysis membranes capable of clearing larger molecules that may include currently unidentified toxins.

Chronic dialysis by infusing and withdrawing special fluids in the abdominal cavity has been shown to provide satisfactory therapy at much lower cost than with an artificial kidney if complications from infections can be controlled (34). Home peritoneal dialysis equipment will be commercially available in the near future. The extent to which this treatment will replace traditional artificial kidneys cannot be predicted. Present indications suggest it is a preferred form of treatment for children, but the restrictions imposed by strict sterility may limit its application outside of medical centers.

Transplantation

At the present time transplantation is becoming an increasingly successful form of therapy for chronic renal failure. The rate of progress in this field can be judged from the survival data—in 1966 the cumulative one-year survival of transplants between siblings was 60%. In January, 1968, one-year graft survival was 70% (35). At the present time (1970) the expectation of one-year survival of transplanted kidneys from siblings is over 90% (36). Cadaveric transplants are improving with similar rapidity and, although they do not yet rival transplants from living related donors, some results are very encouraging. The reasons for the improvement in the short-term survival of renal allografts includes the identification of a suitable regimen for treating the onset of rejection without causing excessive suppression of the white blood cells and antibody mechanisms, and the introduction of antilymphocyte serum (ALS), and globulin (ALG). Considerably more information about ALG therapy is being gathered and will have a significant impact on many aspects of immune response and auto immune disease. The recognition of the HLA system of antigens and the techniques for recognizing both renal and lymphocyte antigens is another decisive factor determining the outcome of renal transplantation.

Additional typing procedures will be required before significant long-term benefits can be demonstrated. The development of organ preservation for kidneys obtained at autopsy would allow time for the antigenic typing both of the donor lymphocytes and of the donor kidney. Computer selection of the best match for the recipient would improve prospects for elective transplantation under optimum conditions.

At the present time the preservation of human cadaveric kidneys up to 50 hours has produced significant improvements in cadaveric transplantation and may lead to significant improvements in cadaveric transplantation with important reductions in the cost of this therapy. There is no evidence at this time that more prolonged preservation of the organ is necessary. Computer techniques for determining the optimum recipient for a cadaver kidney are now in operation both in Europe and the United States. In spite of these developments, renal transplantation remains hazardous mainly due to the need for maintenance immunosuppressive therapy. The future development of the field will undoubtedly involve improved recognition of the significant antigens with improved tissue matching. More specific immunosuppressive or antibody blocking therapy will be developed which might cause less interference with the body defense mechanisms against foreign organisms and malignancy. It can therefore be expected that the morbidity and danger of renal transplantation will be significantly decreased. Thus the highly successful results of 1970 will be considerably improved in the near future.

ANESTHESIOLOGY

Wayne E. Martin and H. F. Stegall

The practice of anesthesiology consists of reversibly altering a patient's neurophysiologic status to produce conditions suitable for operative procedures, yet restoring as soon as possible his normal status at the end of the procedure. "Spinoff" from the anesthesiologist's operating room procedures have broadened the scope of his specialty to include pulmonary therapy, intensive care, and treatment of chronic pain. Obstetrical anesthesia and newborn resuscitation are additional examples of specialized application of these skills.

Anesthesiology is generally practiced only in hospitals. In community hospitals the anesthesiologist gauges his patient's condition from a scanty supply of information. He may monitor heart sounds from precordial or esophageal stethoscopes, periodically determine blood pressure by the classic Korotkoff method, periodically estimate urinary output, and subjectively observe a patient throughout the operative procedure to determine the rate, depth, and quality of spontaneous respiration, appearance of cyanosis or other evidence of inadequate perfusion of the extremities, ocular reflexes, and muscle tone (useful in judging the depth of anesthesia). Information of this sort is largely obtained without the assistance of technical aids, interpreted through the anesthesiologist's training and experience, and used to make vital decisions.

In medical centers, where surgical procedures are likely to be more extensive, an oscilloscopic display of the electrocardiogram and intravascular arterial and central venous pressures are often added to this list. Semiquantitative measures of the effectiveness of peripheral blockade are sometimes obtained by peripheral nerve stimulation. Blood gases—particularly p_{O_2} and p_{CO_2}—can be obtained periodically as desired by blood sampling, body temperature can be monitored via rectal thermistors, and occasionally a selected lead of the electroencephalogram is available.

Several new sources of information, or better ways of obtaining and using old ones, are clearly needed. Continuous, on-line measurement of cardiac output, stroke volume and/or ejection velocity would materially assist the anesthesiologist in assessing the status of his patient, in whom a number of potentially depressant drugs may be circulating at one time. Unfortunately, dye-dilution techniques are slow, cannot be indefinitely repeated, and are inaccurate in the low-flow states where the information is most needed. It seems likely that catheter-tip or transcutaneous measurements of blood velocity or flow in the aorta (or a major branch of it), would yield useful information regarding cardiac performance. Continuous measurement of venous O_2 content in the right atrium offers some promise of assessing metabolic requirements and how well they are being met. The new fiberoptic devices described in Chapter 6 appear stable enough to yield reliable information over several hours. Such devices could also be used on the arterial side of the circulation, but several other variables of interest can be determined from arterial blood and thus arterial p_{O_2} (rather than O_2 saturation), p_{CO_2}, and pH should be included in multifunctional catheter development (see Fig. 6.6). The pH should be determined by intravascular electrodes which are sufficiently stable to remain in place without recalibration for several hours. The arterial oxygen pressure indicates the degree of shunting in the pulmonary vasculature, or other ventilation–perfusion abnormalities. Measurement of p_{CO_2} and pH yield valuable insight into the adequacy of ventilation and perfusion of the body.

Since an anesthetic deliberately changes a patient's neuorphysiologic state, some quantitative measure of central nervous function or other reflexes would be a valuable addition to an anesthesiologist's input for judgment about depth of anesthesia. No general measure of CNS depression is now available, and it is difficult to differentiate from cardiovascular depression; possible contributors to this information might include pupillometry, tensor tympani reflexes (see Chapter 6), or the "H-reflex." Electroencephalographic indication of depth of brain depression can be obtained by wave analysis, or by determination of changes in evoked potentials. In particular, the practice of employing nurse-anesthetists rather than anesthesiologists in small community hospitals could be aided were quantitative measures of anesthetic depth available to supplement the limited clinical experience of such anesthetists. Moreover, such measurements would allow the design of an adequate system for automatically controlling the depth of anesthesia without the continuous attention of an anesthesiologist.

Blood pressure remains a major tool in assessing the cardiovascular status of a patient under an anesthetic agent, yet often becomes unavailable when the blood pressure falls to shocklike levels. A Doppler ultrasonic sensor which functions when a Korotkoff technique fails is commercially available now, but in a machine which is expensive and difficult to use. Modification of this approach might yield a technique which can be automated, easily displayed, and free of observer prejudice. Other measurements of cardiorespiratory function, e.g., pulse rate and respiratory rate, could be recorded continuously rather than intermittently hand-charted and free the anesthesiologist for other duties.

A potentially exciting technique for the future is a replacement for chemical anesthesia such as electronarcosis. At present, this approach appears to produce nearly ideal anesthesia in selected species of primates without troublesome drug side effects, but development in Western medicine has proceeded slowly. It offers the hope of a technique suitable for man which can be employed easily, is rapid in its induction, fully controllable and easily reversed, and without the side effects invariably associated with drug use.

Anesthesiology is a particularly suitable field for assessing the usefulness, accuracy, and feasibility of new measurement techniques. Changes in patient status are easily induced, fairly large in magnitude, and under control. Often surgical exposure of vessels provides a valuable opportunity for comparison with conventional standards (i.e., electromagnetic flowmeters). The requirement for measurement of a large number of variables, the processing of such data into a usable form, and adequate display to aid the anesthesiologist's judgment in handling the conduct of anesthesia justify major efforts in this field. The overlapping interests of anesthesiology with obstetrics, surgery, cardiology, acute medicine, and neurology suggest that greater exchange of information between these areas might accelerate progress.

GENERAL SURGERY

H. F. Stegall

General surgery and its numerous specialties—orthopedic, thoracic-cardiovascular, neuro-, plastic, opthalmological, otorhinolaryngological, urologic, abdominal, and gynecologic-obstetric surgery—share a common interest in removal or restructuring portions of the normal or abnormal anatomy. An object is removed (a fetus, a neoplasm, or part of a normal stomach) and the remainder of the organs are aligned, fixed in place, and allowed to heal; or they may be reshaped (a nose, a breast, an arthritic joint), perhaps with the aid of materials left under the skin.

Aspects of surgery dealing with diagnosis and general care share much in common with other specialties of medicine, and thus share in their techniques as well. The history and physical examination are still the major source of a surgeon's information about disease, but laboratory tests, X-ray examinations, and special tests all add to his ability to anticipate what he will find when he explores a patient surgically. Besides these, community practice of surgery may involve special X-ray techniques (like tomography), isotope uptake studies, cardiac catheterization, and many others in assessing the state of a patient and his suitability for a certain surgical procedure.

Surgery's diagnostic needs are well covered in the discussions of this chapter involving specific organ systems, and include improved methods of visualizing structures—by X-ray, ultrasound, isotopes, or whatever—so that we will focus on therapeutic advances which seem realizable in the next decade. Basic surgical techniques applicable in medical centers soon find their way into community practice; witness the rapid proliferation of heart-lung machines in hospitals throughout the country over the last decade. This probably reflects the great interest of surgeons in technology; they are particularly interested in devices which will enable them to carry out their tasks with a minimum of guesswork and a maximum of manual skill.

The technical problems of cardiac, renal, or hepatic replacement are largely solved now, and the remaining problem is largely a biochemical-genetic one. Surgical replacement of diseased organs with healthy ones depends on improved control of the immune responses which are responsible for rejection of them. The obvious advantages of substituting a living organ include the capacity for self-repair and meeting its energy requirements from circulating blood. Unfortunately there still remain major problems in understanding the immune processes, and controlling or suppressing rejection without long-term side effects. Technical improvements in maintenance of organs in storage until needed would be valuable. Under these conditions, organ banks would be able to supply replacements as needed, rather than as they become available. The potential social, ethical, and moral issues implicit in large-scale transplant programs need to be resolved at an early stage.

Temporary cardiorespiratory support, capable of maintaining a patient for up to two weeks, would tide him over the acute period of many diseases like myocardial infarction. Cardiac support should reduce the incidence of ventricular rupture and improve coronary blood flow. Total and long-term replacement of the human heart by a mechanical device seems many years away.

Improved materials for vascular grafts, patches, valves, and other cardiovascular prostheses are still needed (see also Chapter 8). Some materials are still associated with an unacceptably high incidence of clot formation when in contact with blood, and others (like silicone rubber ball valves) often degenerate with the passage of years.

Devices that can be manipulated at a distance, under direct (endoscopic) or fluoroscopic control, would reduce the problems associated with exposure for repair (see Fig. 6.6). Devices at the end of cardiovascular catheters for valvuloplasty, removal of thrombi or atherosclerotic deposits, and placing a nonthrombogenic lining, are all potential technological developments of the next decade.

Better methods of tissue closure than the time-consuming, traditional suture are still needed. Better devices for stapling or clipping would make repair of minor wounds easy for paramedical personnel, and speed the surgeon's job in more extensive repair. Tissue adhesives offer considerable promise where sutures cannot be used adequately, as in hepatic or renal lacerations, and their use will be expanded to include arteriotomy repair, placement of artificial corneas, skin closure, etc., as they become easier to use and form stronger bonds without excessive tissue reaction.

In orthopedic surgery, exoskeletal splints could avoid the necessity of cast changes whenever some adjustment in bone fixation is required. Such splints would also reduce the skin problems and peripheral edema associated with the application of full plaster casts, and might also be used to apply traction. The major problems in their development include adequate fixation to subcutaneous skeletal structures without abrading the skin. Indeed, materials which bond well to skin are needed in several areas. Wires leading from implanted sensors or to implanted devices (like powered extremity prostheses) are a major source of infection where they penetrate the skin, since the integrity of the surface is broken and bacteria enter around the wire surface. Adhesive bonding to the wire itself, or skin ingrowth into a wire or "skin plug" surface, would avoid the problem.

Simpler methods of controlled tissue destruction without the necessity for full exposure would make the surgeon's task much easier. In particular, the problems associated with exposure of the central nervous system for treatment of Parkinson's disease, Ménière's disease, cordotomy for intractable pain, etc., seem approachable by using focused ultrasound, microwaves, or γ-radiation, since the tissues involved are among the most sensitive to radiation damage.

SUMMARY

This brief review of engineering applications to general surgery serves only to suggest the barest outline of the possibilities. The many different specialties of surgery are focused on direct therapy of diseases or malfunctions in the brain, eyes, ears, nose, throat, heart, blood vessels, gastrointestinal tract, endocrine system, genitourinary system, muscles, bones, and joints, skin. Each of these special areas represent diverse opportunities to develop instruments and special devices which will improve the convenience, safety, reliability and effectiveness of manual surgical techniques. Most surgical instruments and devices in current use have been developed by physicians acting like engineers but without actual collaboration or input from competent engineers. Systematic analysis of the needs, the specifications, optimal design and construction techniques have rarely been utilized for progress toward improved surgical technology.

OBSTETRICS

John T. Conrad

INTRODUCTION

Obstetrics is that branch of medicine dealing with pregnancy and parturation, including events preceding and following them. It extends into normal and abnormal male and female sexual function and ranges into the early life of the newborn infant. It involves not only the prevention of death in the mother, but also the salvage of the infant by preventing stillbirths or the death of the newborn infant. Conversely, it is also concerned about the prevention of conception desired by the individual or because of ecological pressures for population control.

PROBLEMS OF OBSTETRICS

There are two immediate problems in obstetrics. The first is maternal mortality, usually resulting from hemorrhage, infection and the toxemias of pregnancy and accounting for 90% of all maternal deaths. The second is the problem of stillbirths and neonatal mortality resulting from congenital malformations. The causes of prematurity and congenital malformation are multiple and solutions to this problem will present problems to many fields of medical science for years to come.

Other branches of obstetrics will become more important as population pressures increase, particularly reproduction control and human sexuality. The first is a biological problem and the second has overtones that may change our social fabric due to changes in male-female relationships.

The Technology of Obstetrics

The routine practice of obstetrics has demanded little of technology. Once pregnancy is diagnosed by a pregnancy test (once bioassay but now immunological) the main effort consists of routine examination of the mother and later of the fetus (i.e., size, heart rate, etc.). The existence and extent of any pelvic measurements too small to admit the fetus during delivery are detected manually (the Diagonal Conjugate) or more quantitatively by means of X-ray pelvimetry. The X-ray method may use relatively simple techniques for measuring, if factors of film to target distance are known.

The presentation and position of the fetus may be estimated by abdominal palpation, vaginal and rectal examination, or X rays as needed. The heart rate of the fetus is followed by the use of the fetal stethoscope.

Labor, itself, consists of three stages. These are described in the various texts on obstetrics and generally merge one into the other during normal labor. Examination during these periods involves abdominal palpation, rectal palpation of the fetal position and vaginal examination for cervical dilatation and orientation of the fetus.

All of these techniques have required a minimum of technological support during normal labor in a large majority of pregnant women.

Technological Advances Used in Medical Centers

The main use of modern technological innovation in obstetrics have been to ascertain the status of the fetus. The term fetal distress is used to describe the situation when fetal function is so deranged to threaten death or permanent injury. However, there is serious doubt regarding the nature of fetal distress in the human. A large effort is directed toward a search for clinical or laboratory findings which can be coordinated to the known existence of fetal distress.

Fetal Heart Rate and Uterine Contractility. The relation of fetal heart rate to uterine contraction has been shown to be highly predictive of impending fetal distress. Various commercial instruments have been developed to measure this relationship (e.g., the cardiotacograph by Hewlett-Packard). This instrument allows the obstetrician to monitor the cardiological status of the fetus during pregnancy, labor and delivery. It does this by providing a continuous display of the fetal heart frequency (beat-to-beat) as well as a measure of uterine contractility.

The fetal heart rate may be obtained by picking up the heart sounds through the abdominal wall by means of microphone. Likewise, the changes in curvature of the abdominal wall are used as an indirect measure of uterine contraction. Later in labor (e.g., after the membranes have ruptured), it is possible to obtain the fetal ECG from electrodes attached to the scalp of the fetus. This technique requires the use of an additional module built into the instrument as an accessory.

It is also possible to measure intrauterine pressures by means of a pressure transducer and a pressure monitor connected to the cardiotacograph. This requires transabdominal penetration of the uterine wall and the use of a saline filled catheter to measure the pressures. After the membranes have ruptured it is possible to place an open-ended or balloon-ended catheter through the vagina into the uterine cavity to monitor pressure. However, these techniques have limitations because of (1) transabdominal penetration with concommitant infection problems and (2) cluttering of the vaginal orifice with wires and tubing.

Ultrasonic Techniques. In an attempt to reduce the possibility of exposure to X rays, ultrasonic scanning has been used to detect multiple pregnancy, and localize the placenta and diagnose hydatidiform mole in obstetrical work. They have also been used to demonstrate tumors of the uterus and ovaries. The ultrasonic Doppler flowmeter is used routinely in many medical centers to detect fetal life early in pregnancy, to detect the fetal heart beat and to localize the placenta and blood vessels.

Amniotic Fluid Analysis in Rh Sensitization and Intrauterine Transfusion. The technique of intrauterine transfusion first described by A. W. Liley has been employed to salvage a number of infants who would otherwise have died *in utero.* The cause of death in these cases is erythroblastosis fetalis. In most cases from 40–80 ml of blood is injected into the fetal periotoneal cavity one or more times between the twenty-second and thirty-fourth weeks of pregnancy. The need for technology in this case is instrumentation for the detection of hemo-globin in the amniotic fluid.

Maternal Urinary Estriol Estimation. The output of estriol in the maternal urine is a good indicator of fetal health. A fall in this steroid material gives a good indication of fetal distress in chronic maternal diseases, such as toxemia of pregnancy, chronic bleeding, and essential hypertension. It does not correlate too well with acute problems, such as obstetric hemorrhage, prolonged labor, and amnionitis. The techniques used are those of the biochemist with such technological innovations as the gas chromatograph and paper chromatography.

Obstetrical Anesthesia. Only about 15% of all obstetrical anesthesia is conducted by professional anesthesiologists. The infant mortality with registered anesthesiologists in attendance has been found to be one-third that with less skilled people giving anesthesia. One of the major advantages of the medical center is the availability of modern equipment to monitor the vital signs of both mother and fetus.

CONTRIBUTIONS OF TECHNOLOGY IN THE FUTURE

There are many urgent needs in the field of obstetrics.

1. An accurate means of measuring the blood flow in the fetus without surgical manipulation.

2. There needs to be more specific criteria as to the nature of fetal distress. What are the blood flow rates, fetal metabolism, etc., that define a well baby and how can they be measured *in vivo*?

3. Monitoring during labor needs to be expanded to include measurement of cervical dilatation. The stages of this phenomenon are known to correlate with labor. What is needed is a nonobstructive means of measuring the time sequence of cervical dilation with a graphical readout to follow progress and perhaps predict future progress.

4. There needs to be a better means of locating retained products of conception (i.e., the placenta, blood clots, etc.).

5. Better methods for evaluating blood or amniotic fluid steroids. The new methods should be more sensitive and require little or no sample preparation.

QUANTITATIVE DERMATOLOGY

Colin H. Daly

Dermatology is probably the most subjective of the clinical specialties, a fact best illustrated by the fact that in his examination of a patient, the dermatologist rarely if ever performs a measurement that produces a numerical result. The majority of laboratory tests used in diagnosis in this field are qualitative rather than quantitative in nature (e.g., histological assessment of skin biopsies). Whether in his private office or in a large medical center, the dermatologist relies primarily on his senses of sight and touch and uses a minimum of technical aids. Indeed, modern technology has made little impact on this ancient branch of clinical medicine. However, there are several areas in which there exists a large potential for the successful application of this technology.

1. The development of quantitative methods of measuring physical and functional properties of human skin *in vivo* by "nondestructive" techniques. The aim of such development is to extend and complement, not to supplant, the subjective skills of the trained dermatologist. Such methods are not restricted in application to dermatology as much information about generalized disease can be gained from the skin

2. The development of artificial skin substitutes of either a temporary or a permanent nature

3. Methods of controlling the immediate environment of the human body within limits which can be handled by the body's own environmental control system

QUANTITATIVE METHODS

The need for the development of quantitative diagnostic methods is emphasized by the growing interest in the role of computer science in diagnosis because computers require numerical inputs. In view of the ready accessibility of the skin it is surprising to find that only a few isolated attempts have been made to develop such methods.

Most of the work which has been reported has been aimed at the measurement of various mechanical parameters. Indentation of the skin by devices similar to the Schade "elastometer" (37) has been investigated by several authors (38, 39) and has been found to give a measure of skin edema. These devices imitate the clinician's subjective method of pressing on the skin with a finger and watching the recovery of the indentation. They are, however, awkward and time consuming in use and thus have not found clinical acceptance. This is also true of other techniques which have attempted to imitate subjective methods such as measurement of the force required to compress a skin fold (39, 40). A wide variety of properties of the skin are susceptible to study and to quantitative measurement (41) including mechanical properties—(1) at low stresses (elastin dependent) and (2) at high stresses (collagen dependent); surface roughness; color and color changes; thermal properties; water content; water transport; blood flow and content; and sensation of temperature and touch.

An objective measure of elastic properties of skin can be obtained by application of loads in the plane of the skin surface by means of an oscillating disc. Depending on the magnitude and the pattern of loads used, a variety of parameters may be measured. If small amplitude ($\pm 1°$) oscillations of the disc are employed, skin behaves as a linear elastic material, i.e., only one parameter need be measured, this being in the form of a torsional stiffness. Also, as the torque applied to the disc is quite small at these amplitudes, friction forces at the disc–skin interface can be used to transmit this torque. It is, therefore, possible to develop instrumentation which is both convenient and rapid and which can measure this mechanical parameter at any point on the skin. An instrument operating on these principles has been developed and subjected to limited clinical trials (42). Results to date indicate that the greatest value of this device is that it provides the clinician with a "memory" and allows him to follow subtle changes in the skin which occur during the progress of diseases such as sclerodema and during the course of therapy with drugs such as topically applied steroids.

The development of this instrument was based upon *in vitro* tests on human skin which first demonstrated the existence of the linear elastic behavior of skin at low stresses and showed that this was a function of the elastin content (43). At higher stresses, the behavior was determined by the collagen fibers in the dermis and was viscoelastic in nature. Because of this more complex behavior and the higher loads which are necessary it is much more difficult to develop a disc loading technique to make *in vivo* measurement at higher stresses. For example, it becomes necessary to use an adhesive to attach the disc to the skin. However, *in vivo* testing at stresses high enough to load the collagen fibers would open up the possibility of studying changes in the various "collagen" and "ground substance" diseases. In this application, the skin often acts as a very convenient model system for the study of a systemic condition.

It is also possible to measure other relevant physical parameters of the skin by using other energy sources such as electromagnetic radiation covering the spectrum from infrared to ultraviolet. For example, there is currently no method of measuring the water content of the stratum corneum *in vivo*. This water content is important in determining the permeability of the skin and in maintaining the microbiological barrier function of this tissue. In addition, *in vivo* study of the water content may give important information on the progress and treatment of diseases such as ichthyosis, psoriasis, and dermatitis.

One possible method of measuring water content in the stratum corneum is based on a relationship between certain thermal properties and the water content and the adaptation of methodology originally developed for the study of thin films bonded to thick substrates, e.g., ceramic coatings on metal substrates.

If one is to expect the acceptance in private practice of any new techniques of the type discussed above it is essential that certain criteria be met.

1. The measured parameter must be clinically relevant and its measurement must permit significant improvements in diagnosis and/or therapy.

2. The measuring technique must be truly objective, i.e., a number of different observers should be able to obtain the same reading from a given subject.

3. Costs should be reasonably low.

4. Instrumentation should be simple to use and should require minimum set up and measurement time.

In general, if criteria (3) and (4) above are not met, one must expect that use of a given technique would be limited to medical centers. In this case, the benefits obtained must outweigh the additional costs involved.

ARTIFICIAL SKIN

The development of artificial skins, as nonbiological material substituting for some or all of the functions of the skin, would be immensely useful in the reduction of disability and mortality resulting from thermal burns, traumatic avulsion of portions of the skin, diseases characterized by widespread blistering and cracking of the skin and generalized inflammation of the skin with resultant severe thermoregulatory and cardiovascular problems. The problem of recovery from thermal burns is ultimately dependent on the regeneration of an intact integument. Whereas much progress has been made in prolonging the life of people with major third-degree burns by means of controlled sepsis and fluid balance, the absence of a source of intact skin for autograft is sorely felt, particularly in those instances when the burn exceeds 60% of the body surface area. Temporizing devices such as homografts, silver nitrate dressing, etc., are often inadequate in such cases.

It is curious that treatment and research relating to major burns has been largely monopolized by surgeons although there exists a large pool of individuals in the field of dermatology who are qualified to contribute their knowledge of skin biology in this field. However, it is also clear that the development of artificial skin requires a major collaborative research effort involving dermatologists and other skin biologists, plastic surgeons, physiologists, mechanical and chemical engineers, and other physical scientists. The American Academy of Dermatologists (A.A.D.) has proposed a program of this type. Four distinct goals were identified.

First Aid Skin

Intended to replace skin functions essential to life in cases of extensive skin loss, first aid skin should be readily applied by unskilled personnel and should be readily available in emergency situations. This material need only function for short periods (no more than a few days) and should be easily removable or should be biodegradable. Current promising research in this area is centered on the use of the cyanoacrylic adhesives as temporary scabs.

Temporary Skin

Temporary skin is intended as a temporary substitute for vital skin functions from the first-aid stage through an interval long enough to permit the burn surgeon to reestablish epithelial coverage (approximately one year). Again, removal of this material should not result in further trauma and should leave a surface suitable for skin grafting or the mounting of a permanent substitute. It would be highly desirable for such a skin substitute to permit the patient to move around rather than remain bedridden.

Life-Sustaining Permanent Skin Substitute

This is intended as a permanent prosthesis and covers a very wide range of materials and/or devices that would replace some or all of the functions of the skin system. This also implies that the patient will be able to lead an independent and productive but somewhat restricted life. Also included in this category are devices to substitute for the loss or impairment of individual skin functions in case the skin itself remains intact.

Complete Artificial Skin

The ultimate aim is the development of a prosthetic skin which is cosmetically and functionally permissive of a normal life.

It should be obvious that the attainment of any of the above goals would require considerable extension of our present knowledge of the physiology of the skin and of the many facets of engineering science which would be relevant. It must be emphasized that the successful development of such artificial skins should not be limited to a forced application of existing information but depends on a firm committment to support basic research so that rational specifications can be developed for artificial skins. The A.A.D. proposal included a suggested program prepared by the convergence technique, an organizational approach which pinpoints very clearly any problems which are not being met and are causing major bottlenecks in the overall program. Research contracts would be a natural way to encourage investigators to enter these areas.

CONTROL OF THE IMMEDIATE ENVIRONMENT OF THE SKIN

Although the remarkable properties of the skin allow the human body to survive in a wide range of ambient environmental conditions, chronic exposure to some extreme environments can result in a total breakdown of all or part of the integument. For example, chronic immersion in water cannot be tolerated and results in conditions such as the immersion foot syndrome now seen among troops in Viet Nam.

A major problem with external prosthetic devices is the interface between the prosthesis and the skin. Unless a very accurate fit between stump and socket is obtained, local high pressures normal to the skin will result in necrotic sores. Similar problems are encountered by paraplegics and other wheelchair or bedridden patients. In the latter case, the problem is often compounded by a lack of the sensory input which normally warns that the tolerance level of the skin has been exceeded.

There is a very real need for basic research to determine tolerance limits for the skin with respect to various extreme environments. This in turn would allow the rational development of the hardware necessary to protect the skin from such hostile elements.

PHYSICAL MEDICINE: Clinical Applications of Energy

A. W. Guy

Physical medicine and rehabilitation encompass a very broad spectrum of activities with current and potential applications of engineering including prosthetic devices, special tools and facilities for the disabled, occupational therapy, and many other examples. Instead of attempting to cover this multi-faceted field, some applications of wave energies are described to illustrate the variety of opportunities which may be recognized in the future.

Medical applications of wave energies (microwaves or ultrasonic) can be classified into two areas—(a) heating of tissues and (b) diagnostic processes. The former includes the most historic application, diathermy or therapeutic heating of tissues, and newer applications including rewarming of refrigerated whole blood, thawing of frozen human organs, production of differential hyperthermia (elevated body temperature) in connection with cancer treatment, and rapidly reversing a patient's hypothermic state (low body temperature) in connection with open heart surgery. Diagnostic applications include dielectric constant measurements to assess the properties and condition of certain biological tissues and reflectance and transmission measurements to assess significant parameters such as blood or respiratory volume changes. More detailed descriptions of each are given below.

HEATING OF TISSUE

Diathermy

The oldest application of microwaves to medicine is "diathermy," a clinical technique used to achieve "deep heating," the heat induced in tissue beneath the skin and subcutaneous fatty layers. Sufficient deep heating can elevate the temperature to the point where therapeutic benefits are achieved through local increases in metabolic activity and in blood flow by dilation of the blood vessels. The beneficial results are believed to arise from the stimulation of healing and defense reactions of the human body. Diathermy has been used in the treatment of musculoskeletal diseases such as arthritic and rheumatic conditions, fibrositis, myositis, pain, sprains and strains, and many other ailments too numerous to mention here (44). The deep heating cannot be obtained with heating pads or infrared rays, but must be accomplished by the transformation of certain forms of physical energy, such as ultrasound, radio energy, or microwaves, into heat beneath the subcutaneous fat layer. Both microwaves and ultrasound produce the required heating in the deeper tissues though lately ultrasound appears as the most popular due to their deeper penetration. Part of this stems from the historic poor choice of a microwave diathermy frequency of 2450 MHz. Clearly the presently used frequency does not provide a good method for achieving deep heating with minimal surface heating. This problem has been discussed in great detail by Schwan (45).

DIFFERENTIAL HYPOTHERMIA IN CANCER TREATMENT

Research on a new application of heating of body tissues in the treatment of cancer is now being conducted (46). The technique involves the use of microwaves to selectively and uniformly heat the cancer or tumor area while the remainder of the body is maintained in a hypothermic condition at $25°C$ (below normal body temperature). Then a very toxic anticancer drug is administered to the subject. The cooler tissues will absorb very little of the drug while the warmed tumor will have a metabolic rate that allows a significant amount of the drug to be absorbed. Through the combined use of selected frequencies, dielectric loaded waveguide apertures, and surface cooling, controlled heating patterns can be applied to the cancer area. Recent experiments on mice indicate that 75% of the tumors disappear after 4 to 5 hours of treatment.

WARMING OF HUMAN BLOOD

The warming of refrigerated bank blood from its $4°C$ to $6°C$ storage temperature to body temperature prior to transfusion is important to prevent dangerous cardiac and general body hypothermia. This has been done in the past by passing the blood through a long small core plastic tubing that is coiled in a thermostatically controlled water bath. The heat exchanger offers considerable resistance to the blood, slowing down the rate of transfusion which presents problems when rapid blood replacement is necessary. Restall *et al.* (47) have developed a microwave blood warmer which will heat a unit of blood (approximately 500 ml) in its original plastic container from 4° or 6°C to 35°C in one minute. This warmed blood can then be rapidly administered to the patient since both the viscosity has been lowered and the warming coil eliminated. The unit is warmed by rotating it in microwave cavity driven by a 2450 MHz, 1000 W magnetron. Extensive laboratory tests indicate no deleterious effects in the blood. Transfusions to 37 volunteer patients also indicate no abnormal effects.

RAPID ELIMINATION OF HYPOTHERMIA

A standard technique used in open heart surgery is to reduce the body temperature in order to induce a hypothermic state prior to surgery. This is to slow down the metabolic rate of the body cells so that nutrients and oxygen requirements are reduced sufficiently to allow the heart to be stopped for surgery. Surface *cooling* is desired prior to surgery since the peripheral body cells are cooled *before* the body core temperature is reduced to the point where the blood cannot provide nutrients and oxygen. During the rewarming stage after surgery, however, *core heating* is desired so that the blood temperature is sufficient to allow proper metabolism prior to the rewarming of peripheral cells. If surface heating is used, peripheral cells will require oxygen and nutrients at a level that the blood cannot provide due to the lower body core temperature. Core heating can be provided for adults and large children by pumping the blood through heat exchangers. This is a time-consuming process, however, that restricts the total time allocated for surgery. This type of core heating cannot be used for infants below a certain size due to physical limitations of the apparatus. Microwaves do offer a method for rapidly achieving the core heating for any size patient by selectively heating certain portions of the body. Experiments have just begun on the development of such a device.

DIAGNOSTICS AND STUDIES ON BIOLOGICAL SYSTEMS

Moskalenko has formulated methods for assessing the changes in microwave or shortwave reflectance and transmission which are caused by significant parameters such as blood or respiratory volume changes. The author has demonstrated the feasibility of Moskalenko's method by measuring the transmission loss in a 915 MHz microwave beam as it passes through the chest. The variation in microwave loss is proportional to the volume changes in the ventricle of the heart. The potentialities of this approach to cardiac studies appear favorable. Lonngren (49) has been able to determine the degree of pulmonary emphysema in postmortem analysis of a lung by measuring the dielectric properties of the dried diseased portions. This technique can be also applied to study the composition of biomaterials and biological systems through measurements of their electrical properties at microwave frequencies. An example is the strong dependence of the refractive index and absorption characteristics of various tissues on their water content. Thus it is theoretically possible to assess the water content of subcutaneous fat in a living subject without penetrating the skin. Another possibility is the analyses of biological membrane systems with respect to the role of structured or bound water in their function by measuring the refractive index and absorption characteristic (50).

SENSORY SUPPLEMENTS AND SUBSTITUTES

Sam L. Sparks

Nearly all aging humans experience reduced ability to clearly see objects nearby and must depend upon glasses to aid the accommodation of their eyes. Similarly, a substantial fraction of the population have partial loss of hearing and require electronic amplification by a hearing aid. These sensory supplements are very nearly optimal and require little further bioengineering development.

Complete loss of either vision or hearing constitutes a major tragedy for which no solutions are now available. The problems are compounded when a patient is congenitally blind or deaf since they do not appreciate what visual or auditory patterns could be like. The prospects of providing a functional replica of the eye or ear seems particularly remote. There are approximately one million cones in the fovea which is about 1 mm in diameter and responsible for all high-resolution vision. A manufactured fovea with similar resolution capabilities seems unattainable. Perhaps integrated circuit technology might supply tiny components for artificial retinas (see also Chapter 6). But there is still the problem of "connecting" the output of an artificial retina into the person's central nervous system. Long-term chronic implantation of electrodes has not been achieved, particularly with the neurons of the central nervous system. It is physically impossible to fabricate and implant an artificial sensory system which would achieve a one to one correspondence with the biological counterpart, but every effort should be expended to approach this goal as close as possible.

In the development of sensory substitutes, biomedical engineering can effectively integrate the contributions of several disciplines such as psychology, engineering, physiology, and education, to provide workable alternatives to the sensory modalities that have become defective. Two specific examples will be discussed to illustrate the manner in which bioengineering can exercise the necessary leadership and research management to effectively bring forth practical sensory alternatives.

The method by which an auditory substitute might be fabricated and interfaced with the person can be visualized. An artificial cochlea might be used to detect the sound waves and produce an output which can be conveniently conveyed to the person by mechanical or electrical stimulators. The cutaneous receptors, which are excited by the stimuli, would initiate impulses transmitted to the central nervous system. Recognition of the particular incoming sound pattern depends on the person's ability to identify the temporal and spatial aspects of these neural patterns. This latter step must be "learned" and depends upon the degree and pertinence of the particular training program presented for learning the use of the auditory substitute.

Two basic interfaces must be considered in providing an auditory substitute. First, there is the interface between the environment and the auditory substitute. It is not enough to consider only the engineering aspect of designing an artificial cochlea because the psychological aspects are also important. For example, there are certain sounds in the acoustic environment which are important to the person; for instance voices, sounds of music, an oncoming car, etc. There are other sounds which are unimportant or are considered as being noise by the individual. Among these might be echoes or masking sounds such as from a jackhammer. It is important that the engineer be aware of what sounds are desired and what sounds should be disregarded before he can specify the artificial cochlea.

The other interface between the auditory substitute and the skin, is an area in which the physiologist is interested. One concern is personal comfort during the long-term use of the instrument. The applied stimuli must be consistent with the receptor organs but also optimized for interpretation of the patterns. The importance of education cannot be stressed too strongly. This is particularly important in the case of the profoundly deaf, who have never had an opportunity to experience sound and noise, and therefore must learn to appreciate the significance of "sounds" they have never heard.

The successful development of a sensory substitute must involve the integration of many disciplines, as will be illustrated in the second example. Consider next the development of a substitute for vision. This instrument must provide an acceptable contact with the lighted environment for the totally blind. The assumption is that the auditory sense is not the preferred channel of information flow into the central nervous system. The blind rely too heavily on their auditory sense for it to be contaminated with extraneous sound generated by the visual substitute.

The task can be broken down into five phases. Phase one would constitute the development of an acceptable means by which information could be transmitted into the central nervous system. For convenience, let this be called the cutaneous communication system (CCS) in which information is processed through the skin. The second phase, is the determination of the type of light pattern or spatial information which a blind person would desire to obtain from his lighted environment. This particular phase is of extreme importance, as it affects the remaining four. For what is important to the normally sighted person, may be less important to the visually handicapped. Phase three is the development of a training program by which the blind can be taught to use the visual substitutes in identifying not only distant objects but also to read. This phase is of great significance since the blind (particularly the congenitally blind) have little or no concept of such things as perspective, light–dark interaction, and the visual shape of letters. Phase four constitutes the bringing together of all of the previous phases into a workable visual substitute.

CONCLUSION

Several things emerge from the responses from these clinical collaborators when viewed as a whole: first, that there are a number of potentially exciting and important contributions which can be made to the practice of medicine which appear to be within the reach of 1970's technology. Second, we are beginning to see a growing interest in the cost of any such contribution, and a concern with the level of skill required to use it. Third, a wide variety of approaches are needed which will demand multidisciplinary attack, probably by teams of collaborative scientists, and that a number of such teams may be needed to solve the large problems. Fourth, that the explosive expansion of public demands for improved health care delivery at reasonable cost will encourage reductions in the cost of some available treatment (chronic renal dialysis, e.g.), and will receive an increasing share of attention and support.

A final observation, left until last because of its importance, is that useful contributions to medicine will, with rare exception, be made in close collaboration with working physicians. Many diagnostic and therapeutic developments have failed, because they were not useful or too expensive or too trivial. The widely held attribute that large numbers of engineers (perhaps the surplus from aerospace industry), well funded, can by themselves solve significant medical problems has little foundation in experience. The problems faced by physicians are subtle and the solutions complex. The key to diagnosis rests in vague pattern recognition which is poorly understood. Much therapy depends on the manual skills of a physician, skills learned only through years of training, and cannot be circumvented by innovations no matter how impressive. Bioengineering and biotechnology must recognize the vital role the physician plays in their future.

References

1. Tavel, M. E. "Clinical Phonocardiography and External Pulse Recording." Yearbook Publ., Inc., Chicago (1969).
2. Bruce, R. A., and Hornsten, T. R. Exercise stress testing in evaluation of patients with ischemic heart disease. *Progr. Cardiov. Dis.* 11, 371-390 (1969).
3. Zimmerman, H. A. "Intravascular Catheterization," 2nd ed. Thomas, Springfield, Ill. (1968).
4. Kennedy, J. W., Baxley, W. A., Figley, M. M., Dodge, H. T., and Blackmon, J. R. Quantitative angiocardiography: I. The normal left ventricle in man. *Circulation* 34, 272-278 (1966).
5. Johnson, C. C., Palm, R. D., Stewart, D. C., and Martin, W. E. A solid state fiberoptics oximeter, *J. Ass. Advan. Med. Instrumentation* 5, 77-83 (1971).
6. Gabe, I. T., Gault, J. H., Ross, J., Mason, D. T., Mills, C. J., Shillingford, J. P., and Braunwald, E. Measurement of instantaneous blood flow velocity and pressure in conscious man with a catheter-tip velocity probe. *Circulation* 40, 603-614 (1969).
7. Benchimol, A., Stegall, H. F., Maroko, P. R., Gartlan, J. L., and Brener, L. Aortic flow velocity in man during cardiac arrythmias measured with the Doppler catheter-flow-meter system. *Amer. Heart J.* 78, 649-659 (1969).
8. Swan, H. J. C., Ganz, W., Forrester, J., Marcus, H., Diamond, G., and Chonette, D. Catheterization of the heart in man with use of a flow-directed balloon-tipped catheter. *New Eng. J. Med.* 283, 447-451 (1970).
9. Pantridge, J. F. The mobile coronary care unit. *Hosp. Practice* 64-73, August 1969.
10. "Proceedings Artificial Heart Program Conference, Washington, D.C., June 9-13, 1969." U.S. Dept. H. E. W., U.S. Government Printing Office.
11. Popp, R. L., and Harrison, D. C. Ultrasonic cardiac echography for determining stroke volume and valvular regurgitation. *Circulation* 41, 493-502 (1970).
12. Strandness, D. E. Jr. "Peripheral Arterial Disease: A Physiologic Approach." Little, Brown, Boston, Massachusetts (1969).
13. Greenfield, A. D. M., Whitney, R. J., and Mowbray, J. F. Methods for the investigation of peripheral blood flow. *Brit. Med. Bull.* 19, 101 (1963).
14. Strandness, D. E. Jr., Schultz, R. D., Sumner, D. S., and Rushmer, R. F. Ultrasonic flow detection: A useful technique in the evaluation of peripheral vascular disease. *Amer. J. Surg.* 113, 311 (1967).
15. Sumner, D. S., Baker, D. W., and Strandness, D. E. Jr. The ultrasonic velocity detector in a clinical study of venous disease. *Arch. Surg.* 97, 75 (1968).
16. Lassen, H. A. Muscle blood flow in normal man and patients with intermittent claudication by simultaneous ^{133}Xe and ^{24}Ha clearance. *J. Clin. Invest.* 43, 1805 (1964).
17. LoPresti, P. A., Hilmi, A., and Cifarelli, P. The foroblique fiberoptic esophagoscope. *Amer. J. Gastroenterol.* 47, 11-15 (1967).
18. Zeid, S. S., Felson, B., and Schiff, L. Percutaneous splenoportal venography, with additional comments on trans-hepatic venography. *Ann. Int. Med.* 52, 782-805 (1960).

19. Bayly, J. H., and Gonzalez, C. O. Umbilical vein in adult: Diagnosis, treatment and research. *Amer. Surg.* **30**, 56-60 (1964).
20. Laurijssens, M. J., and Galambos, J. T. Selective abdominal angiography. *Gut* **6**, 477-486 (1965).
21. Crosby, W. H., and Kugler, H. W. Intraluminal biopsy of the small intestine. The intestinal biopsy capsule. *Amer. J. Dig. Dis.* **2**, 236 (1957).
22. Krohmer, J. S., and Bonte, F. J. Scintillation scanning of the liver. *Amer. J. Roent.* **88**, 269-288 (1962).
23. Takagi, K., Ikeda, S., Nakagawa, Y., Sakaguchi, N., *et al.* Retrograde pancreatography and cholangiography by fiber duodenoscope. *Gastroenterology* **59**, 445-452 (1970).
24. Flick, A. L., Quinton, W. E., and Rubin, C. E. A peroral hydraulic tube for multiple sampling at any level of the gastro-intestinal tract. *Gastroenterology* **40**, 120 (1961).
25. Farrar, J. T., and Bernstein, J. S. Recording of intraluminal gastrointestinal pressures by a radio telemetering capsule. *Gastroenterology* **35**, 603-612 (1958).
26. James, C., Schedle, H. P., and Clifton, J. A. The basic electrical rhythm of the duodenum in normal human subjects and in patients with thyroid disease. *J. Clin. Invest.* **43**, 1659-1667 (1964).
27. Gault, M. H., and Geggie, P. H. Clinical significance of urinary LDH. alkaline phosphatase and other enzymes. *J. Can. Med. Ass.* **101**, 208 (1969).
28. Taddenham, W. J. (ed.) *Symposium on the perception of the Roentgen image.* Radiologic Clinics of North America, Vol. VII, No. 3., Dec. 1969.
29. Sampson, D. Ultrasonic method for detecting rejection of human renal allotransplants. *Lancet*, 976-978 (1969).
30. Cole, J. J., Fritzen, J. R., Vizzo, J. E., vanPaaschen, W. H., and Grimsrud, L. One year's experience with a central dialysate supply system in a hospital. *Trans. Amer. Soc. Art. Int. Org.* **XI**, 22-23 (1965).
31. Babb, A. L., Bell, R. L., Wilson, W. E. J., Peoples, R. W., and Grimsrud, L. "The Development of a Prototype Single Patient Dialysate Supply System for Home and Hospital Use." Proceedings of the Working Conference on Chronic Dialysis, University of Washington, Seattle, p. 52, 1964.
32. Pendras, J., and Pollard, T. L. Eighth year's experience with a dialysis center. *Trans. Amer. Soc. Art. Int. Org.* **XVI**, 77-84 (1970).
33. Lonergan, E. T., Semar, M., and Lange, K. A dialyzable toxic factor in uraemia. *Trans. Amer. Soc. Art. Int. Org.* **XVI**, 50-57 (1970).
34. Tenchkoff, H., and Schechter, H. A bacteriologically safe peritoneal access device. *Trans. Amer. Soc. Art. Int. Org.* **XIV**, 181-185 (1968).
35. Marchioro, T. L. Current clinical results of renal transplantation. *Arch. Int. Med.* **123**, 485-490 (1969).
36. Russel, P. S., and Winn, H. J. Transplantation. *New Eng. J. Med.* **282**, 786-793; 848-853; 896-906, (1970).
37. Schade, H. Untersuchungen zur Organfunktion des Bindegewebes. *Z. F. Exp. Path. Ther.* **11**, 369 (1912).
38. Kirk, E., and Kvorning, S. A. Quantitative Measurements of the Elastic Properties of the Skin and Subcutaneous Tissue in Young and Old Individuals. *J. Zenontol.* **4**, 273 (1949).
39. Robertson, E. G., Lewis, H. E., Billewicz, W. Z., and Foggett, I. N. Two devices for quantifying the rate of deformition of skin and cutaneous tissue. *J. Scb. Clin. Med.* **73**, 594 (1969).

40. Jochims, J. Elastometric an Kindem bei wechselnder Hautdenung. *Arch. B. Kinder-heilh.* **135**, 228 (1948).

41. Rushmer, R. F., Buettner, K. J. K., Short, J. M., and Odland, G. F. The skin. *Science* **154**, (3748), 343 (1966).

42. Daly, C. H., Tomlinson, M., and Odland, G. F. Clinical applications of mechanical impedance measurements on human skin. *Proc. 23rd ACEMB*, 16.6 (1971).

43. Kenedi, R. M., Gibson, T., and Daly, C. H. The Determination, Significance, and Application of the Biomechanical Characteristics of Human Skin. Digest of 6th Int. Conf. on Med. Elec. and Biol. Eng., Tokyo, August, 1965, pp. 531.

44. Licht, E. "Therapeutic Heat and Cold." Waverly, Baltimore, Maryland, Vol. 2, 1965.

45. Schwan, H. P., and Piersol, G. M. The absorption of electromagnetic energy in body tissues. Part I. *Amer. J. Phys. Med.* **33**, 371-404 (1954), **34**, 425-448 (1955).

46. Zimmer, R. P., Ecker, H. A., and Popovic, V. P. A new application of electromagnetic radiation in cancer treatment. *IEEE Trans. Microwave Theory Tech.* MTT **19**, 238-245 (1971).

47. Restall, C. J., Leonard, P. F., Taswell, H. F., and Holaday, R. E. Warming of human blood by use of microwaves. *4th Annu. IMPI Symp., Sum., 1969*, DB4, pp. 96-99 (1971).

48. Moskalenko, Y. E. Utilization of superhigh frequencies in biological investigations. *Biophysics (USSR)* (English Transl.) **3**, 619-626 (1958).

49. Lonngren, K. E. An application of microwaves to medical research. *4th Annu. IMPI Symp., Sum., 1969* p. 91. (1971).

50. Schwan, H. P., and Vogelhut, P. O. Scientific uses. *In* "Microwave Power Engineering" (E. C. Okress, ed.), Vol. 2, pp. 235-244. Academic Press, New York, 1960.

OVERALL SUMMARY
Future Options

The magnificent accomplishments of medicine in the past are partially obscured by mounting problems and concern for the future. Deficiencies of the American health care system are becoming all too evident, with widespread public dissatisfaction and some 20 million people deprived of adequate health care. Potential solutions to the complex problems in health care delivery cannot be resolved solely by either the health professions or applications of technology in support of their efforts. Increased sophistication is not the sole answer; simplification of procedures and processes are just as essential in many instances. As indicated in Chapter 1, the problems of medicine must be considered in the context of modern society with all its complexities. The pressures and constraints imposed by social, economic, legal, and political influences are absolutely crucial in most aspects of the problems.

The interdisciplinary character of medical problems is clearly evident in comparisons between the health care mechanisms of different countries. The combination of quality, availability, and distribution of health care in several countries of Western Europe exceeds the fondest hopes in the United States for geographical, economic, historical, political, and economic reasons. These programs represent extremely expensive feasibility experiments which hold valuable opportunities to recognize advantages and problems as we address impending

375

health care crises in this country. Modern societies are so complex they are responsive only to intense pressures or crises. The magnitude of current and impending problems in the health field seems fully adequate to stimulate major and basic changes which might otherwise be regarded as totally unattainable. For example, the threat of population explosion has stimulated remarkable adjustments in changes of official and public attitudes toward sensitive issues like birth control and abortion. The deteriorating environment has elicited countermeasures of very broad scope. On this basis it seems appropriate to consider the advantages and consequences of alternative proposals for health care delivery, in seeking optimal approaches, some of which could certainly evoke vigorous opposition.

The imminent prospect of federal expansion of health care support beyond Medicare and Medicaid will further increase the money available to purchase medical services without appropriate provisions for expanding the services. This combination inevitably threatens catastrophic stresses on medical personnel and facilities (see Chapters 2 and 3). Much has been written and stated about the need for increasing the efficiency and effectiveness of the system but most of the suggestions are palliative rather than basic. The rapidly rising costs of medical care are but one consequence of artificially stimulating demand without increasing the supply. Undesirable alternatives are rigid controls or chaos. A nondictatorial system depends on economic restraints to maintain a viable balance between supply and demand. In every country or institution which has rendered health care either artificially cheap or apparently free by reducing or eliminating direct payment by the patients, the result has been an unprecedented deluge of individuals seeking access to the services, whether they really require them or not. The result is a major change in the character of the "patient population" with increasing numbers of individuals who are demonstrably well, chronically concerned, or suffering from mild disturbances which derive no significant benefit in their physical condition by entrance into the health care system. This artificial increase in demand tends to clog the system, impeding access and care of patients who could really benefit significantly from the ministrations of physicians or facilities. Visits to physicians may become hurried, harried, and ineffectual and admission to hospitals may be delayed by long waiting lists. Demand for medical care has no inherent limits and continues to expand. The most important requirement is a mechanism for distinguishing individuals with physical illness which can be favorably affected by physicians from people who are well or "worried well." They may obtain a psychological lift or reassurance but not a tangible benefit from health services. Most countries of Western Europe have large numbers of general practitioners playing the crucial role of directing patients to hospitals, referring them to medical or surgical specialists or sending them to drugstores or home (Chapter 4). These decisions are frequently based on totally inadequate evidence when physicians see as many as 50–100 patients a day. Each visit is so short that the physician can obtain only limited histories and usually cannot perform a physical examination. Snap judgments based on such limited data are contrary to the original objectives of extensive training and experience, particularly when a large proportion of the visits are by people deriving little or no benefit from them. The few studies of medical practice employing industrial engineering techniques have consistently disclosed that physicians devote less than half of their time to management of sick patients for which their exhaustive training and experience prepared them. If the organization and nonprofessional support of health care delivery systems could be improved to fully utilize the unique training of health professionals and adequately distribute their efforts, there would probably be little or no net shortage of health manpower. However, it would be very difficult to accomplish.

The rising costs are inconsistent with the traditional attitude that price is no object when it comes to saving lives. The sanctity of human life has been so dominant that economies in medicine have been specifically avoided by establishing nonprofit hospitals as our fundamental health facility. With the exhaustion of available personnel and resources, it is now necessary to consider methods of assigning priorities for allocating them. One approach is an economic criterion of *added value.*

THE CONCEPT OF ADDED VALUE

One economic gauge of industrial capacity or maturity is the value added to resources or raw materials during processing in commercial concerns. It is generally applied to industrial processes but might be also utilized in assessing the contribution of physicians or health care facilities as they influence the course of illness. Various disease processes can be affected to varying degrees by the best of medical care (see Fig. 2.6). A most effective result of care is the prevention of disease (e.g., vaccination against smallpox, which prevents the development of the disease by a simple, cheap, and safe procedure). The net cost is slight in terms of either resources or hazard while the effectiveness is manifestly high. In decreasing order of added value, one might list the following categories.

1. Prevention of diseases
2. Permanent cure of illness by direct treatment of causative mechanism
3. Supportive treatment of illness for which no direct therapy is available
4. Symptomatic relief without affecting the course of the illness
5. Expensive and/or hazardous diagnosis coupled with ineffectual therapy.

The enormous sale of proprietary drugs directly to patients on the basis of advertisements in communication media demonstrate the existence of symptomatic disorders for which a visit to a physician provides little or no added value in altering the course of the illness (i.e., headaches, colds, "flu," depression, etc.). Similarly, the persistence of medicinal baths, spas, chiropractics, acupuncture and quackery indicate conditions not being effectively handled by standard medical practice. In a sense these constitute examples of disorders for which medical management is inadequate and should be improved or disclaimed as responsibilities of the health care system.

One source of patient dissatisfaction is the fact that physicians charge the same fees for patients experiencing no change or no "added value" as for visits resulting directly in significant improvement. If a physician's fees were related in some way to the extent to which he is able to favorably influence the course of illness, then both the physician and the patient would more generally avoid unnecessary encroachment on his services and resources which are employed with least effectiveness. Such a change in fee structure could conceivably alleviate to a significant degree shortages of physicians and facilities. It would also reduce the current tendencies for health care facilities and services to be clogged with patients with mild disturbances or with "worried well" when health care is rendered artificially cheap by insurance or government support.

The evaluation of cost-benefit or "value added" cannot be based on single arbitrary criteria. On the contrary, a multifactorial approach is required to consider a substantial number of critical variables such as the age of onset of a disease, the expected duration of illness without therapy, the degree of disability, the ultimate effect on life expectancy, and the influence on productivity or quality of life. The important consideration is the effectiveness with which medical intervention favorably influences these various aspects which can be regarded as a benefit or value added. They must include intangibles such as reassurance or relief, which are important even if they cannot be assigned a quantitative value. Because of the range in individual responses to disease there is no real prospect of arriving at a rigid categorization of disease states in terms of cost-benefit in medical care. Indeed there is no real need for this. However, it is equally clear that gross and obvious differences can be readily identified to distinguish between conditions with large positive added value and negative effects. For example, syphilis can be accurately diagnosed by a single test and effectively treated to prevent serious disability in later years. This example also includes a social factor: the protection of others in society from disease. The contrast between the diagnosis and therapy of congenital heart disease and of lung cancer is clear in the sense that the costs are very high in both instances, but one provides the prospects of a greatly increased life expectancy from early childhood, and the other a restricted extension of life complicated by extreme anguish. One would have a strongly positive cost benefit; the other might well be regarded as a questionable or even negative result. Very different effects can be encountered in two patients with sudden severe myocardial infarction and ventricular fibrillation receiving intensive emergency care. One might respond promptly and recover fully from this attack to live a few years longer. The other heart might respond only a few minutes later but ensuing brain damage could produce a human vegetable surviving for several months or even years at enormous cost and suffering to all concerned. It is high time we of the medical profession more critically evaluated the consequences of our objectives, our methods and our results. The positive and negative aspects of cost and benefit must be weighed in a balance representing such factors as those illustrated in Scheme 1.

The increasing requirement for accountability for costs and allocation of limited resources requires a conscientious effort to work toward developing and using criteria such as those illustrated above. Developing a priority list of disease categories indicating the expected range of value added would require interaction of physicians, economists, engineers, sociologists, and many other disciplines, but the effort would seem timely and worth while. Such a combined effort could be oriented toward developing mutually acceptable criteria and a categorization of diseases of various types and organ systems in terms of the cost and benefits derived from diagnosis and therapy. The relative cost-benefit of current practice for a full spectrum of generalized and organ system disease to provide a relative priority list could be most useful is in organizing available health manpower and facilities, and evaluating the areas most in need of research

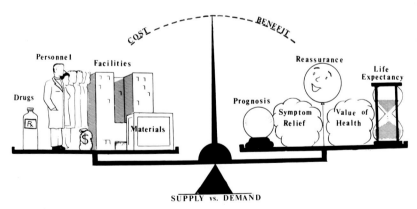

Scheme 1. The relationships between the cost of health care and the intangible benefits derived by the patient are difficult to evaluate. In a free market the value of health care is established by the price clients are willing to pay in accordance with supply and demand. The costs of medicine are being obscured by third party payments (insurance or government). The patient benefit must ultimately be gauged by the value added by medical management compared with the course of disease in untreated patients.

and development. For these purposes there is serious need for effective long-range planning at local, regional and national levels (see Chapter 4). Bioengineers with diverse backgrounds could make substantial contributions to many aspects of this big picture.

DIVERSIFICATION OF HEALTH CARE FACILITIES

Health care in the United States is oriented around private, nonprofit hospitals providing wide spectra of services and diversity of medical and surgical specialties (Chapter 3). A very large proportion of these hospitals have small bed capacity with more than half containing less than 100 beds and 90% containing less than 400 beds (Chapter 4). In Sweden, hospitals of this size are converted into facilities providing very limited services or phased out as inherently uneconomical. Comprehensive medical care is provided in very large hospitals and medical centers ranging from 800 to 2500 beds in many countries with an important difference. In the United States there is little or no integration or cooperation between different hospitals, and little evidence of local, regional, or national planning directed toward organizing existing or planned facilities into any kind of system. It is impractical to either rapidly erect large medical centers or eliminate numbers of existing hospitals despite the fact that many are functioning at occupancy rates well below 80%, some below 60%, even 50%. Hope for the future lies in the remote possibility that groups of smaller nonprofit hospitals will affiliate, amalgamate, and allocate responsibilities for various types of patients to achieve more efficient operation (Chapter 3). For example, one hospital might handle all obstetrical patients for the group; another might assume responsibility for major surgery and intensive care; pediatrics could be confined to one institution; expensive facilities such as cardiac catheterization, radiation therapy, open heart surgery, artificial kidney centers, coronary care units might be confined to one unit each to service the entire complex without undesirable duplication. Similarly, groups of hospitals could share support functions such as automated clinical laboratories, food service, equipment maintenance and pools, computer services, medical records, purchasing, accounting, and laundry. Such consolidated efforts usually fail to work smoothly at first and in some circumstances fail to achieve objectives because of extraneous problems such as personalities, prestige, individuality and tradition. The mounting pressures seem sufficient to require conscientious efforts at feasibility trials and demonstration units. In addition, the experience in Western Europe could be most valuable if properly interpreted for application in North America.

Widespread interest has grown in greatly expanding the availability and use of facilities and mechanisms for proper management of patients at greatly reduced expense through ambulatory clinics, extended care units, nursing homes, home care, and other options. Bioengineering could contribute significantly to the effectiveness of these efforts through design and organization of equipment and facilities for handling patients under these conditions. They will, however, not be used optimally until both the physician and patient have appropriate incentives to utilize less expensive options in place of such inappropriate policies as paying physicians' fees only for patients actually in hospitals. On the contrary, physicians' fees might be substantially reduced for hospitalized patients because the responsibility is shared by the house staff, nurses, and other members of health care teams during a hospital stay. On this basis, physicians' fees for an office visit might be a base which is proportionately increased for a house call, slightly reduced for patients in nursing homes and significantly, reduced for patients in hospitals. This step alone would greatly alleviate the current tendency for overutilization of hospitals, but might elicit strong opposition among the medical profession or hospital administrators.

MALDISTRIBUTION OF HEALTH CARE

Major segments of the country are deprived access to health care on the basis of geography or socioeconomic factors. The solution to such problems is not merely the creation of expanded medical personnel and facilities. It seems paradoxical that three of the greatest technical achievements of the century have not been mobilized to meet some of these requirements—(1) by rapid transportation, (2) by improved communication links between patients and physicians, paramedical personnel and facilities, and (3) through the great advances in data processing of medical records and display techniques.

TRANSPORTATION

Transportation of patients has not yet involved widespread applications of helicopters except in war zones. The location of hospitals at sites convenient to high-speed transportation links has not been featured in current or future urban planning. Emergency vehicles, manned by inadequately trained personnel, are poorly equipped and lack effective communication with health facilities. Transfer of patients between facilities is sufficiently complicated that there is little incentive to transfer recuperating patients to nursing homes or extended care facilities.

COMMUNICATION

Communication between patients and nursing stations in hospitals remains relatively primitive. Access to sources of immediate aid and guidance in health matters remains extremely difficult in most cities. Sophisticated communication media of all sorts have not been mobilized in any organized fashion to improve the level of knowledge of the general public in matters of their own health. For example, it would be entirely possible to rapidly upgrade the knowledge of the general public by a mechanism like that employed to introduce the "new math" some years ago. If a group of experts were assembled to identify information which should be conveyed to the public about the definition, meaning, and significance of signs and symptoms, and regarding therapeutic approaches to common illnesses, the information could be incorporated into books or pamphlets designed specifically for curricula at various levels in our school system. Intensive indoctrination of teachers and parents in preparation for presenting the information to the students would rapidly disseminate vital information to a large and growing segment of our society. This would tap the great mass of American public as supplements to existing health manpower.

Obviously many of the suggestions contained within this monograph are visionary and unlikely to be instituted; however, the magnitude and severity of the impending health care crises demand consideration of possibilities. The rapidly changing times and tides require accelerated responses and longer range planning and forecasting to meet the growing needs of a complex society.

NEED FOR NEW TECHNOLOGIES

The requirements for new diagnostic techniques stems from the fact that traditional signs, symptoms and test results are decidedly nonspecific so that a pattern of changes in several variables or factors is generally required to identify a functional disturbance or disease state. For improved diagnosis, the pressing need is for a progressively expanding array of simple, safe, inexpensive, and effective objective tests which can be performed without hazard by technicians and indicate the existence and severity of specific illness with minimal ambiguity and minimal tests or determinations. When enough is known of the causes and effects of disease that we can detect and evaluate severity of an illness by a single quantitative determination, then we have attained a high level of understanding. An indiscriminate development of new diagnostic technologies would be inappropriate because the total quantity of data currently being collected is very great, with much of it lacking relevance to the individual patient under consideration. Instead there is crucial need for including cost-benefit projections in any decision to engage in new diagnostic or monitoring research and development efforts. To this end we should consider sources of valuable and relevant information, particularly that which can be gleaned from nondestructive, external energy probes as described in more detail in Chapters 6 and 10.

DATA PROCESSING

The enormous quantities of information currently flowing into the medical record defies intelligent handling (see Chapter 5). The objectives, format content and display of data in the medical record at the bedside and in data banks require thorough reconsideration so that modern data-handling techniques can be effectively employed. The current interest in the "problem-oriented" record is extremely important because it signifies a willingness and possibility of introducing major changes in the medical record in hospital and private practice. In addition, such revisions should take into account the amazing pattern recognition ability of the human visual system which is totally wasted on records consisting of narrative accounts or columns of figures. Recognized requirements for new knowledge, techniques, or technology generally call for renewed or reoriented basic research efforts as foundations for improved solutions to practical problems. The priorities established for increased tempo of clinical engineering efforts should be reflected in accentuated emphasis on related fundamental bioengineering science, research and development.

SIMULATION AND MODELING

The analysis of complex structures and functions in engineering is commonly undertaken by means of mathematical simulation or modeling; powerful techniques which are currently being widely employed in many aspects of medicine and biology, from simple control systems in lower forms of life to clinical medicine or to evaluation of whole health care systems. With the growing awareness of complexity in biomedicine in all its ramifications, the utilization of these modern engineering techniques is being called upon with increasing frequency. The most important single contribution of these methods is the identification of missing information so that improvements in data acquisitions and sources are stimulated as suggested by examples presented in Chapter 7.

BIOMECHANICS AND BIOMATERIALS

Among the principal impediments to progress in clinical therapy and monitoring are unmet needs for new biomaterials with stringent requirements for use in contact or conjunction with tissues (see Chapter 8). Physical, chemical and biological compatability can be optimized only by increased understanding of the physical and chemical properties of the tissues and of the materials considered for various applications. Thus, biomechanics and biomaterials are inevitably closely related and serve as basic bioengineering subjects of critical importance for solution of many different problems. A substantial number of comprehensive projects suffer from lack of necessary materials with extreme durability and optimal reactance with tissues and blood; some applications require firm adherence, others require minimal interaction with blood or tissues. The medical profession lags far behind the dental profession in recognition of the needs for new materials and for specifications needed for engineering definitions, isolation or production. Recent years have witnessed a massive convergence of interest by materials scientists and mechanical engineers on these and related problems.

MANPOWER

Manpower requirements for the health care system cannot be met by training more physicians. The composition and scope of health care teams are increasingly diversified, which means that a wide spectrum of training opportunities need to be provided. The specific activities and distribution of time by the various members of health professionals is not known and not being widely explored in most instances. The ultimate requirements for different kinds of health manpower should be determined and based on extensive studies by operations analysis, management engineering, time and motion studies, and other established techniques. In the meantime, a variety of professional career opportunities can be identified including physicians' assistants, health technologies, clinical technicians, and many others. The extent to which engineers and engineering should be included in their training has not been determined but deserves early consideration.

FUTURE CLINICAL REQUIREMENTS FOR NEW TECHNOLOGIES

The training of most physicians is seriously deficient in the areas of quantitative or physical sciences and engineering. For this reason they lack perception concerning what contributions might be expected or achieved from new technologies. When asked what they need in the way of new tools, most clinicians are quite incapable of identifying unmet needs of a technical sort except minor improvements in existing equipment. However, clinicians which have had exposure to engineering and engineers can visualize many needs and requirements as evidenced by the individual responses in Chapter 10. Caution should be exercised in rising to the challenges raised by these suggestions to be certain that optimal cost-benefit relations result. However, the breadth and diversity of opportunities to add technology to the various medical and surgical specialties is extremely impressive.

SUBJECT INDEX

387